Industry 4.0 Technologies in Civil and Mechanical Engineering (ICI4TCME 2020)

IOP Conference Series: Materials Science and Engineering
Volume 1112

Namburu, India
4 - 5 November 2020

Editors:

N. Kumara Swamy
M.K.M. Farooqui

ISBN: 978-1-7138-3401-4
ISSN: 1757-8981

Printed from e-media with permission by:

Curran Associates, Inc.
57 Morehouse Lane
Red Hook, NY 12571

CURRAN ASSOCIATES INC. proceedings.com

Some format issues inherent in the e-media version may also appear in this print version.

Institute of Physics
Dirac House, Temple Back
Bristol BS1 6BE UK

Phone: 44 1 17 929 7481
Fax: 44 1 17 920 0979

techtracking@iop.org

Additional copies of this publication are available from:

Curran Associates, Inc.
57 Morehouse Lane
Red Hook, NY 12571 USA
Phone: 845-758-0400
Fax: 845-758-2633
Email: curran@proceedings.com
Web: www.proceedings.com

TABLE OF CONTENTS

MECHANICAL ENGINEERING

CIVIL ENGINEERING

IOP Conf. Series: Materials Science and Engineering 1112 (2021) 011001 doi:10.1088/1757-899X/1112/1/011001

PREFACE

Vasireddy Venkatadri Institute of Technology was established under the aegis of Social Educational Trust, Guntur in the year 2007.It has become an autonomous college from the academic year 2019-2020. All six departments are accredited by NBA, New Delhi and accredited by NAAC with A Grade. Todayalmost 4300 students, 285 teaching staff and 315 non-teaching staff strive to fulfill the vision

of VVIT. The college has professionally qualified and experienced faculty with 48 doctorates. Siemens Centre of Excellence with 15 high technology laboratories including robotics laboratory has been set up to provide training to the students including neighboring colleges. VVIT has many student chapters like IUCEE (EWB),IE(I),IETE,IEEE. In the Private or self- financed Colleges category, VVIT has been ranked under Band-A (Rank between 6th - 25th) across the entire country by Atal Ranking

of Institutions on Innovation Achievement, MHRD, Govt of India. VVIT stood first in Times of Indian Ranking -2020.

It is the youngest Engineering College in the state to achieve this recognition. MoUs with Google Inc. USA for establishing Google Code labs, Stanford University USA under University Innovative Fellowship (UIF) program and 3D Experience Lab established by Dassault Systems, France signify our commitment to update quality technical education. Realizing the slogan that "Everyone Counts," VVIT has the unique achievement of having placed most of its students in MNCs like Amazon, DBS bank, Tech Mahindra, Infosys, TCS, Wipro, Athena Health Care, Open Text, Genpact, Sutherlands, CSS Corp, Convergys, Apps Associates, etc. (about 95 companies) every year.

About Conference

New Technologies are emerging every day that may radically transform the future of engineering. Vision of this conference is to promote excellence in scientific knowledge and innovations in the field of Civil and Mechanical engineering and related disciplines to motivate young researchers. It also provides a premier interdisciplinary platform for faculty members, students' researchers, practitioners and educators to present and discuss the most recent innovations, trends, and concerns as well as practical challenges encountered and solutions adopted in the practical fields. We welcome technical papers presenting research and experimental results, position papers addressing the pros and cons of specific proposals, such as those being discussed in the standard or in industry consortia, survey papers addressing the key problems and solutions on any of the topics or research papers on work in progress

Conference themes

Civil Engineering

Materials: Advanced building materials and high-performance concretes, Green Buildings and Zero Energy buildings.

IOP Conf. Series: Materials Science and Engineering 1112 (2021) 011001 doi:10.1088/1757-899X/1112/1/011001

Structures: Concrete structures subjected to extreme events, strengthening and repair, Composite structures.

Analysis & Design :Building Information Modelling, Structural safety and reliability, Application of Optimization Techniques, Geo-synthetics and reinforced soil structures, Tunnelling and deep excavations, Pavement Geotechnics.

Advanced Technologies : Big data analytics, Automation, Deep Learning and Machine learning, Smart concrete and Block chain Technology, IOT and Smart citiesDrone Technology.

Mechanical Engineering

Design: Tribology, Composite & Functionally graded materials, Intelligent Robots, Rotor dynamics, Vibrations, Condition monitoring, NDT.

Thermal: Renewable and non-renewable sources of energy, Computation Fluid Dynamics, Energy storage, Refrigeration and Air conditioning, Heat and mass transfer, Aerodynamics, IC engines, Compressible /incompressible flows, Bio-heat transfer, Phase change materials, Alternate Fuels.

Manufacturing: CAD/CAM/CIM, Simulation of manufacturing processes, Non-conventional manufacturing processes, Additive manufacturing, Nano-materials.

Organizing Committee

Chief Patron

Sri Vasireddy Vidya Sagar, Chairman

Patrons

Sri S. Badari Prasad, Secretary

Sri M. Sree Krishna, Joint Secretary (Academics)

Conference Chair

Dr. Y. Mallikarjuna Reddy, Principal

Conference Convenors

Dr. K. Giribabu, Professor in ECE, Dean of Studies

Dr. T. Sreedhar Babu, Professor & HOD, Department of Civil Engineering

Dr. K. Satyanarayana, Professor & HOD, Department of Mechanical Engineering

IOP Conf. Series: Materials Science and Engineering 1112 (2021) 011001 doi:10.1088/1757-899X/1112/1/011001

Conference Coordinators

Dr. N. Kumara Swamy, Professor & Dean of Academics, Department of Civil Engineering

Dr. M. K. M. Farooqui, Professor, Department of Mechanical Engineering

Advisory Committee

Dr. Krishna Vedula Executive Director IUCEE, USA

Dr. Chaitanya Krishna, Professor in CE, University of Tokyo, Japan

Dr. G. Apparao, Professor in CE, IIT Madras

Dr. D. Ramasheshu, Professor in CE, NIT Warangal

Dr. G. Abbaiah, Professor, CE, JNTU Kakinada

Er. K. Srinivasa Rao, Dy. Project Manager, BSCPL Infrastructure Ltd.

Dr. P.N. Rao, Professor, ME, University of Northern Iowa, USA

Dr. Tirupathi. R. Chandraputla, Professor in ME, Rowan University, USA

Dr. N. Mohan Rao, Professor ME, JNTU Kakinada

Dr. Syed Ismail, Professor ME, NIT Warangal

Dr. B. Raghukumar, Professor ME, PVPSIT

Dr. R. Naveen, Professor in ME, Director- COE

Organizing Members:

All Staff members of Department of Civil and Mechanical Engineering

IOP Conf. Series: Materials Science and Engineering 1112 (2021) 011001 doi:10.1088/1757-899X/1112/1/011001

Conference Committees

S. No.	Committee Name	Committee Members (ME)	Committee Members (CE)
1	Correspondence committee	Mr.K. Seetharami Reddy Ms. A.Sirisha Mrs. P.Bharathi Mr. A.Srinivas	Mr. S. Farooq Ahmed Mr. D. Aditya Sai Ram
2	Technical program committee	Mr. D. V. Seshagiri Rao Mr. B. Aditya Mani Sai Pavan Mr. D. Kiran Reddy Mr. E. Saibaba	Mr. D. Sudhakar Mr. G. Leela Suresh
3	Registration, & certificates committees	Dr. K.V.L Somasekhar Mrs. M.L. Vinitha Mr. M. Avinash	Mr. A. Srikanth Mr. D. Jaya Krishna Mr. Abdul Rashid
4	Printing and Publication Committee	Dr. M. K. Mallik Ms. P. Bhagyasree Mr. V.Sivakanna	Mr. Ch. Sukesh Mr. Pavan Sai Durda Reddy
5	Inaugural and valedictory committee	Mr. Shaik Abdul Saleem, Mr. V. Sambasivarao, Mr. Ch. Raghavendra	Ms. K. Krishna Veni Ms. P. Lavanya
6	Finance committee	Dr. P. Venkateswara Babu Ms. K J Padmaja	Ms. K. Adithya Nandini Mr. Mirza Mahboob Baig
7	Press and media committee	Mr. K. Madhusudhan Chowdary Mrs. E. Leela Rani	Mr. V. Rathna Kumar Mrs. K. Sai Ramya
8	Editorial committee	Dr. T. Srinivasa Rao Mr. N. Kishan	Mr. K. S. Vivek Mr. Y. V. Subba Reddy
9	Preparation of Report & sending to higher authorities	Mr. V. Kiran Kumar Mr. SAK Jilani Mr. P. Naga Srinivas	Mr. T. Venkateswara Rao Mr. Sk. Noor Ibrahim

Editors

Dr. N. Kumara Swamy

Dean & Professor

IOP Conf. Series: Materials Science and Engineering 1112 (2021) 011001 doi:10.1088/1757-899X/1112/1/011001

Department of Civil Engineering

VVIT

Guntur

Dr. M. K. M. Farooqui

Professor

Department of Mechanical Engineering

VVIT

Guntur

DISCLAIMER

All the information given about the conference is upto date and correct.

IOP Conf. Series: Materials Science and Engineering 1112 (2021) 011002 doi:10.1088/1757-899X/1112/1/011002

Peer review declaration

All papers published in this volume of *IOP Conference Series: Materials Science and Engineering* have been peer reviewed through processes administered by the Editors. Reviews were conducted by expert referees to the professional and scientific standards expected of a proceedings journal published by IOP Publishing.

- **Type of peer review: Double Blind**
- **Conference submission management system:** Online Mode
- **Number of submissions received:** 45
- **Number of submissions sent for review:** 40
- **Number of submissions accepted:**35
- **Acceptance Rate (Number of Submissions Accepted / Number of Submissions Received X 100):**77.7
- **Average number of reviews per paper:** 3
- **Total number of reviewers involved:** 7
- **Any additional info on review process:** No
- **Contact person for queries:** **Dr.M.K.M Farooqui,**
Professor,
Department of Mechanical Engineering, VVIT,
Nambur, Guntur-522001
9885060475
farooqui.mm@rediffmail.com

IOP Conf. Series: Materials Science and Engineering 1112 (2021) 012001 doi:10.1088/1757-899X/1112/1/012001

Analysis of composite raft mount for vibration reduction of a centrifugal pump

Dr. G.Vijay Kumar[1*] and Dr.P.Phani Prasanthi[2]

[1]Professor, Department of Mechanical Engineering, Prasad V Potluri Siddhartha Institute of Technology, Kanuru, Vijayawada, Andhra Pradesh, India.

[2]Associate Professor, Department of Mechanical Engineering, Prasad V Potluri Siddhartha Institute of Technology, Kanuru, Vijayawada, Andhra Pradesh, India.

*Corresponding author:gvkumar@gmail.com

Abstract: The noise induced by centrifugal pump is one of the main concerns in the design and optimization of onboard equipment for ships and submarines. In this work, an attempt has been made to estimate the natural frequency of centrifugal pump with an ordinary conventional steel material foundation. Later, the conventional foundation is replaced with composite material with carbon reinforcement. The vibration studies are accomplished by using FE software ANSYS 18.1. The natural frequencies of centrifugal pump with steel foundation are compared with natural frequencies of centrifugal pump with composite raft foundation. It is observed that the natural frequencies are enhanced by providing composite material foundation.

Keywords: centrifugal pump, Natural frequency, Composite raft foundation, ANSYS

1. Introduction

The possible reason of centrifugal pump and its associated system vibration is unbalance of mechanical rotating parts of pump, driver natural frequency, resonance, flow induced vibrations, hydraulic resonance in piping. The centrifugal pumps are mostly used by many industries and the problem encountered with the centrifugal pump is the generation of sounds during the working. The Rotating parts, unbalanced mass, flowing medium vibrations and resonance in supported piping's are main attributing factors for vibrations in this category.

Many authors showed a path to estimate the natural frequency of the centrifugal pump. Ahmed Ramadhan Al-Obaid [1] showed one solution to the present problem. The authors detected and diagnosed the cavitation of centrifugal pump by applying vibration studies. Speed and flow rate are the important parameters while characterizing the caviatation phenomenon. The natural frequency of the centrifugal pump can be estimated with Finite element method. This is proved by estimating natural frequency of the pump by treating above mentioned parameters [2]. A multistage centrifugal pump includes many parameters that control the performance. An optimized design is required to get good performance. Similar attempts were made to optimize the design using Energy Loss Model(ELM) and Computational Fluid Dynamics (CFD).

Different grid numbers, turbulence models, convergence precisions, and surface roughness are calculated for a typical multistage centrifugal pump was proposed [3]. The demand of reducing vibration and underwater noise caused by the air compressor, the design of an isolation device with the floating raft is described based on the comprehension of the excitation source. FEA models of the vibration isolation system in case of test bank status and installation in ship were established separately by means of the software ANSYS. The vibration modes, the transmission characteristics of the system as well as the shock response were analyzed [4] Flow-induced noise is a significant concern for the design and operation of centrifugal pumps. The negative impacts of flow-induced noise on operating stability have been shown in many cases [5]. With these insights an analysis is performed to understand the effect of composite foundation for centrifugal pump vibrations.

2. Finite Element Modelling

2.1 Geometrical Model

Centrifugal pump is modelled in Pro-e software and the IGS file is imported into ANSYS Workbench 18.1. Fig.2.1 depicts the model geometry considered for the present work. Finite element model of the present study is presented in Fig.2.2. No of elements on the FE model are 1,84,433.

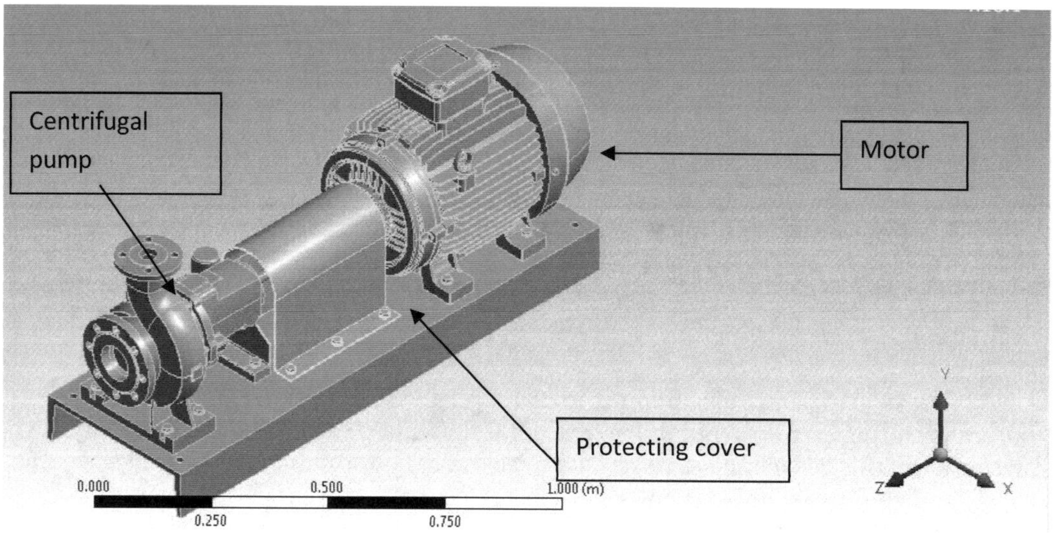

Figure 2.1. Centrifugal Pump Geometrical Model with conventional mount

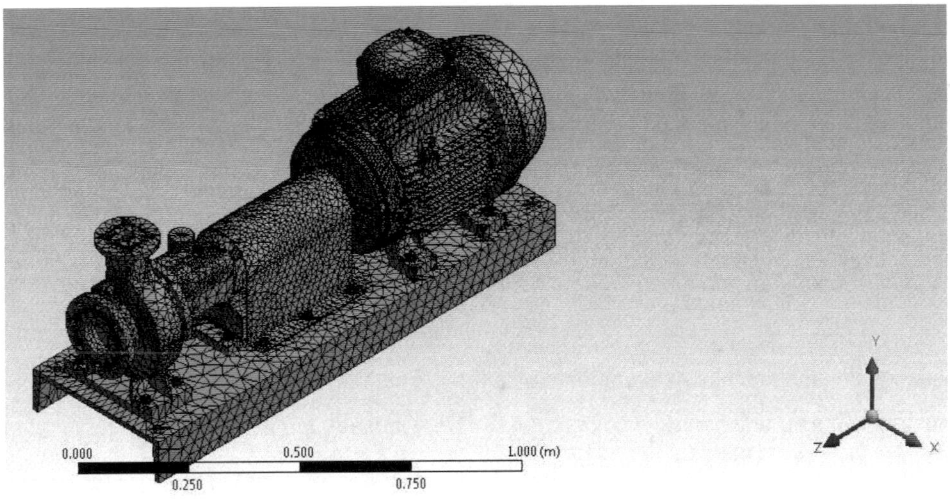

Figure 2.2. Finite element model of centrifugal pump with conventional mount

2.2 Modal analysis

To predict the natural frequencies of centrifugal pump, the bottom part of the supporting member is fixed and mode shapes are extracted. The analysis is performed by considering the all the parts with the stainless steel and protecting cover is made with aluminium alloy. The stainless steel Young's modulus is 210GPa, Poisson's ratio is 0.3.

Loading and boundary conditions: The bottom supporting structure is constrained in all aspects. As the lowest part of the pump is fixed to the ground, the FE model bottom displacement is arrested in all directions i.e X,Y and Z. The fine mesh option is used to perform the present analysis.

Table 2.1. Natural frequency of the centrifugal pump

Mode No	Natural Frequency (Hz)
1	57.471
2	142.91
3	172.18
4	203.14
5	206.85
6	240.76

IOP Conf. Series: Materials Science and Engineering 1112 (2021) 012001 doi:10.1088/1757-899X/1112/1/012001

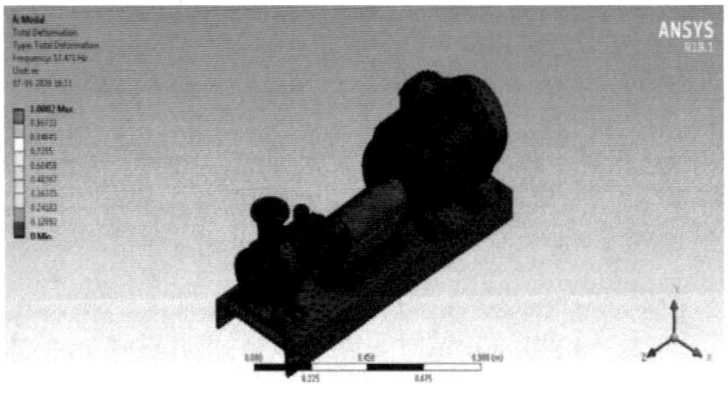

Figure 2.3. Total deformation in mode-1

Figure 2.4. Total deformation in mode-2

Figure 2.5. Total deformation in mode-3

Figure 2.6. Total deformation in mode-4

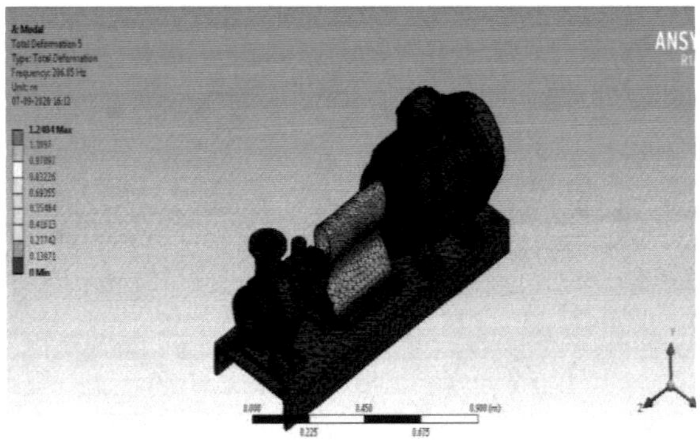

Figure 2.7. Total deformation in mode-5

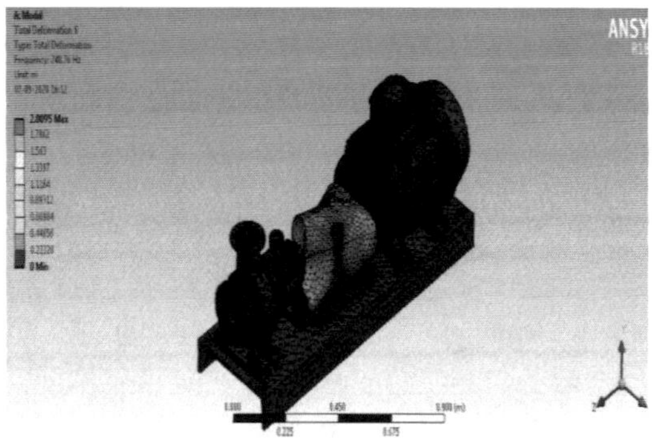

Figure 2.8. Total deformation in mode-6

3. Composite material support for base of the Centrifugal pump:

The present model considered for the study is analysis for composite base. The composite base with 5mm is modelled and carbon unidirectional fiber reinforced epoxy composite properties are attributed and the fiber direction is parallel to the global Z-axis of the pump. The model analysis is performed and the stiffness of the centrifugal pump is estimated from FE results.

3.1 Composite material properties

The carbon epoxy composite [5] is an orthotropic composite in the X-direction

The modulus of elasticity is 61340Mpa,

Ey=61340Mpa and Ez=6900Mpa.

μxy=0.04, μyz=μxy=0.3

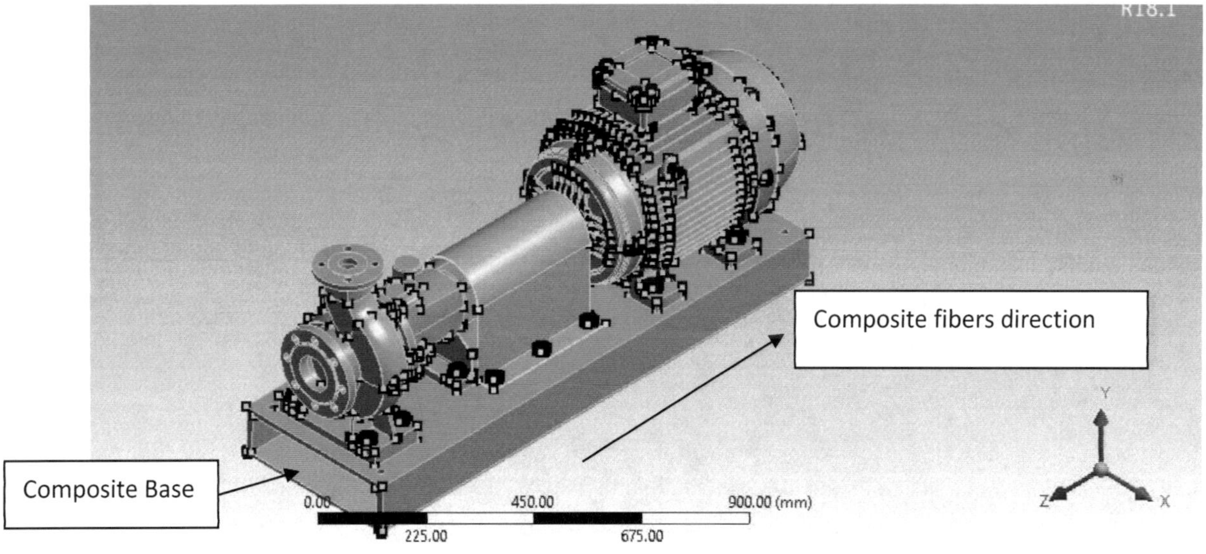

Figure 3.1. Centrifugal pump with composite base mount

Form the study it can be observed that by utilizing composite raft base, the Natural Frequency is increased. The strength to weight and stiffness to weight ratio is more fro composite material as a result, the natural frequency is more for composite materials foundation then conventional materials.

Table 3.1. Natural frequency of the pump with composite base

Mode No	Frequency (Hz)
1	204.54
2	324.18
3	476.74
4	654.7
5	679.41
6	724.16

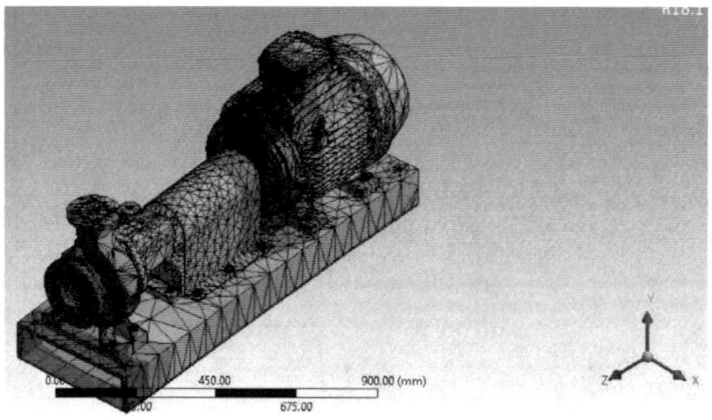

Figure 3.2. Total deformation in mode-1

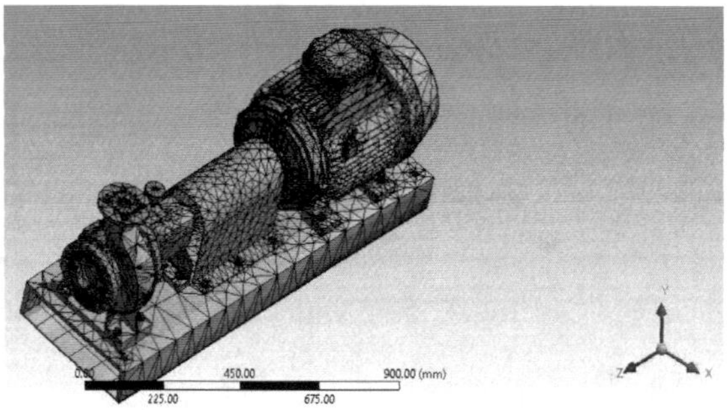

Figure 3.3. Total deformation in mode-6

IOP Conf. Series: Materials Science and Engineering 1112 (2021) 012001 doi:10.1088/1757-899X/1112/1/012001

4. Conclusions

Analysis on the centrifugal pump natural frequencies was conducted out by using ANSYS. The following are the conclusions derived from the present study

- The natural frequency of centrifugal pump with composite foundation is improved when compared to steel foundation
- Composite mount is suggested for reducing the vibration of centrifugal pump
- The composite mount can be used in ships and submarines to reduce the vibration signature and improve stealth

References

[1] Ahmed RamadhanAl-Obaidi, "Investigation of effect of pump rotational speed on performance and detection of cavitation within a centrifugal pump using vibration analysis", Heliyon, Volume 5, Issue 6, June 2019, pp 1-19

[2] Dalia M. El-Gazzar, "Finite element analysis for structural modification and control resonance of a vertical pump", Alexandria Engineering Journal , 2017, 56, pp 695-707

[3] ChuanWang, WeidongShi· XikunWang, XiaopingJiang, YangYang, WeiLiLingZhou, "Optimal design of multistage centrifugal pump based on the combined energy loss model and computational fluid dynamics", Applied Energy, Volume 187, 1 February 2017, pp 10-26

[4] WANG Guo-zhi,FANG Yuan-yuan, "Design of Vibration Isolation System for Air Compressors in Ship and Analysis of Its Anti-Shock Properties", Noise and Vibration Control, 2005

[5] Chang Guo, Ming Gao and Suoying He, A Review of the Flow-Induced Noise Study for Centrifugal Pumps, Appl. Sci. 2020, 10, 1022; doi:10.3390/app10031022.

Investigation of Electro-chemical performance of AZ31 Magnesium alloy – $CaSiO_3$ Metal Matrix Composites in different environments

Penugonda Suresh Babu[1], Y. Pratapa Reddy[1], K. L Narayana[2], M. Kedar Mallik[3]

1,2 – Department of Mechanical Engineering, Koneru Lakshmaiah Education Foundation, Vaddeswaram, Guntur District-522 502.Andhra Pradesh, India
3- Department of Mechanical Engineering, Vasireddy Venkatadri Institute of Technology, Nambur, Guntur District-522 508, Andhra Pradesh, India

Abstract:
The electrochemical behaviour of AZ31 magnesium alloy as matrix made as metal matrix composites (MMC) by adding the reinforcement with various weight percentages of Calcium Silicate ($CaSiO_3$) powder was investigated under the influence of 3.5 percent Sodium Chloride (NaCl) and Simulated Body Fluid (SBF) environments using electrochemical workstation. After experimentation, the results exhibit that the pure magnesium alloy sample with 0% calcium silicate exhibits the corrosion rate higher value (40.562mmpy) compared to the other samples when 4 % of calcium silicate is added as additive which exhibits the lower corrosion rate (25.510mmpy). Similarly, the polarisation resistance is also calculated for the samples and the average value of polarisation resistance for the pure magnesium sample is 153.813mA and the other for other samples it lies in between 140.5 mA to 185.72mA respectively. By experimentation it concludes that addition of reinforcement with various weight percentages plays a significant role in the reduction of corrosion rate of AZ31 magnesium alloy rather than using it in a pure form.

Key Words: AZ31 Magnesium Alloy, Simulated Body Fluid, Electrochemical Behaviour, MMC, NaCl

1. Introduction: Magnesium is the sixth abundant element available in earth from different sources like Magnesite, dolomite, carnallite and sea water, and similarly it is fourth abundant cation in human body. Magnesium has an exclusive and inherent corrosion behaviour and kinetics properties compared to other metals. The degradation of metals generally occurs by three well known reasons like wear, corrosion and fracture. When pure magnesium is exposed to environments like oxygen or water, its surface undergoes chemical reaction by forming an oxide surface film. With its inherent limitations including vulnerability to corrosion resistance, low creep resistance, many research works was focussed on coatings of magnesium to control corrosion rate and addition of reinforcements, fabrication process using conventional methods like casting, forming, process and other fabrication methods like stir casting process, advanced manufacturing methods like additive manufacturing , Wire Arc Additive Manufacturing (WAAM) where deposition of metal is done in layer by layer form. Adding of other elements as additives to pure magnesium alloys series like AZ31, AZ91,AZ91D,AZ92 etc also effects the corrosion rate [1]. Conventional corrosion test methods require vast time and specific environment conditions, where the corrosion test is analysed by both mass and volume loss methods. By technology development, new techniques are introduced to presage the corrosion rate of metal in defined environment which are familiar as electrochemical techniques that consumes less time and cost. These electro-chemical techniques offer an easy study of kinetics of corrosion which is an added advantage and mostly preferred in many labs at present. Corrosion of metals is mainly occurred due to an irreversible oxidation and reduction reactions due to the influence of oxidizing agent. The electrochemical reaction which is a chemical process where the charge transport occurs at the interface of metallic conductor (electrode) and ionic conductor(electrolyte).

By adding different additives as reinforcement in pure magnesium alloy to make them as metal matrix composites will also influence its improvement of specific strength, stiffness, damping behaviour, wear rate, creep and fatigue properties compared by conventional magnesium alloy. The microstructure and mechanical properties like tensile strength, yield strength, ductility, hardness is also improved by adding reinforcement particles [2].

The AZ31 magnesium alloy of commercial grade exhibits much corrosion rate which is main drawback to use the alloy for different applications. The corrosion rate of these alloys is predicted in different environments like various percentages of NaCl , KOH, H_2So_4,KCl, Simulated Body Fluid etc. The pH value of these environments is to be considered while working with different solutions, because by chemical reactions layer is formed on the surface of metals can be observed which acts as protecting layer against corrosion and also material reaction with environment also initiates and enhances the corrosion rate of metal in particular environment[3].

Surface improvements like coatings with other materials can be done which are preferred in various applications where the coatings are done by different process like hot spray process, electroplating, sol-gel coatings, synthetic coatings, other natural and artificial polymer coatings which influences the corrosion resistance of magnesium alloy. Addition of elements like Al, Mn, Ca, Zn, and other rare earth elements shows better results related to corrosion resistance, wear resistance and mechanical properties etc.. the silicon carbide particles added as reinforcement with magnesium alloy enhances the corrosion resistance with the improvement of volume fractions. Calcium phosphate [4,5].

As these magnesium alloys are preferred as bioimplants which are implanted for Orthopaedic, dental applications due to its low density, high strength to weight ratio and easy formability properties. But its biodegradability in some environments controls its usage in vast manner which has high grade demand in designing a controllable in vivo corrosion rate and also improvement in its mechanical properties at present which attracts many researchers in this field. The metal matrix composites from these magnesium alloys cover some percentage of its demand but not able to cover all fields like high temperature conditions [6].

The in vivo corrosion is predicted by experimental analysis for different materials uing the apparatus like potentiostat where alloys , commercial available materials like Al, MS, Copper, SS and other coupons fabricated by different process also can be done by these polarisation techniques like Tafel curves, and impedance spectroscopy analysis, cyclic voltammetry analysis etc with the help of electrochemical workstation under the influence of various environments [7,8]

The prediction the electro-chemical behaviour and tribo-corrosion behaviour of coupons deposited with advanced technologies of additive manufacturing like LENS technology, or WAAM process, where samples are deposited by selected process parameters in a controlled atmosphere to restrict the oxidation of samples from surrounding environment which alters the results related to mechanical, microstructural, and electro-chemical behaviour depend upon process parameters and properties of selected material to fabricate the sample and also selected fabrication process . [9,10,12,14-17]

.The biocorrosion behaviour is generally predicted by using Potentiostat / Galvanostat equipment in defined environments as simulated body fluid (SBF), Hank's solution, etc where the immersion of coupon or exposure of coupon surface area in the media will exhibit the related corrosion rate. The ratio of surface volume to surface area of coupon will effect the corrosion rate of metal in different solutions [11].The post processing technique like heat treatment also effect the properties of coupons made of magnesium alloys, austenitic steels, Co-Cr-W alloys, Co-Cr-Mo alloys, Aluminium alloys etc which are utilised for different applications in various fields. [18-23].

Beyond metals, polymers, ceramics, the fibre material also plays a role which acs as a substrate material in copper coated bio-implantable wide network antennas. These antenna acts as a detecting device and to detect the object by networks [24].

2. Materials and Equipment:

IOP Conf. Series: Materials Science and Engineering 1112 (2021) 012002 doi:10.1088/1757-899X/1112/1/012002

2.1 Specimen Preparation:

AZ31 Magnesium alloy samples are first fabricated by stir casting process by adding various weight percentages of calcium silicate $CaSiO_3$ which are mentioned in Table-2. In AZ31 magnesium alloy, the Aluminum content is at maximum of 3% and Zinc as 1% and the miscellaneous elements contributes at a maximum of 0.40% and the balance is Magnesium. Here in the present work, the samples are fabricated by addition of calcium silicate from 0%, 2%,4%,6% and 8% by altering the magnesium percentage in the alloy, without altering the other elements, Aluminum and Zinc. A general pure AZ31 magnesium alloy chemical composition is mentioned in Table-1. The compositions of calcium silicate added to prepare the specimens and the composition details are mentioned in Table-2. The specimen samples are shown in Fig-1, the samples are S-1 to S-5 from left to right.

Table-1 Chemical Composition of AZ31 Pure Magnesium Alloy (in weight%)

Element	**Mg**	**Al**	**Zn**	**Mn**	**Si**	**Cu**	**Ca**	**Fe**	**Ni**
Min.	---	2.50	0.60	---	---	---	---	---	---
Max.	97	3.50	1.40	0.20	0.10	0.050	0.040	0.005 0	0.005 0

Table-2 Composition of Calcium Silicate in Specimens

Sl. N o	**Sample Numbe r**	**Magnesiu m Percentag e**	**Calcium Silicate Percentag e**	**Aluminiu m Percentag e**	**Zinc Percentag e**	**Total Percentag e of Other Elements**	**Total Percentag e**
1.	S-1	95.6	0				
2.	S-2	93.6	2				
3.	S-3	91.6	4	3	1	0.4	100
4.	S-4	89.6	6				
5.	S-5	87.6	8				

Fig-1 AZ31 Magnesium Samples

From the Table-2, the S-1 sample is a pure magnesium sample with 0% Calcium Silicate added as additive. Similarly, S-2 is added with 2% of Calcium Silicate, S-3 is added by 4%, S-4 has 6% and lastly S-5 is added with 8% Calcium Silicate which are fabricated using Stir casting process.

2.2 Electrolytes:

The media or electrolyte always plays a critical role when the metal coupon is surrounded by a media which initiates the corrosion degradation of metal by chemical process occurred between metal and media. Materials always will be in contact with anyone of the media either a dry atmosphere or a liquid solution.

In the present work, various electrolytes are selected to evaluate the corrosion rate of AZ31 Magnesium Alloy which are Simulated Body Fluid (SBF), Distilled Water, 3.5 percent NaCl Solution.

2.2.1 *Simulated Body Fluid (SBF):* The body fluid with an ion concentration which has nearer composition of human blood plasma with similar human body temperature conditions which is Simulated Body Fluid. The various ions present in SBF are listed in Table-3. The SBF is useful to predict the in-vivo bone bio-activity of material and examination of formation of apatite formation on material by the influence of SBF.

Table-3 Ions present in Simulated Body Fluid

Sl. No	Chemical Name	Abbreviation	Percentage in mmol/L
1.	Sodium Chloride	NaCl	136.8
2.	Sodium Bicarbonate	$NaHCO_3$	4.2
3.	Potassium Chloride	KCl	3.0
4.	Potassium Phosphate	K_2HPO_4	1.0
5.	Magnesium Chloride	$MgCl_2$	1.5
6.	Calcium Chloride	$CaCl_2$	2.5
7.	Sodium Sulphate	Na_2SO_4	0.5
8.	Sodium Nitride	Na_3N	3.08

2.2.2 *NaCl Solution:* Sodium chloride solution with 3.5% M is used to predict the corrosion rate of AZ31 magnesium alloy, where the NaCl solution is prepared by adding 35gms of pure Sodium Chloride form to double time distilled water of 1 litre.

2.3 Equipment:

2.3.1 *Electrochemical Workstation:* The instant corrosion rate prediction is done using an electrochemical workstation (Make: Biologic SAS, France, Model No: SP-300) which is shown in Fig-2 attached with a flat corrosion cell. A flat corrosion cell shown in Fig-3 is filled with electrolyte of approximately 270ml. Specimen and Electrodes (Working Electrode, WE; Counter electrode, CE; and Reference Electrode, RE) are fixed in corrosion cell and the terminals are connected to electrochemical workstation. The specified terminal colours attached to electrodes are Blue colour to Reference Electrode (RE) which is made of platinum, the White colour terminal is connected to Counter Electrode (CE) which is a Ag/AgI_2 and the red colour terminal is connected to Working Electrode(WE) which is the specimen(here, AZ31 Magnesium Alloy). The terminal connections to electrodes are shown in Fig-4.

EC-Lab software (supplied by Biologic SAS, France) is used to predict the corrosion of magnesium samples fabricated by stir casting process added with various compositions of Calcium Silicate ($CaSiO_3$). This software provides a facility to predict Corrosion Rate (CR), polarisation resistance (PR), Cyclic Voltammetry (CV), Potentio-static Impedance Spectroscopy (PEIS) and many more.

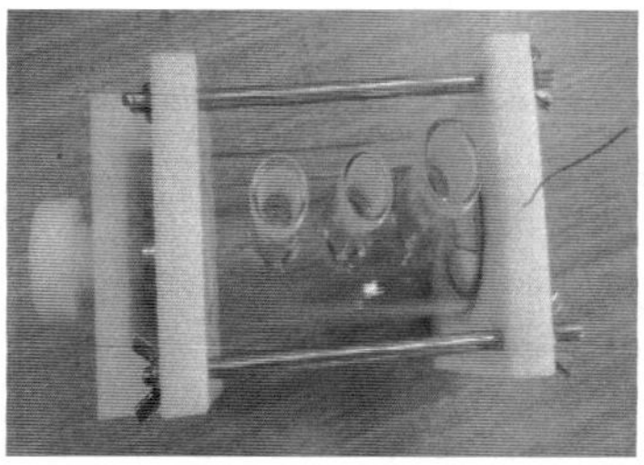

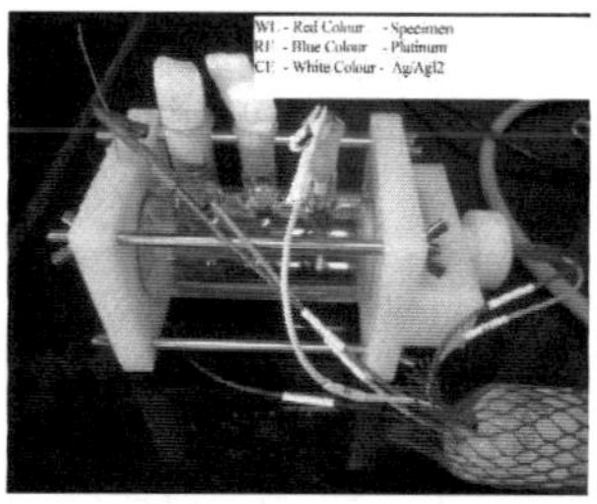

IOP Conf. Series: Materials Science and Engineering 1112 (2021) 012002 doi:10.1088/1757-899X/1112/1/012002

Fig-2 Electrochemical Workstation(SP-300)

Fig-3 Flat Corrosion Cell

Fig-4 Terminal Connections of Electrodes

3. Methodology:

The NaCl Solution is prepared by adding 35 grams of Sodium Chloride powder to double filtered distilled water of 1 litre, where the weight of NaCl powder is measured with weighing balance (MAKE: OHUS, JAPAN) which has 0.0001g accuracy. A quantity of 270ml of NaCl solution, three electrodes (platinum, Ag/ AgI_2, Specimen) are placed in specified places and made the flat corrosion cell ready, and the whole setup of cell is attached to terminals of electrochemical workstation (Potentiostat / Galvanostat) which are shown in Fig-4.

In the present work, corrosion rate in mmpy and polarisation resistance in ohms are predicted for the magnesium samples under the influence of both 3.5%M NaCl and Simulated Body Fluid environments.

The interface of Tafel plot is shown in Fig-5 and the interface of polarisation resistance is shown in Fig-6.

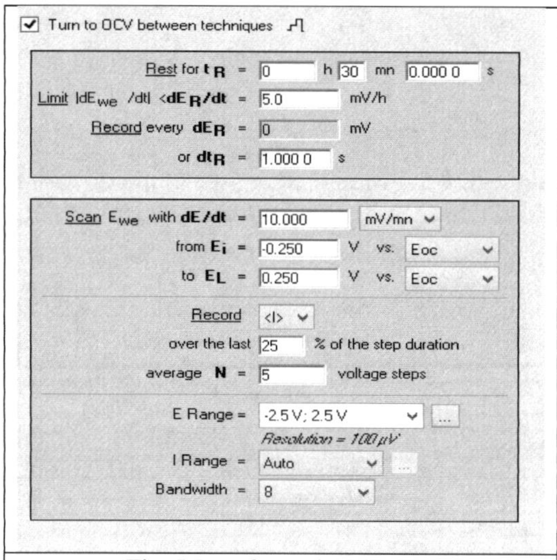

Fig-5 Interface of Tafel plot

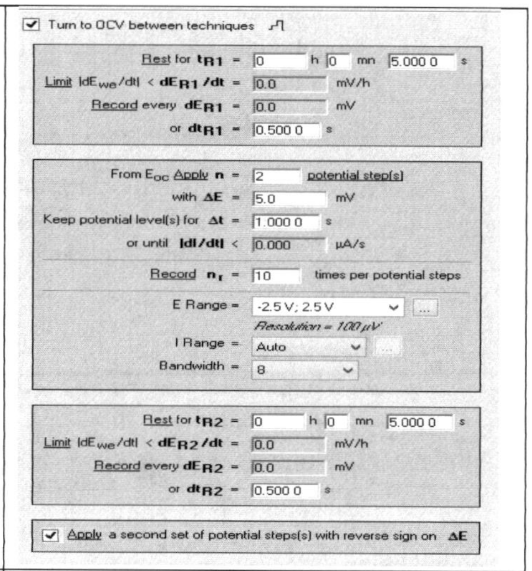

Fig-6 Interface of Polarisation Resistance

3.1 Tafel Plot Curve:

In the interface of Tafel plot curve, Rest for t_R indicates the fixation of time in hours, minutes and seconds, by default it shows 30min of time for t_R, which means the recording of potential stays in rest for 30minutes of time. The Scan Ewe with dE/dt should be mv/sec , otherwise the scan rate of experiment will become slow and takes time to complete, if it is in mV/sec implies that per every second 10 milliVolts (mV) of potential is applied as the external source. This Tafel plot consists oxidation and reduction process which shows as positive and negative regions or anodic and cathodic regions. The beta (β) coefficients for both anodic and cathodic regions are denoted by β_a and β_c which is the slopes of Tafel curve. A typical Tafel plot curve is shown in Fig-7 and the flow chart of Tafel curve is shown in Fig-8.

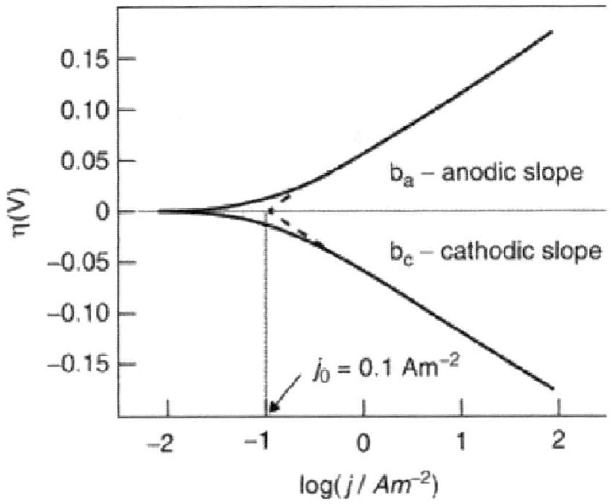

Fig-7 General Tafel Plot Curve

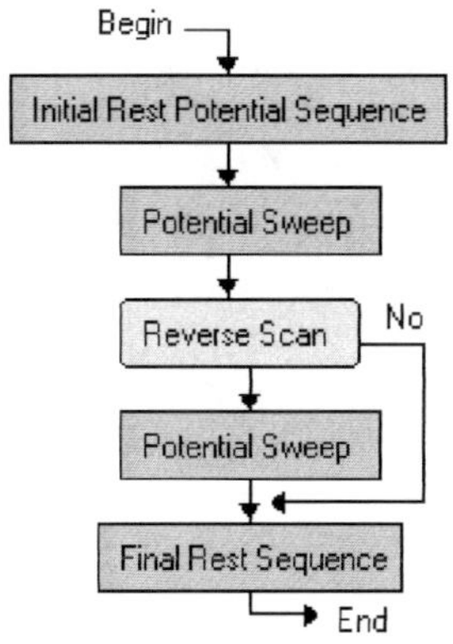

Fig-8 Flow chart of Tafel Plot Curve

3.2 Polarisation Resistance:

It is defined as the slope of the potential- current density for a free corrsion potential indicated by Eq-1.It is also used to find the corrosion rate and beta coefficients of Tafel plot curve.

$$R_p = \frac{dE}{dI} \text{ , dE→0} \quad (1)$$

In this the polarisation resistance is determined by only three or four potential steps, where in general the four step method is used to exhibits maximum accuracy. The general R_p procedure is indicated in flow chart shown in Fig-9 and the four step process is sh own in Fig-10 and three step is mentioned in Fig-11.

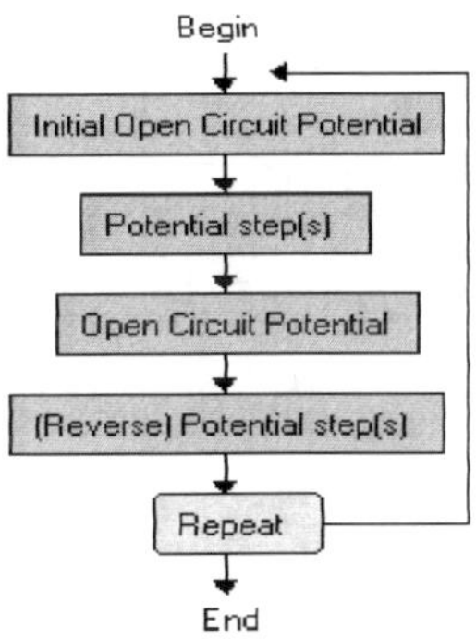

Fig-9 Flow chart of Polarisation R Resistance

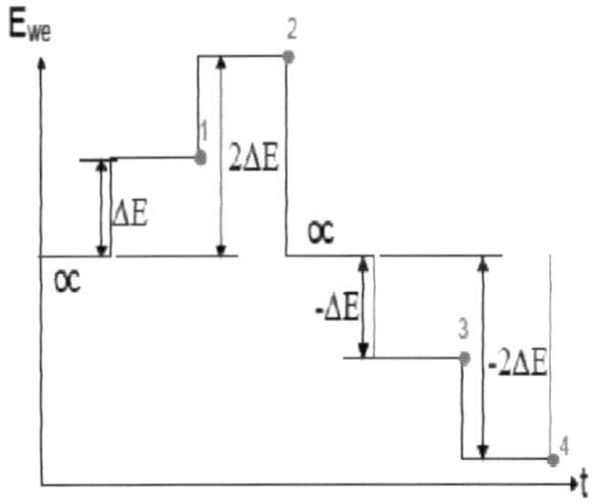

Fig-10 Four Step Method

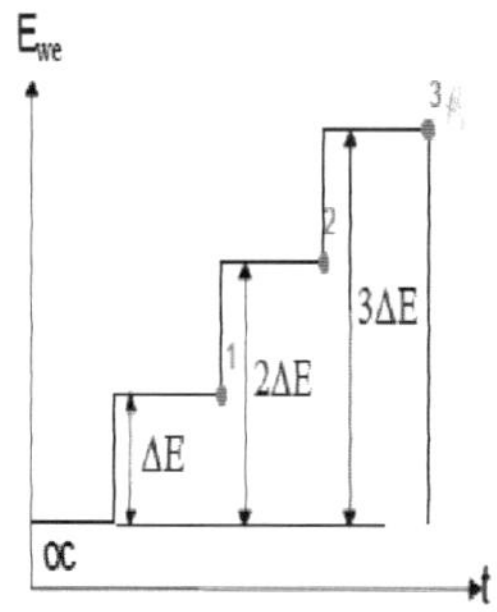

Fig-11 Three Step Method

4. Results and Discussions:

4.1 Curves in 3.5% NaCl Solution

4.1.1 Tafel Curves

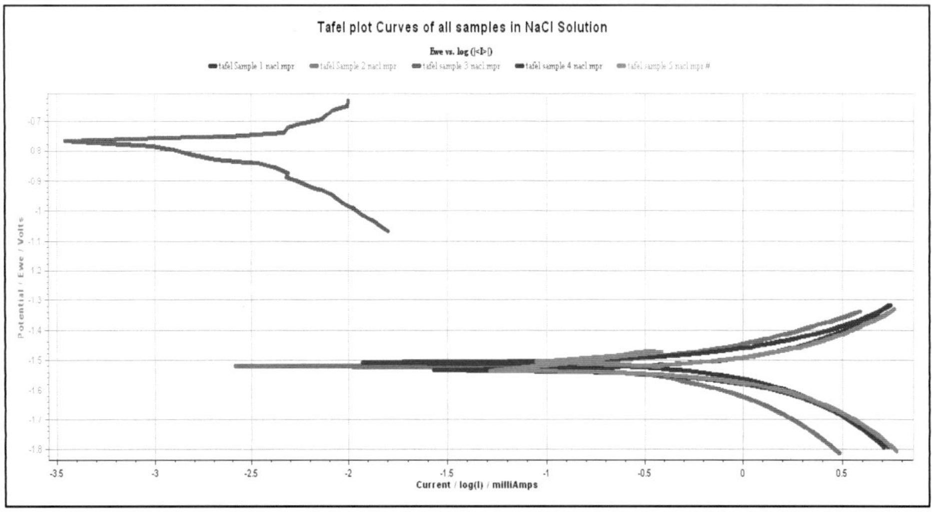

Fig-12 Tafel Plot Curves of All Samples in 3.5% NaCl Solution

4.1.2 Polarisation Resistance:

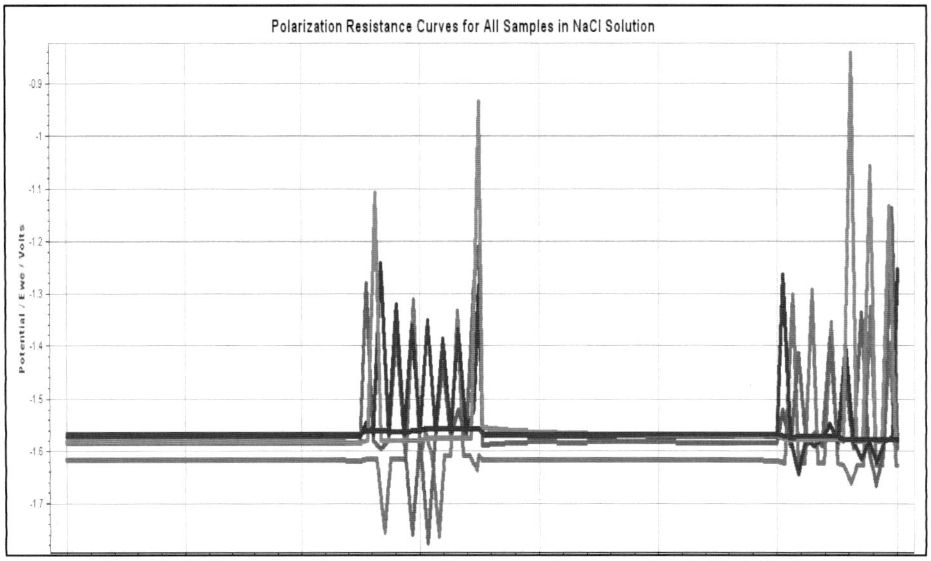

Fig-13 Polarisation Resistance Curve of all sample in NaCl Solution

4.2 Curves in Simulated Body Fluid Solution

4.2.1 Tafel Plot Curves:

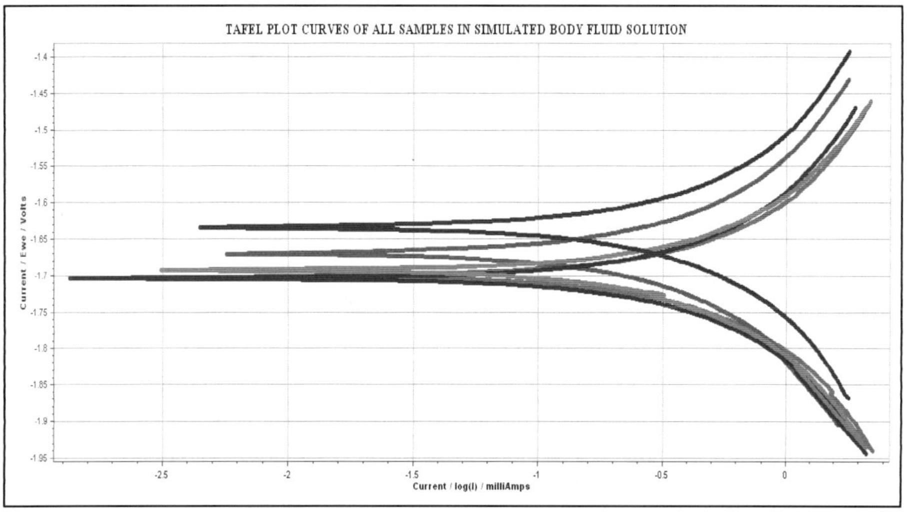

Fig-14 Tafel Plot Curves of All Samples in SBF solution

4.2.2 Polarisation Resistance Curves

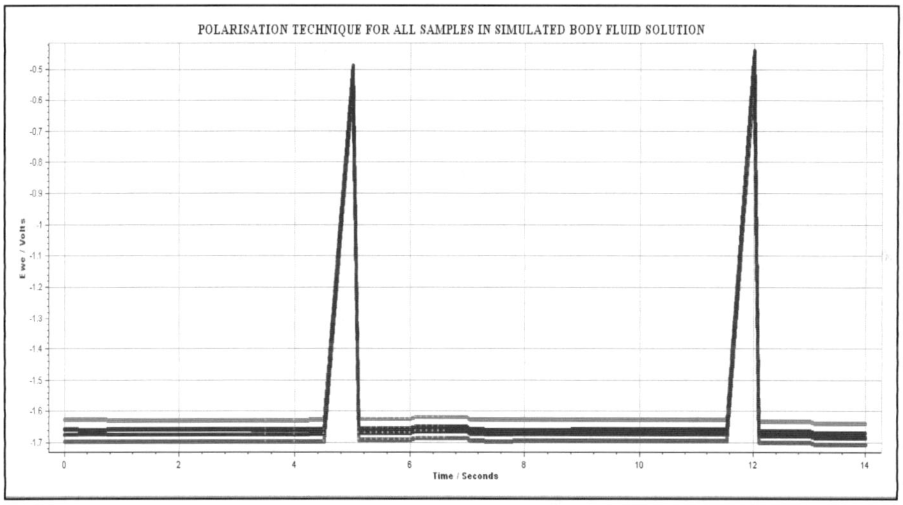

Fig-15 Polarisation Resistance Curves of all samples in SBF Fluid

Table-4 Polarisation resistance Values in SBF fluid

Sl. No	Specimen Number	R_p anodic Value	R_p Cathodic Value	R_p Average Value	I_{corr} Value

1.	S-1	159.1	148.53	153.81	-0.056.669
2.	S-2	162.5	141.56	152.03	-0.053.125
3.	S-3	160.8	150.09	155.44	-0.062.258
4.	S-4	155.7	163.36	159.53	-0.083.255
5.	S-5	148.9	175.47	162.18	-0.096.892

Table -5 Tafel Curve Values in SBF Fluid

Sl. No	Sample Number	Corrosion Potential (E_{corr}) Value	Corrosion current Density (I_{corr}) Value	Tafel Slopes		Corrosion Rate	
				Beta Anodic Coefficient	Beta Cathodic Coefficient	in millimetres per year	in mils per year
		mV	µA	(β_a) in mV	(βc) in mV	(mmpy)	(mpy)
1.	S-1	-1641.326	283.753	245.9	278.1	40.562	1486.236
2.	S-2	-1702.501	482.886	317.0	328.9	31.569	1102.354
3.	S-3	-1669.949	360.874	324.2	310.8	25.510	895.365
4.	S-4	-1701.036	416.036	310.8	315.8	29.881	1059.368
5.	S-5	-1703.622	431.555	299.2	314.8	35.654	1384.495

5 Conclusions:

After completion of experiment , the following conclusions are drawn

- The sample(S-1) exhibits less value of beta anodic and beta cathodic where these ar the slopes of anodic and cathodic values of Tafel curve.
- The sharp point in the curve is actually the point where the current reverses polarity as the reaction changes from cathodic to anodic.
- The log values in current are provided which are used to plot the Tafel plot curve because of wide of input current values.
- The Sample S-2 Exhibits more I_{corr} value than other samples, where this I_{corr} value is needed to calculate the corrosion rate of metal in mmpy or in mpy
- The corrosion rate of sample S-1 which is a pure AZ31 magnesium alloy sample exhibits more corrosion rate than other samples.
- In general the cathodic process the metal receives more electrons from electrodes or from other source in electrolyte by the supply of external current which initites the release of electrons and shifts the potential of metal or slow down anodic process and speeds up the cathodic or vice-versa.
- The average R_p value is more for sample S-5 thsn other samples.

References:

[1] M. Esmaily, J. E. Svensson, S. Fajardo, N. Birbilis, G.S. Frankel, S. Virtanen, R. Arrabal, S. Thomas, L.G. Johansson, (2017), Fundamentals and advances in magnesium alloy corrosion, *Progress in Materials Science*, 89, 92–193.

[2] Abhijit Dey and Krishna Murari Pandey, (2015) Magnesium Metal Matrix Composites - A Review, *Revisions of Advanced Material Science*,42,58-67.

[3] Somi Doja, Lukas Bichler, Simon Fan, (2017), Corrosion Behavior of AZ31 Magnesium Alloy in Highly Alkaline Environment, Acta Metall. Sin. (Engl. Lett.),22 Feb.1-9.
[4] R. Arrabal *et al.*, "Corrosion behaviour of a magnesium matrix composite with a silicate plasma electrolytic oxidation coating," *Corrosion Science*, vol. 52, no. 11, pp. 3738–3749, 2010.
[5] S. Agarwal, J. Curtin, B. Duffy, and S. Jaiswal, "Biodegradable magnesium alloys for orthopaedic applications: A review on corrosion, biocompatibility and surface modifications," *Materials Science and Engineering C*, vol. 68, pp. 948–963, 2016.
[6] M. Basista, "Recent advances in research on magnesium alloys and magnesium – calcium phosphate composites as biodegradable implant materials," vol. 0, no. 0, pp. 1–23, 2016.
[7] Y. Pratapa Reddy, Dr. K. L Narayana, Dr. M. Kedar Mallik, (2019), Electro-chemical behaviour of different metals in Sodium Chloride solution, International Journal of Emerging Trends in Engineering Research, 7(11), 715-719.
[8] P. S. Babu, K. L. Narayana, and V. V. Kondaiah, (2020). "Corrosion behaviour of stir casted AZ31 magnesium alloy-$CaSiO_3$ composites in an aqueous solution," *International Journal of Advanced Science and Technology*, Vol. 29, No. 5, 2333–2345,
[9] G. Suresh, K. L. Narayana, and M. K. Mallik, (2019). "Hardness, wear and corrosion properties of Co-Cr-W alloy deposited with laser engineered net shaping in medical applications," *International Journal of Innovative Technology and Exploring Engineering*, 8(8), 1951–1955,
[10] M Kedar Mallik, K L Narayana, (2019) "Impact Strength and Fracture Analysis of Co-Cr-Mo Alloy Deposited with Laser Engineered Net Shaping - An Additive Manufacturing Technology," World Congress on Engineering, Imperial College, London,
[11] M. S. Uddin, C. Hall, and P. Murphy, "Surface treatments for controlling corrosion rate of biodegradable Mg and Mg-based alloy implants," *Science and Technology of Advanced Materials*, vol. 16, no. 5, p. 53501.
[12] N. Nassif and I. Ghayad, "Corrosion Protection and Surface Treatment of Magnesium Alloys Used for Orthopedic Applications," vol. 2013, 2013.
[13] A. Dey and K. M. Pandey, "Magnesium metal matrix composites-a review," (2015).*Reviews on Advanced Materials Science*, vol. 42, no. 1, pp. 58–67,
[14] Ganzi Suresh, K L Narayana, M Kedar Mallik, (2018), "Electro-Chemical Behaviour of LENS™ Deposited Co-Cr-W Alloy for Bio-Medical Applications" International Journal of Mechanical and Production Engineering Research and Development, Special Issue, , 41-52
[15] Ganzi Suresh, K L Narayana, M Kedar Mallik, (2018), "Processing & Characterization of LENS™ Deposited Co-Cr-W Alloy for Bio-Medical Applications" International Journal of Pharmaceutical Research, 1(10), , 276-285
[16] Ganzi Suresh, K L Narayana, M Kedar Mallik, (2018) "Bio-Compatible Processing of LENS™ Deposited Co-Cr-W alloy for Medical Applications" International Journal of Engineering & Technology, 7(2), 362-366
[17] Ganzi Suresh, K L Narayana, M Kedar Mallik, (2019), "Hardness, Wear and Corrosion Properties of Co-Cr-W Alloy Deposited with Laser Engineered Net Shaping in Medical Applications", International Journal of Innovative Technology and Exploring Engineering, 8(8), 2951-2955.
[18] M Kedar Mallik, Mitun Das, Vamsi Krishna Balla, Ch Srinivasa Rao, V. V. S. Kesava Rao, "Laser deposited Co-Cr-Mo alloy, microstructure, wear and electrochemical properties" Journal of Material Research (Material research society) 17(29), 2021-2027.
[19] M Kedar Mallik, Ch Srinivasa Rao, V. V. S. Kesava Rao "Effect of Process Parameters on the Tensile Properties of LENS deposited Co-Cr-Mo alloy" International Journal of Manufacturing Technology and Management, 6(32), 509-516.

[20] K L Narayana, M Kedar Mallik,(2019) "Comparison of Properties of Laser Deposited Co-Cr-Mo and C-Cr-W alloys, World Congress on Engineering, Imperial College, London,

[21] M Kedar Mallik, Mitun Das, Vamsi Krishna Balla, Ch Srinivasa Rao, V. V. S. Kesava Rao (2015), "Additive Manufacturing of Co-Cr-Mo alloy influence of heat treatment on microstructure, tribological and electrochemical properties", Computer-aided and digital manufacturing technologies , Frontiers in mechanical engineering, 2(1).

[22] M Kedar Mallik, Ch Srinivasa Rao, V. V. S. Kesava Rao, "Microstructure and tribological properties of weld deposited Co-Cr-Mo alloy" Journal of Manufacturing Science and Production (De Gruyter), 2(15), 239-241.

[23] M Kedar Mallik, Ch. Srinivasa Rao, V. V. S. Kesava Rao, Abdul Munaf Sk., "Effect of Heat Treatment on Corrosion Behaviour of Weld deposited Co-Cr-Mo Alloy" ARPN Journal of Engineering and Applied Sciences, 20(11), 12188-12191.

[24] D. Ram Sandeep, N. Prabakaran, B. T. P. Madhav, K. L. Narayana, Y. Pratapa Reddy, (2020), Semicircular shape hybrid reconfigurable antenna on Jute textile for ISM, Wi-Fi, Wi-MAX, and W-LAN applications, International Journal of RF and Microwave Computer Aided Engineering, Wiley, July,1-14.

Optimization of dry sliding wear parameters of Al4Mg system reinforced with high strength alloy particulate (HSAp)

Ch. Kishore Reddy[1], M. Gopi Krishna[2], Rahul[3], V.S.N. Venkata Ramana[4], K. Sri Ram Vikas[5],

[1,5]Department of Mechanical Engineering, Prasad V. Potluri Siddhartha Institute of Technology, Vijayawada, 520007, Andhra Pradesh, India.
[2]Department of Mechanical Engineering, ANU College of Engineering & Technology, Acharya Nagarjuna University, Guntur, 522 510, Andhra Pradesh, India.
[3]Department of Mechanical Engineering, Chaitanya Bharathi Institute of Technology(A), Gandipet, Hyderabad,500075, Telangana, India.
[4]Department of Mechanical Engineering, GITAM Institute of Technology, GITAM (Deemed to be University), Visakhapatnam, Andhra Pradesh, 530045, India.
*Corresponding author: kishorereddymtech@gmail.com

Abstract: The dry sliding wear behaviour of Al4Mg binary alloy and Al4Mg reinforced with Al20Cu20Mg composites was examined using pin on disc apparatus. The composites were synthesized by the reinforcement of different weight % (5, 10 and 15) of the high strength alloy particulate (HSAp) in the base binary alloy Al4Mg using stir casting technique followed by hot extrusion. In this way, composites were prepared with metal particulate reinforcement which are termed as metal-metal composites which is a novel idea instead of reinforcement of ceramic particulate. These extrudates of non-reinforced base alloy and composites were characterized to optimise dry sliding wear parameters, wear resistance and coefficient of friction under the load conditions of 0.5, 1 and 1.5 kgf at the sliding velocities is of 100, 120 and 140 metre per second and sliding times of 15, 30 and 45 seconds respectively. The wear parameters were optimised using Grey relational analysis (GRA) and ANOVA techniques and obtained the optimal combination of input parameters.

Keywords: High strength alloy particulate (HSAp), metal-metal composites, stir casting, hot extrusion, GRA, ANOVA.

1. Introduction

'Wear' is the property that acts on both ends positive and negative when considered in engineering applications. For anecdotal reasons the wear property intentionally induced or reduced depending on the application. Wear may be because of corrosion, adhesion, and abrasion. Each of these has their own advantages and disadvantages that lead to benefits and troubles depending on the type of application for a particular purpose. The benefits or problems may be influenced by the factors type of material, type of lubrication and amount of surface finish. Among Metal matrix composites (MMCs), Aluminium matrix composites (AMCs) hold more than 69% by weight for industrial functions as they holds excellent mechanical and thermal properties couples with better tribological properties [1-2]. Under dry lubrication conditions, the Aluminium based matrix composites are exhibiting poor wear resistance. These are confined to very limited applications due to this inadequacy [3]. It is examined that the strength of matrix alloy AA7075 was enhanced by 6%, on addition of hybrid particulate (SiC+ Al2O3) [4]. Among the hybrid reinforcement particulates RHA (Rice Husk Ash), Al2O3 and graphite in AA6063, the graphite influenced more on the wear rate. As graphite composition increases from 0% to 1.5% the rate of decreases drastically, as the graphite particulates are soft in nature [5]. The AMMCs with inclusion of hard metallic or ceramic particulates are greatly influenced by wear mechanisms like abrasive and adhesive [6-8]. In the global scenario, due to increasing competition for manufacturability with reduced weight to strength ratio regarding engine parts of automobiles, the research in the AMMCs area is still running after new and innovative techniques for better performance [9]. The need is still in take away situation and the lacunae are to be bridged for the emerging applications and needs in all the related industries.

The wear properties of AMMCs are greatly influenced type of reinforcement particulates and volume and size of the particulate as well. Depending on the type of application and requirement of level of attainment of particular outcome from the material the type of reinforcement and its level of composition may be tailored. In the case of Al6061 matrix the hybrid composite is examined with inclusion of Silica and alumina particulates and the rate of wear was substantially decreased with gradual increase of volume of content of reinforcement particulate due to formation of high hard phases [10]. The friction coefficient is increasing with increase in the percentage of particulate in hybrid composite, whereas, the wear rate is decreasing. The higher values of permanent strain (plastic strain) due to TiO2 particulate which is adhesive in nature leads to higher values of frictional coefficient [11]. The distribution fashion as well as the volume and size of the reinforcement particulates are the determinants of stirring speed, impeller size, and holding temperature [12]. Depending on the type of matrix and its properties like viscosity and wettability and chemical composition and the type of reinforcement material the stirring speed and stirring time vary instant to instant. These parameters govern the vortex which results in required richness of mixing of matrix and reinforcements. We may also use the optimization techniques like ANOVA and Artificial neural networks for a better combination of stirring parameters to obtain the required output properties for the prepared composite. Using the ANOVA technique, it is concluded that at 550 rpm stirring speed and 12% weight percentage of SiC+Al2O3 particulate in AA7075 matrix will result in optimal or better strength properties [13]. Among industrial, automobile, aviation, and shipping, the wear property plays a vital role for the selection and design of materials [14]. The rate of wear for matrix or base alloy is uniform till 200N, later suddenly raised above the load 200N. But, for the composites the transition load point is higher than that of 200N as there is no sudden raise in rate of wear is observed [15]. Improved hardness is the main reason for enhancement of wear resistance with the inclusion of TiC particles, which was illustrated in the figure 5 for various percentage compositions of TiO2 [16]. The improvement of volume of wear of the material with load particularly for composite and this is because of the heat generated between the sliding surface and composites surface as heat softens materials which in turn causes easy aberration of material surface [17]. Speed increases volume loss also increases from the compositions. At higher speeds degenerated heat leads to soften the bond between the matrix and reinforcement which is the cause for higher laws of where volume at higher speeds sliding [18]. At the beginning of sliding the surface gruffness is having higher sharpness for both pin and disc which results in temperature rise which leads to adhesion welding open material with disc and hence increase in rate of wear. But as the sliding distance further increases the effluent debris which is formed due to the wear of pin which causes skidding of pin in the air track and hence decrease in the rate of wear.

2. Materials and Methods

In this investigation Al4Mg is matrix material, acknowledged for high compatibility for work hardening. When mixed with copper the alloy has most significance in aircraft industry, auto mobile industries and in many other load applications [19]. From the work done by M.Gopikrishna et al., the ternary alloy Al20Cu20Mg which is high dense and high strength material was chosen as the reinforcement [20]. Both binary and ternary alloys were casted and the particulate from Al20Cu20Mg was prepared from the conventional filing technique and obtained HSAp of average size of particle 200 micron. The composites were casted into billets of 60mm diameter using stir casting technique. The casted billets of composites were homogenized for 24 hours at a temperature of 1500C. The Billets were then extruded in to rods of 16mm diameter using a hot extrusion press of 200T capacity. The temperature of billet and the pressure pad both were maintained 4000C during extrusion process. The extrudates were tested for wear characteristics.

Dry sliding wear test was conducted using DUCOM TR-20 WEAR TESTING MACHINE pin-on-disc apparatus in ASTM: G99 standards. Standard wear specimens of diameter 4mm and 30mm length were cut from both base alloy and composites using wire EDM process. The cut specimens were shown in figure 1. The friction coefficient is the output parameter observed by considering the input parameters load, speed and sliding time along with the percentage of reinforcement.

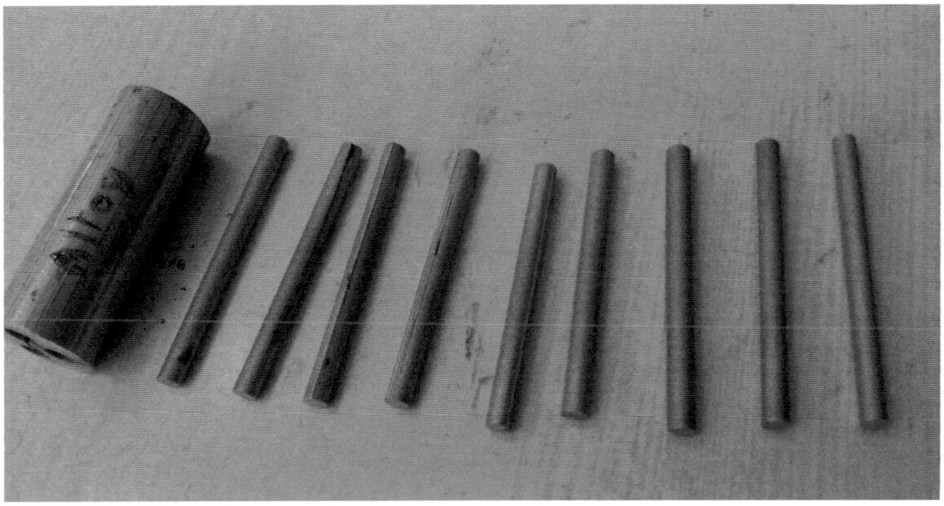

Fig. 1 Wear samples cut using EDM

3. Results and Discussions

3.1 Grey Relational Analysis (GRA):

The coefficient of friction is optimization through GRE relational analysis. The experiment is designed as per the orthogonality, the three factors are selected at three levels as shown in Table 1. Table 2 shows the analytical responses.

Table 1: Input levels of the process parameters

Parameter	Level1	Level2	Level3
Velocity (m/min)	100	120	140
Load (N)	4.92	9.81	14.72
Time (min)	15	30	45

Table 2: Analytical table of responses

Exp	VEL	LOAD	TIME	Coefficient of Friction - COF			
				Alloy	5%P	10%P	15%P
1	1	1	1	0.026	0.042	0.05	0.011
2	1	2	2	0.023	0.016	0.023	0.023
3	1	3	3	0.056	0.079	0.12	0.033
4	2	1	2	0.049	0.263	0.127	0.014
5	2	2	3	0.034	0.158	0.096	0.011
6	2	3	1	0.029	0.032	0.04	0.043
7	3	1	3	0.018	0.178	0.055	0.228
8	3	2	1	0.019	0.035	0.136	0.051
9	3	3	2	0.018	0.086	0.067	0.173

The grey rational coefficients (GRF) calculated form S/N ratio of each response parameter, grey rational grades (GRG) also calculated for each experiment and the ranks are assigned.

$$\eta = -10\log_{10}\left(\frac{1}{n}\sum_{i=1}^{n} y_{ij}^{2}\right) - (1)$$

$$\eta = -10\log_{10}\left(\frac{1}{n}\sum_{i=1}^{n} \frac{1}{y_{ij}^{2}}\right) - (2)$$

3.1.1 Grey Relational Analysis (GRA) for coefficient of friction:

In GRA, grey relations connected to the test information identified with quality attributes, the aftereffects of which used to acquire the grey relational grades consequently to rank every information arrangement. In the grey relations, the S/N ratio normalized as N-S/N ratio ($0\leq \geq 1$) by utilizing the following equation to keep away from the impact of receiving diverse units and to decrease the variability.

$$\text{S/N ratio} = \frac{(\eta_i) - \min(\eta_i)}{max(\eta_i) - min(\eta_i)} \;\ldots\ldots\; (3)$$

$$f = \max\left(\eta_{(\eta_i)} - (\eta_{\eta_i})\right) \;\ldots\ldots\ldots\; (4)$$

$$\text{GRC} = \frac{min(fDc_i) + f * max(Dc_i)}{(Dc_i) + f * max(Dc_i)} \;\ldots\; (5)$$

$$\text{GRG} = \frac{1}{m}\sum_{k=1}^{m} GRC_{ik} \;\ldots\ldots\; (6)$$

Where m= number of output parameters

In the Grey relations, the quality loss function 'f 'calculated by using the N-S/N ratio, Computed the grey relational coefficient (GRC) form the quality loss function f, Where Dc = Distinguishing coefficient, which defined in the range $0\leq Dc\leq 1$. Table 3 shows the S/N ratios. In table 4 and table 5 GRC and GRG were calculated and tabulated respectively.

Table 3: Calculation of S/N ratio

Exp	Coefficient of Friction - COF			
	Alloy	5%P	10%P	15%P
1	4.293	5.680	8.874	9.897
2	4.883	8.404	9.897	10.458
3	5.352	9.897	10.458	10.458
4	5.193	6.745	7.744	7.535
5	6.558	8.874	10.458	9.897
6	7.535	9.897	11.057	13.979
7	5.352	6.745	7.959	8.874
8	5.514	9.119	9.897	10.458
9	5.680	10.458	10.458	13.979

Table 4: Grey relational Coefficient (GRC)

	COF			
Exp	Alloy	5%P	10%P	15%P
1	1.000	1.000	0.595	0.577
2	0.733	0.467	0.435	0.524
3	0.605	0.362	0.379	0.524
4	0.643	0.692	1.000	1.000
5	0.417	0.428	0.379	0.577
6	0.333	0.362	0.333	0.333
7	0.605	0.692	0.885	0.706
8	0.570	0.410	0.435	0.524
9	0.539	0.333	0.379	0.333

Table 5: Grey relational grade (GRG)

	GRG			
Exp	Alloy	5%P	10%P	15%P
1	0.781	0.667	0.479	0.455
2	0.751	0.508	0.472	0.459
3	0.802	0.681	0.467	0.484
4	0.491	0.641	0.743	0.716
5	0.385	0.549	0.656	0.742
6	0.551	0.590	0.667	0.667
7	0.469	0.536	0.609	0.591
8	0.489	0.427	0.417	0.463
9	0.528	0.407	0.389	0.462

The obtained level means are tabulated in table 6. In grey relational analysis, higher value of level means better quality. So, optimal conditions using GRA for Alloy A1, B3, C1, for 5% A1, B3, C1 for 10% A2, B1, C3 for 15% A2, B1, C3 respectively.

Table 6: Level means of GRG

For Alloy:						
Parameter	Level1	Level2	Level3	Delta	Rank	Optimum level
Velocity (A)	**0.778**	0.476	0.496	0.302	1	L1
Load (B)	0.58	0.542	**0.627**	0.085	2	L3
Time (C)	**0.607**	0.59	0.552	0.055	3	L1
For 5%:						
Parameter	Level1	Level2	Level3	Delta	Rank	Optimum level
Vel (A)	**0.619**	0.593	0.457	0.162	1	L1
Load (B)	0.614	0.542	**0.627**	0.085	2	L3
Time (C)	0.561	0.519	**0.589**	0.07	5	L1
For 10%						
Parameter	Level1	Level2	Level3	Delta	Rank	Optimum level
Vel (A)	0.473	**0.689**	0.472	0.217	1	L2
Load (B)	**0.611**	0.515	0.508	0.103	2	L1
Time (C)	0.521	0.535	**0.578**	0.057	3	L3
For 15%:						
Parameter	Level1	Level2	Level3	Delta	Rank	Optimum level
Vel (A)	0.466	**0.708**	0.505	0.242	1	L2
Load (B)	**0.588**	0.554	0.538	0.05	3	L1
Time (C)	0.528	0.546	**0.606**	0.078	2	L3

3.2 ANOVA

3.2.1 Check for adequacy of model:

The normality of the data is shown in figure 2. The contribution of the three process parameters velocity, load and time on the responses, i.e., coefficient of friction has been calculated with the aid of ANOVA.

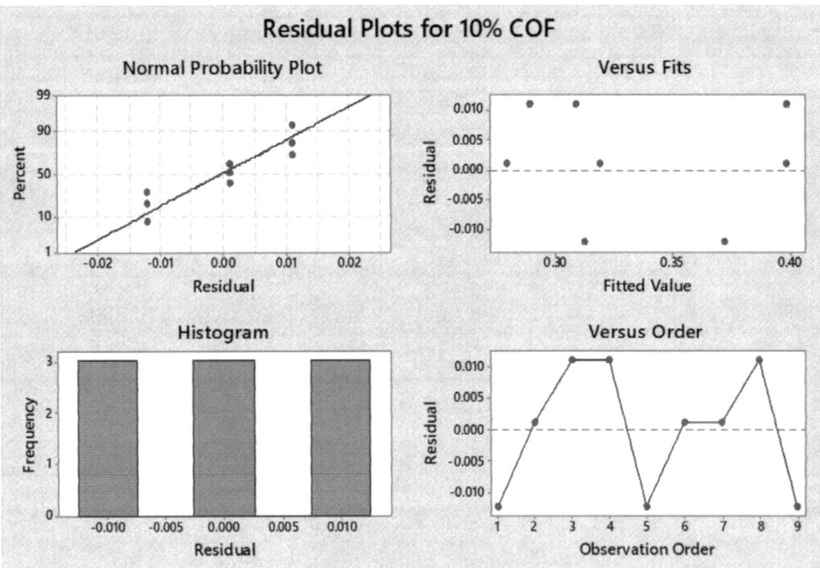

Fig.2 Normality plot for input data analysis.

Table 7: ANOVA table for GRG Values

			Alloy:			
Source	DF	SS	MS	F value	P Value	Contribution
Velocity	2	0.17151	0.08575	109.27000	0.00900	3.72835
Load	2	0.01098	0.00549	7.00000	0.12500	0.23874
Time	2	0.00472	0.00236	3.01000	0.24900	0.10270
Error	2	0.00157	0.00079			0.03413
Total	8	0.18879	0.09439			
			5%			
Source	DF	SS	MS	F value	P Value	Contribution
Velocity	2	0.04542	0.02271	14.14000	0.06600	58.48483
Load	2	0.02163	0.01082	6.73000	0.12900	27.85188
Time	2	0.00740	0.00370	2.30000	0.30300	9.52773
Error	2	0.00321	0.00161			4.13555
Total	8	0.07767	0.03883			
			10%			
Source	DF	SS	MS	F value	P Value	Contribution
Velocity	2	0.093807	0.046904	11.32	0.081	73.76350512
Load	2	0.019837	0.009918	2.39	0.295	15.5975278
Time	2	0.005242	0.002621	0.63	0.613	4.121911712
Error	2	0.008289	0.004144			6.517055373
Total	8	0.127175	0.063587			
			15%			
Source	DF	SS	MS	F value	P Value	Contribution
Velocity	2	0.101168	0.050584	144.22	0.007	87.48605574
Load	2	0.003853	0.001926	5.49	0.154	3.331056132
Time	2	0.00997	0.004959	14.21	0.066	8.575826495
Error	2	0.000702	0.000351			0.607061631
Total	8	0.115692	0.05782			

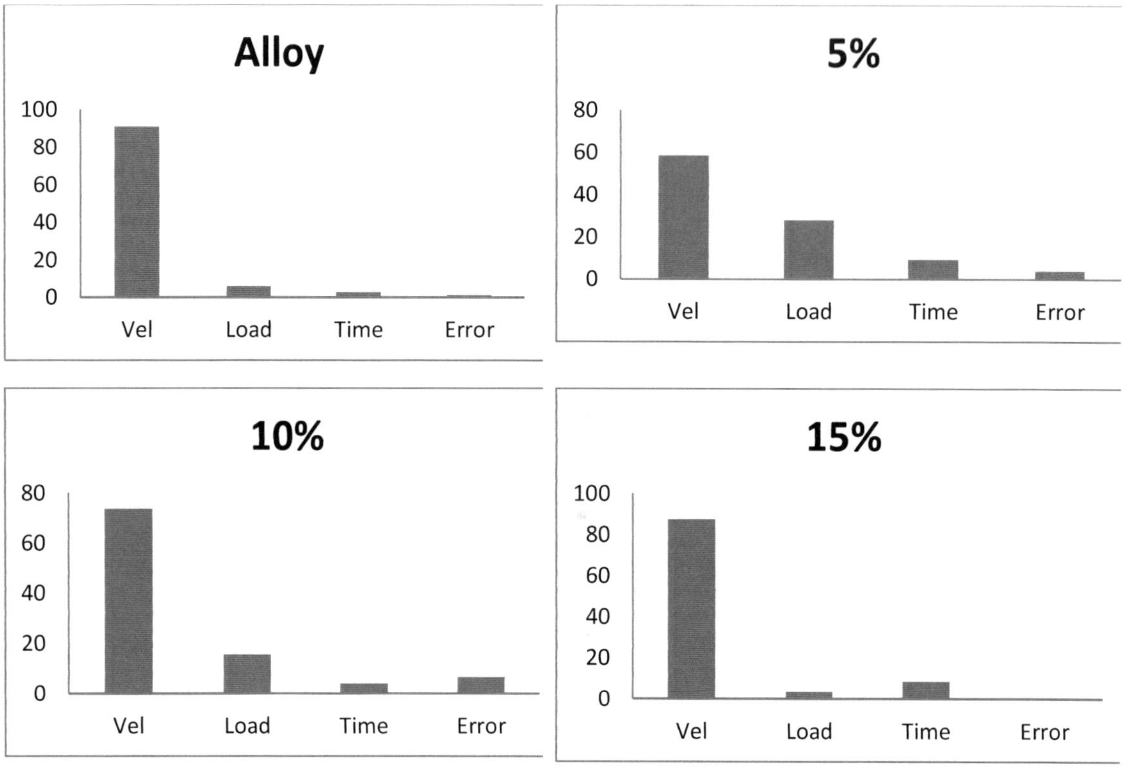

Fig.3 Percentage contributions of input parameters for GRA method.

The ANOVA results are tabulated in Table 7. The p-value is less than 5%. The confidence level for the prepared model has 95%. Also, the p-value is less than 0.05 compared to the F-value. It confirms that the developed models are adequate, and the predicted values are in good agreement with the measured data which can be depicted in figure 3.

4. Conclusions

- Optimal conditions using GRA for Alloy A1,B3,C1, for 5% A1,B3,C1 for 10% A2,B1,C3 for 15% A2,B1,C3 respectively.
- The contribution of each input parameters i.e. Velocity (V), Load (L) and Time (T), on Coefficient of friction (COF) has been calculated by using Analysis of Variance (ANOVA) and GRA method.
- For alloy, it is found that velocity (V) has the highest contribution 3.7 % followed by Time (T) and Load (L) having contribution of 0.23% and 0.10 % respectively.
- For 5% composite, it is found that velocity (V) has the highest contribution 58 % followed by Time (T) and Load (L) having contribution of 27% and 9.5 % respectively.
- For 10% it is found that velocity (V) has the highest contribution 73% followed by Time (T) and Load (L) having contribution of 15% and 3.3 % respectively.
- For 15% it is found that velocity (V) has the highest contribution 87.48 % followed by Time (T) and Load (L) having contribution of 8.5% and 3.3 % respectively.

IOP Conf. Series: Materials Science and Engineering 1112 (2021) 012003 doi:10.1088/1757-899X/1112/1/012003

References

[1] Prakash, J. Udaya, et al. "Multi-objective optimization of wear parameters for aluminium matrix composites (413/B4C) using grey relational analysis." *Materials Today: Proceedings* 5.2 (2018): 7207-7216.
[2] Kumar, S. Mohan, R. Pramod, and H. K. Govindaraju. "Evaluation of mechanical and wear properties of aluminium AA430 reinforced with SiC and Mgo." *Materials Today: Proceedings* 4.2 (2017): 509-518.
[3] Surappa, M. K. "Aluminium matrix composites: Challenges and opportunities." *Sadhana* 28.1-2 (2003): 319-334.
[4] Rajeswari, B., K. S. Amirthagadeswaran, and K. G. Anbarasu. "Investigation on mechanical properties of aluminium 7075-silicon carbide-alumina hybrid composite using Taguchi method." *Australian journal of mechanical engineering* 13.2 (2015): 127-135.
[5] Alaneme, Kenneth Kanayo, and Kazeem Oladiti Sanusi. "Microstructural characteristics, mechanical and wear behaviour of aluminium matrix hybrid composites reinforced with alumina, rice husk ash and graphite." *Engineering Science and Technology, an International Journal* 18.3 (2015): 416-422.
[6] Walczak, M., D. Pieniak, and M. Zwierzchowski. "The tribological characteristics of SiC particle reinforced aluminium composites." *Archives of Civil and Mechanical Engineering* 15.1 (2015): 116-123.
[7] Lee, Stuart M., ed. *Handbook of composite reinforcements*. John Wiley & Sons, 1992.
[8] Mavhungu, S. T., et al. "Aluminum matrix composites for industrial use: advances and trends." *Procedia Manufacturing* 7 (2017): 178-182.
[9] Kevorkijan, Varuzan M. "Aluminum composites for automotive applications: a global perspective." *Jom* 51.11 (1999): 54-58.
[10] Bahmani, E., et al. "Investigation on microstructure, wear behavior and microhardness of Al– Si/SiC nanocomposite." *Surface Engineering and Applied Electrochemistry* 54.4 (2018): 350-358.
[11] Kumar, C. Antony Vasantha, and J. Selwin Rajadurai. "Influence of rutile (TiO2) content on wear and microhardness characteristics of aluminium-based hybrid composites synthesized by powder metallurgy." *Transactions of Nonferrous Metals Society of China* 26.1 (2016): 63-73.
[12] Hashim, J., L. Looney, and M. S. J. Hashmi. "Metal matrix composites: production by the stir casting method." *Journal of materials processing technology* 92 (1999): 1-7.
[13] Rajeswari, B., K. S. Amirthagadeswaran, and K. G. Anbarasu. "Investigation on mechanical properties of aluminium 7075-silicon carbide-alumina hybrid composite using Taguchi method." *Australian journal of mechanical engineering* 13.2 (2015): 127-135.
[14] Vedrtnam, Ajitanshu, and Anuj Kumar. "Fabrication and wear characterization of silicon carbide and copper reinforced aluminium matrix composite." *Materials discovery* 9 (2017): 16-22.
[15] Mistry, Jitendra M., and Piyush P. Gohil. "Experimental investigations on wear and friction behaviour of Si3N4p reinforced heat-treated aluminium matrix composites produced using electromagnetic stir casting process." *Composites Part B: Engineering* 161 (2019): 190-204.
[16] Yigezu, Belete Sirahbizu, P. K. Jha, and M. M. Mahapatra. "Effect of sliding distance, applied load, and weight percentage of reinforcement on the abrasive wear properties of in situ synthesized Al–12% Si/TiC composites." *Tribology Transactions* 56.4 (2013): 546-554.
[17] Sharma, Sandeep, Tarun Nanda, and O. P. Pandey. "Effect of particle size on dry sliding wear behaviour of sillimanite reinforced aluminium matrix composites." *ceramics international* 44.1 (2018): 104-114.
[18] Mistry, Jitendra M., and Piyush P. Gohil. "Experimental investigations on wear and friction behaviour of Si3N4p reinforced heat-treated aluminium matrix composites produced using electromagnetic stir casting process." *Composites Part B: Engineering* 161 (2019): 190-204.
[19] Nishida, Yoshinori. *Introduction to Metal Matrix Composites: Fabrication and Recycling*. Springer Science & Business Media, 2013.
[20] Krishna, M. Gopi, et al. "Metal-metal composites-An innovative way for multiple strengthening." *Materials Today: Proceedings* 4.8 (2017): 8085-8095.

Studies on Machining Characteristics and Microbiological Growth over Stir Casted A356-Graphite Metal Matrix Composites
(A Comparison between Pure Metals and Composite)

Suneel Donthamsetty[1,*] and Penugonda Suresh Babu[2]

[1]Dean (Academics) & Head of the Dept., Department of Mechanical Engineering, Narasaraopeta Engineering College (Autonomous), Narasaraopet, Guntur District, Andhra Pradesh, India, 522601. Mobile No: 91-9441895535
[*]Corresponding Author, Email id: suneeldonthamsetty@gmail.com

[2]Associate Professor, Department of Mechanical Engineering, Narasaraopeta Engineering College (Autonomous), Narasaraopet, Guntur District, Andhra Pradesh, India, 522601

Keywords: A356; Graphite; Composites; Machining; Bacterial Count

Abstract. The need of Metal Matrix Composites (MMCs) are increasing day by day due to their good properties like light weight, high strength, durability, corrosion resistance etc. In the present work A356 is taken as main base material due to its excellent mechanical properties and graphite as fortifying material because of high conductive property to form MMC. Machining is done with and without cutting oil by using automatic feed lathe machine by varying the speed and depth of cuts to find out the cutting forces, cutting tool temperatures and surface roughness and found that these values are gradually increased in many cases. Also done microbiological check and found that the colony count is gradually increased from fresh oil usage to used cutting oil after machining. These tests also done on MMCs, pure brass and pure aluminum and compared the results with MMCs. Novelty focused in present research work is regarding Microbial Organism Growth and its pattern over a period of time since these organism are cancerogenic for the operator and in health and hygiene point of view, lot of research work is going on over this organism growth on combination of several cutting fluids and materials. With the evolvement of composite materials, effect of cutting fluids on the above phenomenon is focused in present work. As a routine work, effect of cutting forces also established by changing machining parameters and presented.

1. Introduction
The MMC is composite material with in any event two constituent parts, one being a metal essentially, the other material might be an alternate metal or another material, for example, an artistic or natural compound. When at any rate three materials are available, it is known as a hybrid composite. A MMC is correlative to a cermet. MMCs are made by scattering a fortifying

IOP Conf. Series: Materials Science and Engineering 1112 (2021) 012004 doi:10.1088/1757-899X/1112/1/012004

material into a base material or matrix which is a monolithic material and is completely continuous. The fortifying material sometimes coated to avoid any reactions with the base material. In structural uses the matrix is generally a lighter metal such as magnesium, aluminum etc., which provides the support to the fortifying material [1].

The strength, stiffness, and density of the composite is depends on its constituent materials properties, the fortifying material's size, shape, quantity & distribution and the bond between base and fortifying material. The base material forces sharing the load among all the fibers. The main types of fortifying materials are fibers of continuous, discontinuous (short fibers), whiskers, and particulates, which are shown in Figure 1 [2].

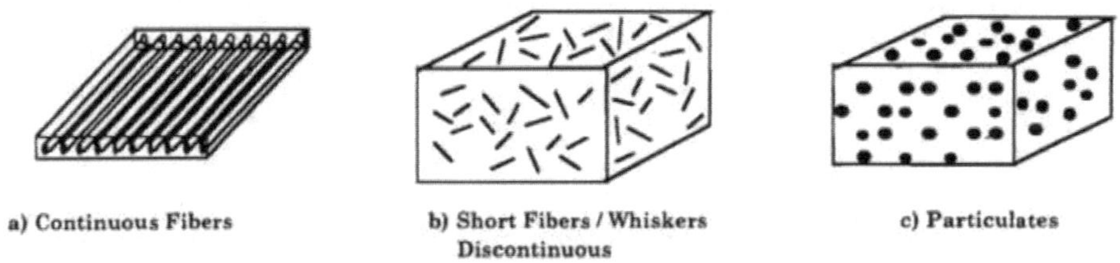

Figure 1. Types of reinforcements

The composite materials are classified into Metal Matrix Composites (MMC) of metals based, Ceramic Matrix Composites (CMC) of ceramic materials based and Polymer Matrix Composites (PMC) of plastic materials based [3].

Aluminum metal matrix composites are getting much importance especially for aerospace, automobile, agriculture farm machinery industries etc., due to their good properties such as high strength, low density, good wear resistance compared to any other metal [4].

The role of cutting fluid is increasing day by day due to finding and innovating new materials in engineering and to use the high speed machining. These cutting fluids reduces the cutting temperature by minimizing the friction between work piece and cutting tool, so that the life of both cutting tool and machine apart from improving the efficiency of the machine, lubrication, cleaning and corrosion protection [5].

In the present work the Graphite powder is used as fortifying material and Al 356 as base material synthesized by using stir casting machine

2. Materials

A356 Alloy

Al 356 alloy is used as a base material or matrix, used in the fields of automotive and avionics industries, has excellent wear resistance and good mechanical properties [6] and selected as matrix material owing to good and readily castable [7, 8]. The chemical composition of Al 356 alloy is given in Table 1.

Table 1. Chemical composition of Al 356 Alloy

Element	Si	Cu	Mg	Mn	Fe	Zn	Ni	Ti	Al
Wt. (%)	7.20	0.02	0.29	0.01	0.18	0.01	0.02	0.11	Balance

Fortifying material (Graphite)

Graphite, taken as a fortifying material, is a crystalline form of the element carbon with its atoms arranged in a hexagonal structure, arises naturally and most stable form of carbon under standard conditions, which is used in pencils and lubricants. It converts to diamond under high pressures and temperatures. Due to its high conductivity of heat and electricity, it is most useful in electronic products such as electrodes, batteries, and solar panel. Also used the components made by Al MMCs with graphite in the fields of aerospace and automobile industries due to its anti-friction nature and high melting point [9].

Cutting fluid

Despite the advantages of cutting fluids of cooling, lubrication, flushing away the chips, reduce the adhesion wear, extended tool life with surface finish to enhance the efficiency of machining process causes increase in productivity, there are some precautions to be followed while using these cutting fluids as these are degrading its quality gradually while in use, its disposal and consequently the pollution to the environment, mist generation etc. Especially, the water based cutting oils are fully contaminated with the microorganisms, which deteriorate the cutting fluids properties causes corrosion of work pieces, choking of fluid flow lines etc. And also chances of getting skin deceases and health issues to the workers who are exposed to these contaminated oil [10]. Used Power fist Cutting Oil/Coolant in this work, which is shown in Figure 2. This water-soluble oil provides protection against rust and corrosion, while also reducing surface tension and improving cooling in cutting areas. Emulsions can be prepared at water-to-oil ratios ranging from 5:1 to 100:1 [11]. In this work, taken this ratio as 20:1 and used for machining as shown in Figure 3.

Figure 2. Cutting Oil Tin

Figure 3. Machining of MMC bar with cutting oil

After machining, the used oils were taken and stored in sterilized bottles and tested to study the microbial contamination.

3. Equipment

Surface Roughness Tester

The Surftest SJ-210 Portable Surface Roughness Tester shown in Figure 4, is used for testing the surface roughness of the work pieces in microns (μm) after machining on the lathe machine.

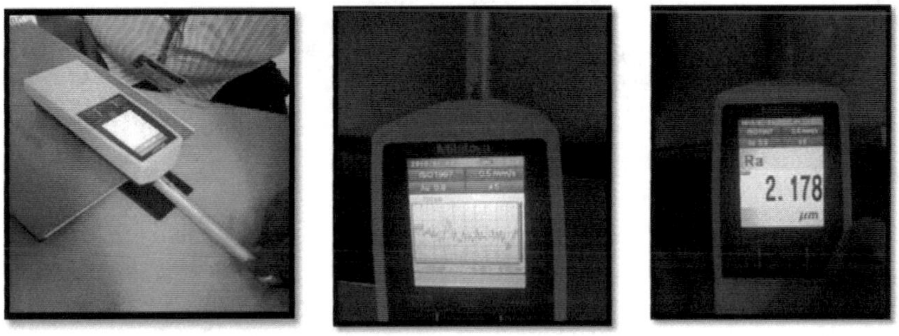

Figure 4. Surface Roughness Tester with work piece

Automatic feed lathe machine with Coolant Pump Facility

The machining process is done by using Banka 35 Automatic feed lathe machine, shown in Figure 5, along with a coolant pump of 40 LPM capacity with 2900 rpm used to circulate the coolant oil.

Figure 5. Automatic feed lathe machine with coolant pump

Tool dynamometer

IEICOS (India) make electronic machine tool dynamometer, shown in Figure 6, used to find out the cutting forces while machining the MMC bar samples on the lathe machine.

Figure 6. Tool dynamometer set up

Microbial Growth

Cutting fluid samples are collected in the sterilized bottles shown in Figure 7 after machining Al-Graphite MMC bars, Pure Al and Pure Brass each with 5 hours machining per day to find the microbial growth.

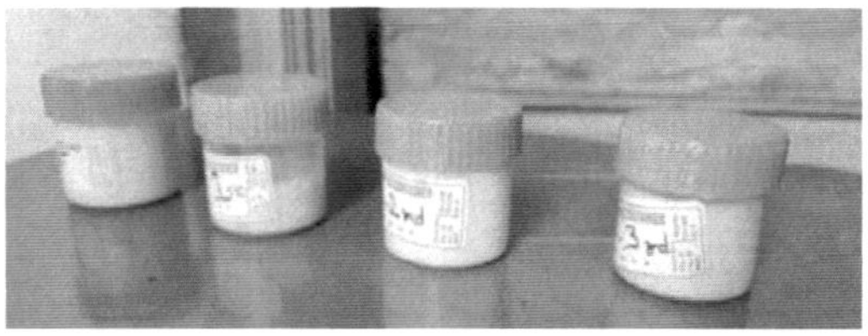

Figure 7. Different oils in the sterilized bottles for microbial growth

Petri Plates

A Petri dish or petri plate is a transparent lidded dish used to find bacteria, fungi or small mosses. It is the most common type of culture plate [12, 13]. Standard plate count which is conventional method for monitoring microbial levels in Cutting fluids is used. In the present work, Petri plates duly sterilized were taken and filled with nutrient Agar which is liquid and the temperature is above 45°C. Upon on cooling of this to the room temperature, it will be dried and hardened like a gel.

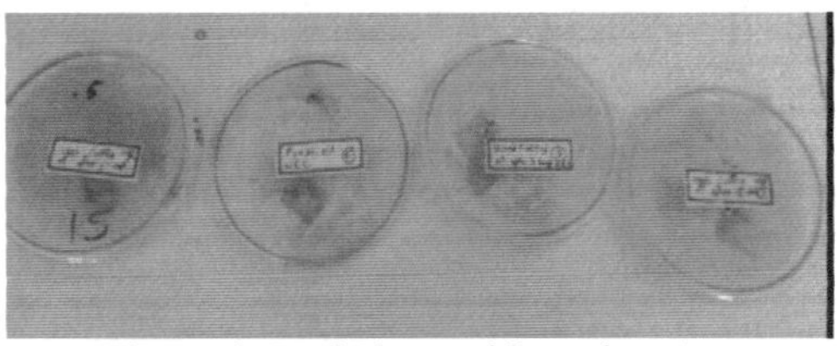

Figure 8. Petri Plates with nutrient Agar

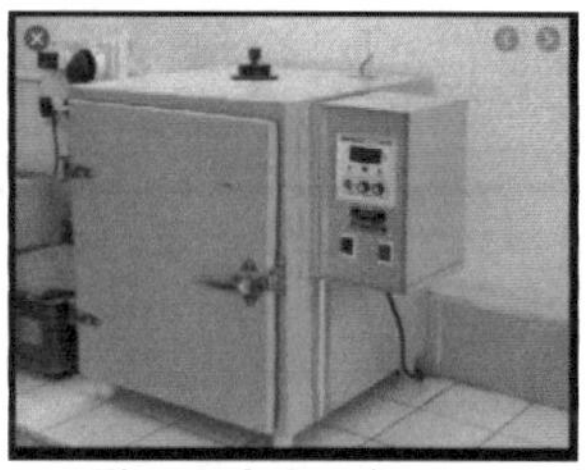

Figure 9. Incubator

IOP Conf. Series: Materials Science and Engineering 1112 (2021) 012004 doi:10.1088/1757-899X/1112/1/012004

Incubation

An incubator shown in Figure 9 is a device used to grow and maintain microbiological cultures or cell cultures. The incubator maintains optimal temperature, humidity and other conditions such as the CO_2 and oxygen content of the atmosphere inside. As both agar media and fluids are easily attacked by the bacteria present in the air, in order to prevent the contamination, the Laminar Air Flow chamber is used to transfer the fluid samples to the Petri plates in the presence of blue flame. The samples were then kept in the Incubator (incubation kept at 37°C for 24 hours, the bacteria formed is known as colonies shown in Figure 10. These colonies (figure naked eye and counted). A colony-forming unit (CFU, cfu, Cfu) is used

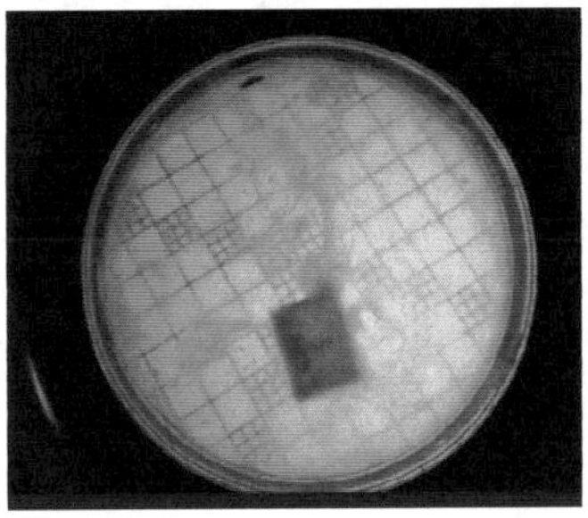

Figure 10. Bacteria developed in the petri Plates

4. Experimentation

To prepare the MMC with long durability, strength and less weight, selected the A356 alloy material as base, graphite powder of 50 microns size as fortifying material and stir casting route for melting these materials due to lengthy interaction between base material and fortifying material causes good uniform spreading, which gives MMC with good mechanical properties. Chosen the Bottom Pouring Type Stir Casting Machine (Make: M/s. Swam Equip, Chennai, India) to melt the both the raw materials.

Initially, dropped the pieces of A356 material into the furnace. After melting the same, poured the preheated graphite powder of 5% by weight into the furnace through the feeder and use the stirrer for even mixing of both the A356 and graphite powder. After that poured the liquid material into the die and allowed for solidification. Machined that solidified rough A356-Graphite MMC bar into the required 30 mm dia, which is further tested to find the Microbial Organism Growth apart from establishing the cutting parameters by using Banka 35 Automatic feed lathe machine, with a coolant pump to circulate the Power fist Cutting Oil (in the ratio of 20:1 of water: oil) for improving cooling in cutting areas. In parallel, found the surface roughness values by using the Surftest SJ-210 Portable Surface Roughness Tester and cutting forces by using IEICOS electronic machine tool dynamometer. And, after that machined both the pure aluminum and brass bars and done the similar tests of MMC bar, compared all the results.

5. Results and Discussions

The prepared samples were tested and the results are tabulated

Forces in X, Y & Z Directions with & without coolant, Temperature and Surface roughness studies

Al 356 / Graphite MMC with variable D.O.C and at Fixed speed

The Table 2 gives the tested values of forces in X, Y & Z directions, cutting tool temperature and roughness values of Al 356 / Graphite MMC with variable D.O.C and at fixed speed and found that all the parameters of cutting Forces, cutting tool temperatures and surface roughness values are gradually increased.

Table 2 Al 356 / Graphite MMC with variable D.O.C and at Fixed speed

Variable D.O.C (mm)	Fixed speed in RPM	F_x (N)		F_y (N)		F_z (N)		Temperature °C		Surface Roughness in Micro Meters			
										Without Coolant		With Coolant	
		Without Coolant	With Coolant	Without Coolant	With Coolant	Without Coolant	With Coolant	Without Coolant	With Coolant	Individual Values	Avg. value	Individual Values	Avg. value
1.5	685	5	15	6	17	9	22	36	34	4.965 4.856 4.765	4.862	5.97 5.819 5.719	5.836
2	685	7	19	8	21	10	23	45	40	6.95 6.85 6.65	6.817	6.477 6.317 6.312	6.369
2.5	685	7	22	-9	23	11	26	60	47	7.105 7.205 7.235	7.182	7.005 7.015 7.015	7.012

Al 356 / Graphite MMC with variable speed and at Fixed D.O.C

The Table 3 gives the tested values of forces in X, Y & Z directions, cutting tool temperature and roughness values of Al 356 / Graphite MMC with variable speed and at fixed D.O.C and found that parameters of cutting Forces of Fx and cutting tool temperatures are gradually increased, Fy & Fz are varied and surface roughness values are gradually decreased.

Table 3 Graphite MMC with variable speed and at Fixed D.O.C

Variable Speed (RPM)	Fixed D.O.C (mm)	F_x (N)		F_y (N)		F_z (N)		Temperature °C		Surface Roughness in Micro Meters			
										Without Coolant		With Coolant	
		Without Coolant	With Coolant	Without Coolant	With Coolant	Without Coolant	With Coolant	Without Coolant	With Coolant	Individual Values	Avg. value	Individual Values	Avg. value
1025	1.5	12	18	14	21	17	32	38	36	5.165 5.215 5.319	5.233	9.505 9.105 9.201	9.270
685	1.5	15	21	16	27	18	31	40	38	4.192 4.135 4.012	4.113	8.704 8.105 7.996	8.268
303	1.5	19	24	21	28	20	33	45	41	3.492 3.393 3.195	3.360	4.665 4.325 4.105	4.365

Pure Al with variable D.O.C and at fixed speed

Table 4 Pure Al with variable D.O.C and at fixed speed

Variable D.O.C (mm)	Fixed speed in RPM	F_x (N)		F_y (N)		F_z (N)		Temperature °C		Surface Roughness in Micro Meters			
										Without Coolant		With Coolant	
		Without Coolant	With Coolant	Without Coolant	With Coolant	Without Coolant	With Coolant	Without Coolant	With Coolant	Individual Values	Avg. value	Individual Values	Avg. value
1.8	303	4	3	5	5	27	18	37	34	4.811 4.269 4.44	4.507	23.1 23.7 22.9	23.233
2	303	4	12	7	15	30	21	43	39	4.128 4.058 4.845	4.344	23.9 24.6 22	23.500
2.8	303	7	14	12	17	28	20	53	44	3.925 3.814 3.736	3.825	24.24 24.05 27	25.097

The Table 4 gives the tested values of forces in X, Y & Z directions, cutting tool temperature and roughness values of Pure Al with variable D.O.C and at fixed speed and found that the parameters of cutting Forces of Fx & Fy, cutting tool temperatures and surface roughness values with coolant are gradually increased. But, Fz and surface roughness values without coolant are varied.

5.1.4. Pure Al with variable speed and at Fixed D.O.C

Table 5. Pure Al with variable speed and at Fixed D.O.C

Variable Speed (RPM)	Fixed D.O.C (mm)	F_x (N)		F_y (N)		F_z (N)		Temperature °C		Surface Roughness in Micro Meters			
										Without Coolant		With Coolant	
		Without Coolant	With Coolant	Without Coolant	With Coolant	Without Coolant	With Coolant	Without Coolant	With Coolant	Individual Values	Avg. value	Individual Values	Avg. value
1025	2	3	3	4	5	4	3	37.6	35.2	7.2	7.567	40.9	42.000
										7.9		44	
										7.6		41.1	
685	2	4	4	7	5	20	9	41.2	41.2	2.2	2.467	24.5	24.700
										2.5		25.7	
										2.7		23.9	
303	2	6	9	9	13	27	22	44	44	3.1	3.467	14.5	14.563
										3.6		15.99	
										3.7		13.2	

The Table 5 gives the tested values of forces in X, Y & Z directions, cutting tool temperature and roughness values of Pure Al with variable speed and at fixed D.O.C and found that the parameters of cutting Forces of Fx & Fy, cutting tool temperatures and surface roughness values with coolant are gradually increased. But, Fz and surface roughness values without coolant are varied.

Pure Brass with variable D.O.C and at fixed speed

The Table 6 gives the tested values of forces in X, Y & Z directions, cutting tool temperature and roughness values of Pure Brass with variable D.O.C and at fixed speed and found that all the parameters of cutting Forces, cutting tool temperatures and surface roughness values are gradually increased.

IOP Conf. Series: Materials Science and Engineering 1112 (2021) 012004 doi:10.1088/1757-899X/1112/1/012004

Table 6 Pure Brass with variable D.O.C and at fixed speed

Variable D.O.C (mm)	Fixed speed in RPM	F_x (N)		F_y (N)		F_z (N)		Temperature °C		Surface Roughness in Micro Meters			
										Without Coolant		With Coolant	
		Without Coolant	With Coolant	Without Coolant	With Coolant	Without Coolant	With Coolant	Without Coolant	With Coolant	Individual Values	Avg. value	Individual Values	Avg. value
1.5	303	0	8	3	5	6	4	34	31	7.179	5.595	9.1	9.917
										5.212		10.85	
										4.394		9.8	
2	303	3	12	3	8	8	6	36	33	8.138	7.452	13.3	13.767
										7.0000		14.5	
										7.217		13.5	
2.5	303	5	17	3	13	11	10	53	36	10.366	10.380	14.9	15.667
										11.643		16.6	
										9.132		15.5	

Pure Brass with variable speed and at Fixed D.O.C

The Table 7 gives the tested values of forces in X, Y & Z directions, cutting tool temperature and roughness values of Pure Brass with variable speed and at fixed D.O.C and found that the parameters of cutting Forces of Fx & Fy, cutting tool temperatures are gradually increased. But, Fz and surface roughness values are varied.

Table 7 Pure Brass with variable speed and at fixed D.O.C

Variable Speed (RPM)	Fixed D.O.C (mm)	F_x (N)		F_y (N)		F_z (N)		Temperature °C		Surface Roughness in Micro Meters			
										Without Coolant		With Coolant	
		Without Coolant	With Coolant	Without Coolant	With Coolant	Without Coolant	With Coolant	Without Coolant	With Coolant	Individual Values	Avg. value	Individual Values	Avg. value
1025	2	3	3	0	8	11	3	31.2	30	2.839	2.635	7.976	7.607
										2.611		7.526	
										2.456		7.318	
685	2	3	8	3	13	15	10	33.5	31.3	6.704	6.547	7.016	7.181
										6.612		7.202	
										6.325		7.325	
303	2	4	11	4	17	9	15	32.6	32	4.225	4.354	5.985	5.608
										4.315		5.625	
										4.522		5.213	

Bacterial Count:

Each cutting Fluid was tested for Bacterial Count after using for 5 hrs. of machining and waited for 24 hrs. of each machining and results are tabulated in Table 8 and Figure 11. It is found that the colony count is gradually increased from fresh oil to Cutting Oil after machining Al 356 + Graphite MMC bar, Pure Al bar and Pure Brass bars.

Table 8 Bacterial Count

Sl. No.	Cutting Fluid Details	Colony Count
1	Fresh Cutting Oil	10
2	Cutting Oil after machining Al 356 + Graphite MMC	20
3	Cutting Oil after machining Pure Al	31
4	Cutting Oil after machining Pure Brass	50

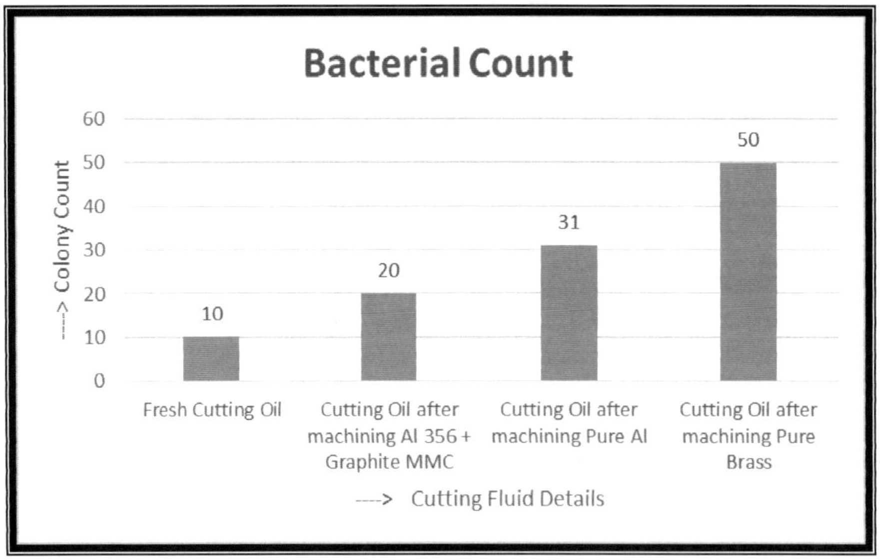

Figure 11. Bacterial Count

CONCLUSIONS

The various testings were done on Al 356- Graphite Metal Matrix Composite, Pure Al & pure Brass and below is the summary

- In majority of cases, all the parameters of cutting Forces of Fx, Fy & Fz, Cutting tool temperatures and Surface roughness values are gradually increased

- The colony count is gradually increased from fresh oil to used Cutting Oil after machining. Found colony count of 10 in fresh oil, 20 in Cutting Oil after machining Al 356 + Graphite MMC, 31 in Cutting Oil after machining Pure Al and 50 in Cutting Oil after machining Pure Brass.

References

[1] Deborah D. L. Chung 2010 Composite Materials Science and Applications *Springer-Verlag London Limited, second edition*
[2] https://www.asminternational.org/documents/10192/1942084/composite.pdf/0eb5cca8-4613-482c-811c-993c58b42e92
[3] https://www.substech.com/dokuwiki/doku.php?id=classification_of_composites
[4] Rohit Sharma, Saurabh Jha P, Khushboo Kakkar, Kushal Kamboj and Pardeep Sharma 2017 A Review of the Aluminium Metal Matrix Composite and its Properties *International Research Journal of Engineering and Technology (IRJET) Volume: **04** Issue: 02*
[5] Pei Yan, Yiming Rong and Gang Wang 2015 The effect of cutting fluids applied in metal cutting process *Proc IMechE Part B:J Engineering Manufacture 1–19, DOI: 10.1177/0954405415590993*
[6] Z. Mišković, I. Bobić, S. Tripković, A. Rac and A. Vencl 2006 The Structure and Mechanical Properties of an Aluminium A356 Alloy Base Composite With Al_2O_3 Particle Additions *Tribology in industry, Volume **28**, No. 3&4*
[7] Suneel Donthamsetty 2013 Investigations on mechanical properties of A356 nano composites reinforced with high energy ball milled SiC nanoparticles with ultrasonic assisted cavitation (with a comparison of micro composite) *International Journal of Nanoparticles (IJNP), Inderscience Publishers, Vol.**6**, No.1, pp.38-49*
[8] Suneel Donthamsetty and Penugonda Suresh Babu 2017 Experiments on the wear characteristics of A356 MMNCs fabricated using ultrasonic cavitation *International Journal of Automotive and Mechanical Engineering, Volume **14**, Issue 4 pp. 4589-4602*
[9] Hazim Faleh, Muna Noori and Ştefănescu 2015 Florin Properties and Applications of Aluminium-Graphite Composites *Advanced Materials Research Vol. **1128** pp 134-143.*
[10] Prasad M.M.S and Srikant R.R 2014 Evaluation of Microbial Contamination of Cutting Fluids with Nano Graphite Inclusions *Research Journal of Engineering Sciences, Vol. **3(2)**, 12-16*
[11] https://www.princessauto.com/en/detail/cutting-oil-coolant/A-p8034012e
[12] Frederick J. Passman 1988 Microbial problems in metalworking fluid *Tribology and Lubrication Technology*
[13] Marc Veillette, Peter S. Thorne, Terry Gordon and Caroline Duchaine 2004 Six Month Tracking of Microbial Growth in a Metalworking Fluid After System Cleaning and Recharging *Ann. occup. Hyg., Vol. **48**, No. 6, pp. 541–546*

An experimental evaluation on revamping the productivity of solar PV panel using wind tunnel as an optimizer

R Rajasekar[1]*, R Ramkumar[2]* and P Prasannaa[3]
[13]Research Scholar, [2]Associate Professor
[123]Department of Mechanical Engineering, Annamalai University,
***Corresponding Author: rajeee.mech@gmail.com, rrramkumarhai@gmail.com**

Abstract At high working temperatures, the performance of the solar photovoltaic (PV) cells will be affected. With the increase in temperature of the PV cells, the conversion of solar radiation into electrical output will be decreased. The operating temperature of the PV cells is one of the prime factors for determining its electrical conversion efficiency. In order to get the maximum electrical output from the solar PV cells, the temperature should be maintained properly. In this experimental study, the temperature of the solar PV cells was maintained by cooling both sides of the PV panel simultaneously. The front surface of the PV panel is sprayed with a thin film of pot water and the rear side is attached with a fabricated duct that was connected into a wind tunnel blower. The mass flow rate of the moist air from the wind tunnel to the rear side of the PV is ranged between 0.1m/s to 0.5m/s. The performance characteristics of both the commercially available PV panel and the experimental PV panel (PVT Clay pipe panel) were compared. The results show that the PVT clay pipe panel has an electrical efficiency of about 11.72%, whereas the commercially available PV panel has only 9.77%. Also, the overall efficiency of the experimental PVT panel is 65.55% respectively.

Key words: solar PV panel, clay pipe, wind tunnel, pot water, moist air, electrical efficiency

1. Introduction

Photovoltaic system is the utmost important device of energy converter, which is directly converts the solar energy into electrical energy without any mechanism [1,2,3]. However, The PV panel is generally produced low amount of electricity, the majority of solar irradiation is converted into heat [4]. Hence, only 13-20% of the solar irradiation converted into electricity, the residual of solar energy gets as a heat [5]. finally, the PV results get elevates the panel temperature [6]. Increase in cell temperature, their efficiency will decrease around 0.4% - 0.5% [7]. In order to overcome this problem, either to reduce PV panel temperature or maintaining operating cell temperature. When the cell temperature decreases, the efficiency of PV panel will increase [8]. Therefore, following researchers have accomplished many methods to improving PV panel efficiency like, water cooling, air cooling, pcm etc.

[9] George Popovic et al. Presents the improved on photovoltaic panel efficiency by the way of temperature reduction. The panel temperature was reduced by the air-cooled heat sinks, it was high thermal conductivity material. The efficiency of the cooling was examined for different configuration of the heat sink. The result shows that, a maintained average temperature of solar PV panel.

[10] Zeyad A. Haidar an experimental investigates that, the rear surface of PV panel heat was absorbed by evaporative cooling effect. The rear surface of PV panel was maintained on wet condition by the water, it was collected from the water tank by the gravitational force. This method was very effective and also reduced more than 20% panel temperature, hereby, the electrical efficiency was achieved around 14% compared than the reference PV panel.

[11] Dhritiman Adhya, experimentally proved that, the PV panel was cooled by the distilled water as a coolant fluid. However, an aluminum pipe was mounted on rear surface of the PV panel. The coolant fluid flow through the mounted pipe line, it was extracting the accumulate the heat at rear side of the PV panel. This phenomenon would decrease the temperature level up to 33.1% and an efficiency increased about 18% of power output.

[12] Farag Mahel Mohammed proposed, an enhancing the efficiency of PV panel with the combination of water flow double glazing system and dual axis tracker has been implemented at hot climatic condition. However, the combined PV panel temperature was maintained around 40°C when ambient temperature exceeds up to 50°C, also increased a maximum efficiency about 12.7%.
[13] Haitham M. Bahaidarah experimentally investigates that, a hybrid PV system was designed to flow active water cooling on back side of PV surface. This active cooling method has been significantly reduced operating temperature around 20%. Furthermore, this water-cooling method is simultaneously produced hot water and electrical energy. The results show that an electrical efficiency achieved about 9% at 900W/m2 irradiation level.
[14] Calebe Abrenhosa Matias examined the performance of power generation equipped with cooling and without cooling systems. The experiment shows that, the cooling of panel was decreased operating temperature during the peak hour. The water is supplied on over surface of the PV panel. The supplied water flow increased power output about 77Wh and also gained 24% total energy.
[15] DjamilaNebbali investigates an increasing the efficiency of solar PV panel by the use of pulsed ambient air through the fan. An electrical efficiency achieved around 71% under the radiation of 2000W/m2 and 50°C air temperature.
[16] Azadeh Kordzadeh have examined different techniques for reduce solar PV panel temperature. finally, they concluded that, the water-cooling method have to be reduced cell temperature through the pump motor. It is most effective method for increasing efficiency solar PV and also reduced cell temperature.
[17] Jie Ji designed PVT water heating system with help of natural circulation methods has been implemented. The authors have used to different mass of water and different initial temperature of water on outdoor condition. Thereby, an increasing electrical efficiency and thermal efficiency to be 10.15% and 45% respectively. The total efficiency was achieved about 52% and also recovered primary energy saving up to 65%.
[18] M. Chandrasekar improving an efficiency of solar PV panel combination with heat spreaders and cotton wicks used for controlling the cell temperature during peak hour. An electrical efficiency and thermal efficiency of PVT system results were compared with the nominal flat PV panel. The thermally regulated PV panel temperature was reduced about 12% also be increased an electrical efficiency up to 14% through this cooling methods.
In this experimental study is focused on enhancing the efficiency of solar PV panel by using pot water to reduce front surface panel temperature. In addition, the duct was designed at rear side of the PV panel to create a way to air flow on it. It was connected with the wind tunnel. By using moist air to control the released heat. Finally, the electrical efficiency and thermal efficiency were analyzed both PV and PVT panels.

Nomenclature

A_p	PV panel surface area (m^2)
Cp_{water}	Specific heat of water (kJ/kg-K)
A_{duct}	Area of the duct
m_{water}	Mass of water flow(kg/sec)
FF	Fill Factor
I_G	Global solar radiation (W/m^2)
P_{max}	Maximum Power (W)
I_{max}	Maximum Current (A)
V_{max}	Maximum Voltage (V)
V_{oc}	Open Circuit Voltage (V)
I_{sc}	Short circuit current (A)
T_{ref}	Reference Temperature (°C)
Q	Heat Transfer rate (W)
T_A	Temperature of ambient air(°C)
T_M	Temperature absorber of PV module(°C)
$T_{moist\ air}$	Temperature of wind tunnel moist air(°C)
β	Temperature coefficient (°C)
T_g	Temperature of glass cover(°C)
T_{sky}	Sky temperature(°C)
U_L	Overall heat loss coefficient (W/m^2 K)
h_w	Convective heat transfer coefficient due to wind (W/m^2K)
h_{air}	heat transfer coefficient in air duct (W/m^2 k)
T_{inlet}	Inlet water & air temperature (°C)
T_{outlet}	Outlet water & air temperature (°C)
η_{ele}	Electrical efficiency (%)
η_{th}	Thermal efficiency (%)
η_o	Overall efficiency (%)
ε	Emissivity of the glass cover
σ	Stefan-Boltzmaan constant (W/m^2K^4)

2. Experimental setup

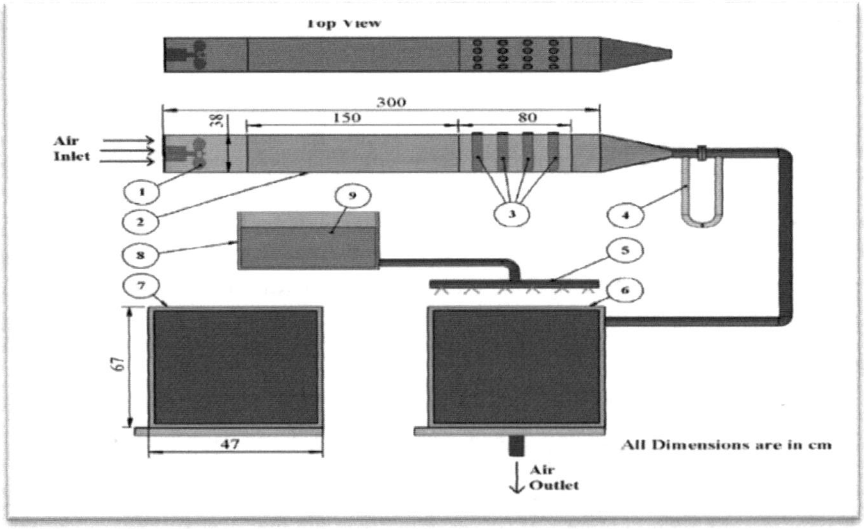

Figure 1. Shows the schematic view of experimental setup.

The Experimental setup consists of two identical 40W solar PV panel. Among these, one PV panel used as reference panel for comparing the electrical efficiency and another one is working panel equipped with cooling system as an evaporative cooling moist air. Here, the evaporative cooling system comprised a rectangular wind tunnel made by plexi glass with size of 3.14 m length, 0.38 m X 0.38 m square cross section and 9 mm thickness. Hence, the following apparatus has been used in this experiment work like, 1. wind blower 2. rectangular wind tunnel 3. clay pipes 4. Utube monometer 5. Water shower 6. PVT panel 7. reference PV panel 8. clay pot water tank 9. Pot water. Fig 1. Shows the schematic view of experimental setup.

Fig 1 clears that; the wind tunnel has been separated into two parts. The initial part of conical adaptor is built with blower and second part is made with a clay pipe with the size of 40cm length and width is 8cm. As per the reference [Ramkumar et al] the clay pipes are settled in different arrangement like align position and staggered position in 16 and 14 tubes respectively. Moreover, the clay pipes are placed at vertical position in the rectangular wind tunnel besides, one end of the wind tunnel has been directly connected into the solar PV panel. This experiment was conducted from 9 am to 5 pm during the month of February to June 2019 under the climatic condition of Chidambaram. A thin film of pot water supplied on over the front surface of the PV panel, Whereas the rear surface was made as duct to make a way to air flow. An ambient air is allowed to flow over the rectangular wind tunnel with the help of an induced air blower. The supplied air flow cross over the clay pipe to the solar PV panel. The clay pipes were filled with water and arranged in two different position. A set of sixteen clay pipes were arranged in an aligned position and another set of fourteen clay pipes where are arranged in staggered position as shown in the fig 2. Both the type of arranged pipes were tested for three pitch distances (3cm, 6cm & 9cm) and five different velocities (0.1m/s, 0.2m/s, 0.3m/s, 0.4m/s,0.5m/s) individually.

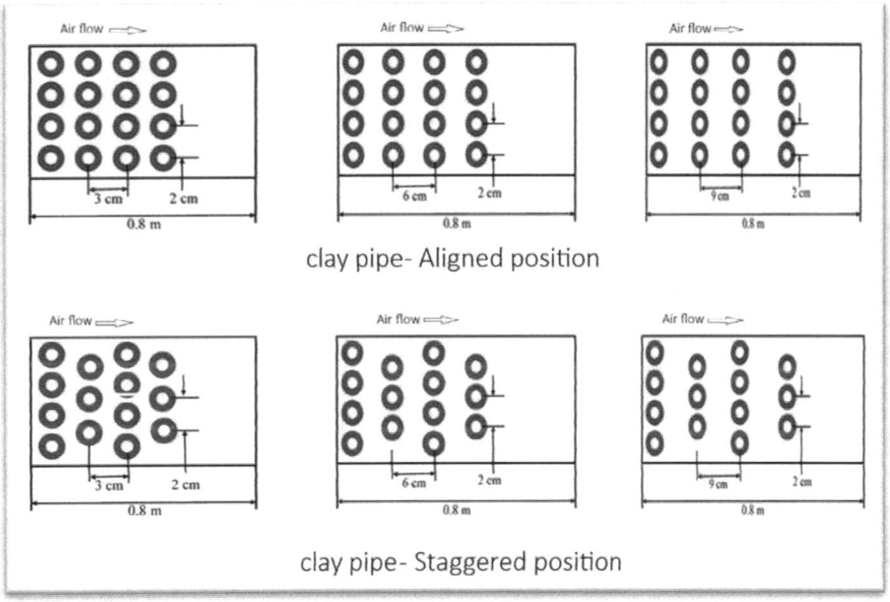

Figure 2. shows the aligned and staggered position of clay pipe with different pitch distance.

3. Mathematical formulation of energy balance equation for PV & PVT panel

The energy balance equation was given below adapted from the equation [19]

The following equation describes the total energy absorbed, the top surface supplied pot water and moist air is bottom side:

$$Q_{overall} = Q_{front}\,(convection + radition) + Q_{rear}(convection)$$

$$U_L A(T_M - T_A) = h_W A_P(T_g - T_A) + \sigma\varepsilon(T_g^4 - T_{sky}^4)A_P(T_M - T_A)$$

$$+ h_{air}A_{duc}(T_M - T_{moist\ air}) \quad (1)$$

The following parameters were measured influence of cooling technique like, maximum power P_{max}, short circuit current I_{sc} and open circuit voltage V_{oc}. This equation is given by the examine the effect of temperature on the electrical performance of PV panel:

$$P_{max} = I_{max}\,V_{max} = FF*I_{SC}*V_{OC} \quad (2)$$

The Eq. (2) shows the FF, as it is called the fill factor, which describes the ratio of maximum power obtainable of the PV module's short circuit current I_{sc} and open circuit voltage $V_{oc.}$ The electrical efficiency η_{ele} is can be determined with various incident solar irradiation. It is given by:

$$\eta_{ele} = \frac{I_M*V_M}{IG} \quad (3)$$

The effective electrical efficiency is can be determined below equation:

$$\eta_{ele} = \eta_{re}[1 - \beta_{ref}(T - T_{ref})] \quad (4)$$

Here, η_{ele} denotes the electrical efficiency, η_{ref} indicates the reference efficiency of Photovoltaic (PV) module. The co efficient of cell temperature is (β=0.0045°C) [20].

The total heat energy gained from the PVT system:

$$Q = q_1 + q_2$$

In Eq. (5) defines the rate of heat energy absorbed by the water:

$$q_1 = m_{water}\,Cp_{Wate}(T_{outlet} - T_{inlet}) \quad (5)$$

Where, m_{water} denotes the mass flow rate of water and $Cp_{\ water}$ denotes the specific heat of the water. The below equation gives the rate of heat energy absorbed by the air [21].

$$q_2 = m_{air}\,Cp_{ai}(T_{outlet} - T_{inlet}) \quad (6)$$

The thermal efficiency can be calculated as the below equation; Q is denoting the gained heat energy, 'G' is solar irradiation and 'A' is the area of PVT module system.

$$\eta_{th} = \frac{Q}{GA} \quad (7)$$

Addition of Both the thermal and electrical energy determined the overall efficiency; below equation could be solve it:

$$\eta_O = \eta_{th} + \eta_{ele} \quad (8)$$

4. Result and Discussion

This research work was conducted to increase the efficiency of PV panel, and wind tunnel clay pipe and pot water were used as coolant medium. These, techniques a way to increase in heat transfer rate between PV panel and atmospheres. The combined pot water and wind tunnel moist air has been used to improving the efficiency of PV panel. Fig.3 represents the temperature difference between ambient temperature, wind tunnel air and pot water temperature. It was observed that, an ambient temperature gradually increased upto 41°C at peak hour at the same time a wind tunnel air and pot water temperature inversely decreased to be 32°C and 28°C respectively. A reason behind that, generally clay pot have evaporative cooling effect, which makes better performance at peak hour due to the pore holes of clay pot. And of clay pot water always been 5°C to 8°C lower than the ambient temperature [20].

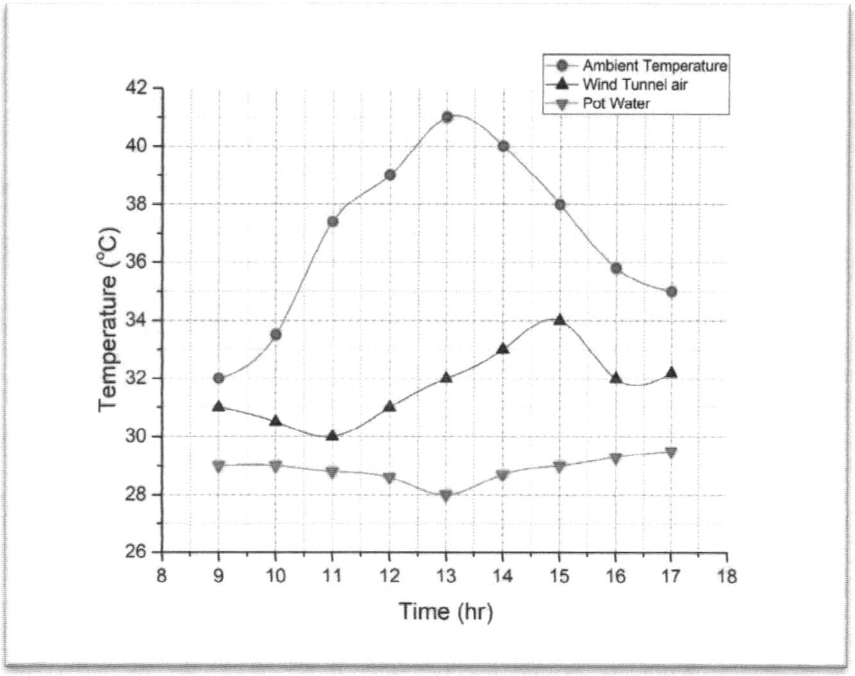

Figure 3. Temperature difference of ambient temperature, Wind tunnel air and pot water

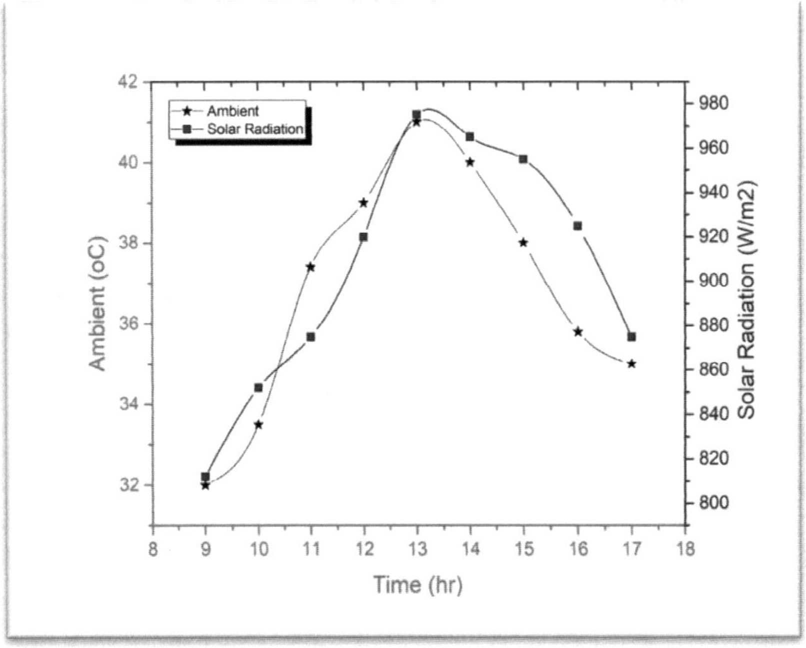

Figure 4. shows the variation of ambient temperature and solar radiation

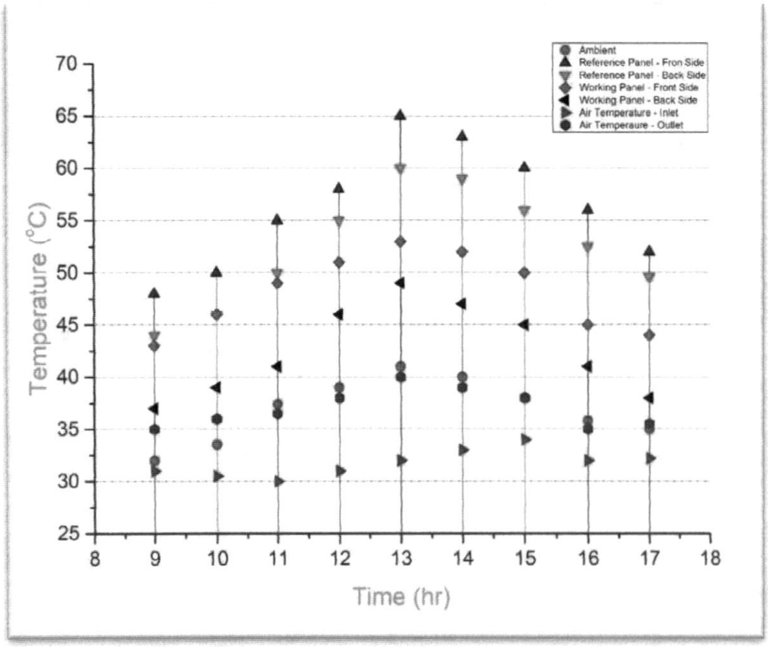

Figure 5. shows the temperature difference between PV and PVT modules

Figure 4 shows that, the variation of ambient temperature and irradiance level range is found to be 41°C and 975W/m2 respectively. At 9am to 5pm the ambient temperature and intensity of solar radiation is achieved about 32°C to 35°C and 812W/m2 to 875W/m2 respectively.

Hence, the experimental study is carried out, during the month of February to June 2019. Figure 5 represents the comparison between surface of reference PV panel and Working PVT panel temperature. All the temperatures were recorded by the data logger. From the initiating work the temperature difference between reference PV and working PVT panel surface is to be about 48°C and 43°C. At peak hour the reference PV and PVT panel temperature reaches up to 65°C and 53°C respectively. The maximum temperature difference between the reference PV panel surface and ambient temperature is 24°C, similarly; for the working PVT panel is about 12°C. Moreover, the back side of PVT panel surface is 8°C lower than the ambient temperature due to the wind tunnel moist airflow on it. From the graph 6,7 represents the power production and electrical efficiency of PV and PVT panel. Here, both the power output and electrical efficiency has been increased with the decrease of PVT module temperature.

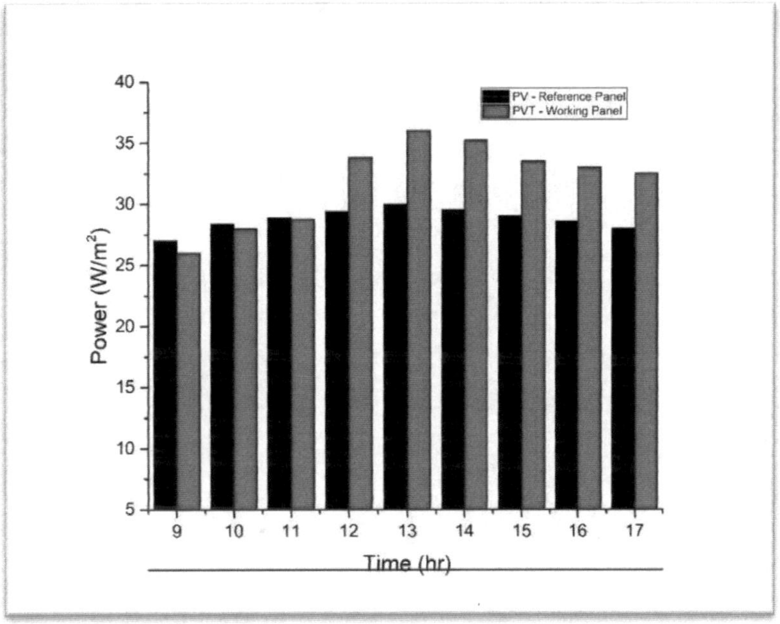

Figure 6. Shows the variation of power production between reference PV and PVT clay pipe panel

Figure 7 clears that, electrical efficiency were compared between reference PV and PVT clay pipe panel. Both the panels should be produce same amount of energy but from the begning time of morning 8 am to 11 am, the PVT clay pipe panel electrical efficiency has been continuesly degereased due to the cooling technics on it. The reason behind that, both the sides of panel were cooled before it recives the required heat to generate electricity.

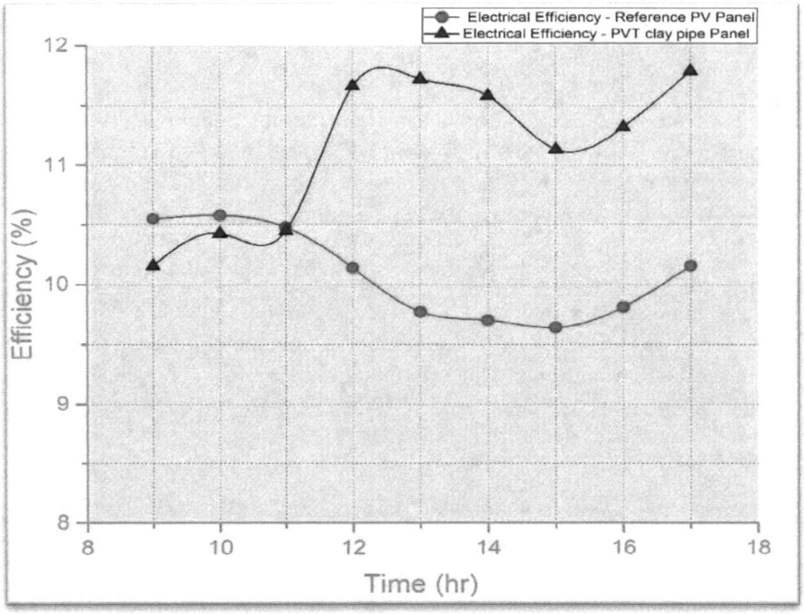

Figure 7. Shows the electrical efficiency of reference PV and PVT clay pipe panel

The overall efficiency can be determined by the absorbed solar radiation of PVT module, Figure 8 shows the electrical efficiency, thermal efficiency and overall efficiency of both PV and PVT module.

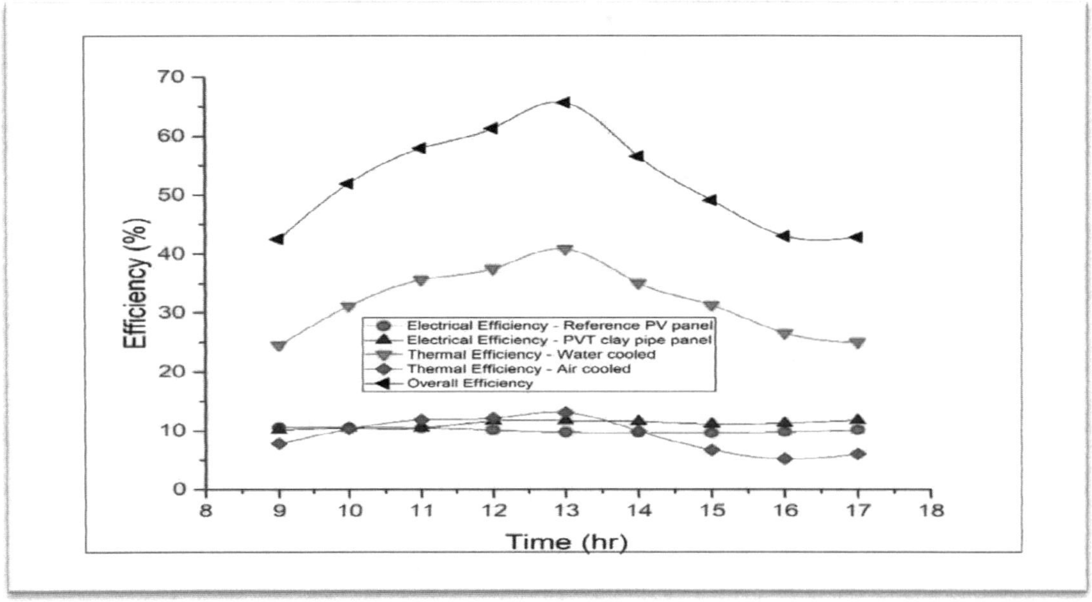

Figure 8. Overall efficiency compared with electrical and thermal efficiency of PV and PVT clay pipe panel

Hence, the reference PV system can be generating only electrical energy, but the working PVT system to yield both electrical and thermal energy. Due to the sensible heat was absorbed by the pot water on PVT front surface, additionally the rear side heat was captured by the wind tunnel moist air. Hereby, the dissipated heat was recovered on front and rear side of PVT system, whereas water and moist air is used as coolant medium.

Figure 8 clears that, the overall efficiency of PVT achieved about 65.5% total energy output. Which is better efficiency of compared with reference PV panel. A reason behind that both the side of PVT panel were simultaneously cooled continuously by the water and air as coolant medium. The maximum thermal efficiency of water and moist air is to be 40.82% and 13.10% respectively, the air efficiency is lower than the water efficiency due to the specific heat (Cp) of air and thermal conductivity. Which makes better cooling effect both the side of PVT panel surface.

Conclusion

The experimental study has discussed about electrical and thermal performance of PVT module system. From the obtained result shows that, the temperature of reference PV reaches upto 65°C and its electrical efficiency dropped significantly to 1.95%. similarly, the working PVT panel temperature is absorbed around 53oC only, due to the evaporative cooling effect, the temperature drop will be significantly reduced around 10°C to 12°C. At peak our, the efficiency of PV and PVT panel achieved about 9.77% and 11.72% respectively. Although, both the side of PVT panel were cooled continuously, front surface was cooled thin film of water and rear surface is cooled by wind tunnel blower moist air thereby the thermal efficiency would be reached upto 53.92%. due to the better cooling effect the overall efficiency of PVT panel is achieved about 65.55%. it is concluded that, the PVT clay pipe panel has achieved better electrical and thermal efficiency compared than the reference PV panel.

Acknowledgement

The authors would thanks to the all authorities of Annamalai University, Annamalai Nagar, Tamil Nadu, India, for the facilities provided to conduct the experiment in the heat transfer laboratory in the Department of Mechanical Engineering for the research work

Reference

[1] Mazón-Hernández, R., García-Cascales, J. R., Vera-García, F., Káiser, A. S., & Zamora, B. (2013) Improving the electrical parameters of a photovoltaic panel by means of an induced or forced air stream. *International Journal of Photoenergy, 2013.*

[2] Chen, H., Riffat, S. B., & Fu, Y. (2011) Experimental study on a hybrid photovoltaic/heat pump system. *Applied Thermal Engineering, 31*(17-18), 4132-4138.

[3] Rasham, A. M., Jobair, H. K., & Alkhazzar, A. A. A. (2015) Experimental and Numerical Investigation of Photo-Voltaic Module Performance Via Continuous and Intermittent Water-Cooling Techniques. *Technology, 6*(7), 85-96.

[4] van Helden, W. G., van Zolingen, R. J. C., & Zondag, H. A. (2004) PV thermal systems: PV panels supplying renewable electricity and heat. *Progress in Photovoltaics: Research and Applications, 12*(6), 415-426.

[5] Tonui, J. K., & Tripanagnostopoulos, Y. (2007) Improved PV/T solar collectors with heat extraction by forced or natural air circulation. *Renewable energy, 32*(4), 623-637.

[6] Aly, S. P., Ahzi, S., Barth, N., & Abdallah, A. (2018) Using energy balance method to study the thermal behavior of PV panels under time-varying field conditions. *Energy Conversion and Management, 175*, 246-262.

[7] Dehra, H. (2017) An investigation on energy performance assessment of a photovoltaic solar wall under buoyancy-induced and fan-assisted ventilation system. *Applied energy*, *191*, 55-74.

[8] Skoplaki, E., & Palyvos, J. A. (2009) On the temperature dependence of photovoltaic module electrical performance: A review of efficiency/power correlations. *Solar energy*, *83*(5), 614-624.

[9] Popovici, C. G., Hudişteanu, S. V., Mateescu, T. D., & Chereches, N. C. (2016) Efficiency improvement of photovoltaic panels by using air cooled heat sinks. *Energy Procedia*, *85*(2016), 425-432.

[10] Haidar, Z. A., Orfi, J., & Kaneesamkandi, Z. (2018) Experimental investigation of evaporative cooling for enhancing photovoltaic panels efficiency. *Results in Physics*, *11*, 690-697.

[11] Adhya, D., Bhattacharjee, S., & Acharya, S. (2016, March). Back surface cooling of photovoltaic panel-an experimental investigation. In *2016 IEEE 6th International Conference on Power Systems (ICPS)* (pp. 1-6). IEEE.

[12] Mohammed, F. M., Mohammed, J. A. K., & Nouri, R. A. (2018). Efficiency Enhancement of a Dual-axis Solar PV Panel Tracker Using Water-Flow Double Glazing Technique. *Al-Khwarizmi Engineering Journal*, *14*(3), 32-47.

[13 Bahaidarah, H. M., Rehman, S., Gandhidasan, P., & Tanweer, B. (2013, June). Experimental evaluation of the performance of a photovoltaic panel with water cooling. In *2013 IEEE 39th Photovoltaic Specialists Conference (Pvsc)* (pp. 2987-2991). IEEE.

[14] Matias, C. A., Santos, L. M., Alves, A. J., & Calixto, W. P. (2016, June) Electrical performance evaluation of PV panel through water cooling technique. In *2016 IEEE 16th International Conference on Environment and Electrical Engineering (EEEIC)* (pp. 1-5). IEEE.

[15] Nebbali, D., Nebbali, R., & Ouibrahim, A. (2015, December) Enhanced efficiency of a solar photovoltaic concentration system cooled by a pulsed air. In *2015 3rd International Renewable and Sustainable Energy Conference (IRSEC)* (pp. 1-4). IEEE.

[16] Kordzadeh, A. (2010) The effects of nominal power of array and system head on the operation of photovoltaic water pumping set with array surface covered by a film of water. *Renewable energy*, *35*(5), 1098-1102.

[17] He, W., Chow, T. T., Ji, J., Lu, J., Pei, G., & Chan, L. S. (2006) Hybrid photovoltaic and thermal solar-collector designed for natural circulation of water. *Applied energy*, *83*(3), 199-210.

[18] Chandrasekar, M., & Senthilkumar, T. (2015) Experimental demonstration of enhanced solar energy utilization in flat PV (photovoltaic) modules cooled by heat spreaders in conjunction with cotton wick structures. *Energy*, *90*, 1401-1410.

[19] Ramkumar, R. (2017) Experimental investigation of indirect evaporative cooler using clay pipe. *Journal of Thermal Engineering*, *3*(2), 1163-1180.

[20] Rajasekar, R., Prasannaa, P., & Ramkumar, R. (2019) Efficiency of solar PV panel by the application of coconut fibres saturated by earthen clay pot water. *Environmental Technology*, 1-8.

[21] Teo, H. G., Lee, P. S., & Hawlader, M. N. A. (2012) An active cooling system for photovoltaic modules. *applied energy*, *90*(1), 309-315.

Multi-response optimization of machining factors in pocket milling of AISI304 using grey relational analysis

M Rajyalakshmi[1,2]* M Venkateswara Rao[3]

1. Research Scholar, Acharya Nagarjuna University, Andhra Pradesh, India

2. Department of Mechanical Engineering, PVP Siddhartha Institute of Technology, Andhra Pradesh, India.

3. Department of Mechanical Engineering, Bapatla Engineering College, Bapatla, Andhra Pradesh, India.

Email: j.rajyalakshmi@gmail.com, +91 9440387264

Abstract. Pocket milling finds applications in ship building and aerospace industries. To cut the required profile of the pocket on the material various tool trajectories may be used. Surface roughness is one of the quality parameters to accept products at the same time production time must be reduced to reduce the production cost. This shows the need of selecting the best parameters machining parameters to get good surface finish with less production time. Tool trajectories also play a vital role in pocket generation and production time. In the current study, first, optimum machining parameters are identified for AISI304 using Taguchi L9 Orthogonal Array and Grey Relational Analysis. Experiments are done with Speed (S), Feed (F) and Stepover (SO) as the influencing parameters. Secondly, suitable tool path is to be identified to improve the surface quality (SR) and Material Removal Rate (MRR). Two tool paths viz, follow periphery and zigzag are used to generate pockets. For each tool trajectory, from the obtained results, Grey Relational Grade is calculated. The combination of parameters with highest Grey Relational Grade is identified as the optimal parameters. Confirmation experiments indicate that predicted responses are closer to the optimal values.

Keywords: Multi-Response Optimization, Pocket Milling, AISI304, Taguchi, GRA,

1. Introduction

In the present scenario, quality of the object with less production cost and high production volume is forcing the manufacturing industry to use advanced technologies. Pocket milling is one of the basic machining processes in the manufacture of dies and molds. The quality of the pocket is directly assessed with its surface roughness Characteristic from the functional properties of the product.

AISI304 is the largely used austenitic stainless steel in various fields, where corrosion resistance and formability is given more importance. Some applications of AISI 304 include piping, heat exchangers, dish washers, fasteners etc., It is also one of the difficult to cut material. Hence its surface roughness plays important role in the quality of the end products. Productivity depends on production time and material removal rate. Surface roughness and material removal rate are affected by many factors like feed, Speed, Depth of cut, stepover, Number of passes, tool geometry etc., To obtain better surface finish the tool path can be modeled using various algorithms(1). In pocket milling, the path followed by tool to cut the given profile also important(2),(3). P.E. Romero et al.(4)studied the impact of pocket geometry and tool movement on surface roughness and cutting forces. From the study, they concluded that each tool path has its own influence on the responses based on pocket geometry selected. H. Perez, et al.(5)analyzed the effect of tool path on surface roughness and cutting force by developing a new mathematical model. A spiral tool path was produced using an algorithm in` `POWERApx" package by Martin Held et al.(6) to generate pocket in high speed machining. They have identified good matching of the tool path data with real world data. A constrained based optimization technique is applied by Bouard et al.(7)for tool path computation in pocket milling, to study the effect on machining time. J Medina et al.(8) have tested different tool movements through simulation for calculating cutting time. In the study, they tried to check the influence of interpolation, distance between those points and toolpath curvature on cutting time. A. Ghani et al. (9) applied Taguchi optimization to study the impact of cutting factors on surface roughness. The influence of cttting parameters on surface roughness in end milling is studied by Mohammed T. Hayajneh et al.(10). They used random experiments at all possible level combinations of the factors. Milon D. Selvam et al.(11)applied Taguchi and Genetic Algorithm to study the effect of process parameters on surface roughness in face milling of Mild Steel. B.Jabbaripour et al.(12) used four cutter directions and speed variations to study their influence on surface finish and cutting force.

A. Noorul Haq, et al.(13) applied Taguchi OA with grey relational analysis to analyze multi response optimization in drilling Al/SiC metal matrix composite. M Maiyar, et al.(14)analyzed multi-responses in milling of Inconel 718Super Alloy using Taguchi and grey relational analysis. Through experimentation they showed that through this approach, they are able to improve machining performance. Nayak et al.(15)analyzed multi response criteria in turning of AISI304 using Grey relational analysis based on Taguchi method. They observed 88% improvement in the responses, surface finish, MRR and cutting force through test experiments. Taguchi method coupled grey relational analysis is applied by Santha kumar et al.(16) to optimize one direction tool movement in pocketing of Ti-6Al-4V. By doing confirmation experiments, they observed that MRR, radial tool deflection and surface roughness are optimized by this method. Rajyalakshmi and Babu (17)optimized surface roughness in pocket milling of SAE304 using Taughi analysis. In the experimental study, they have applied two different tool paths and compared the surface roughness in both criteria. M. Ay (18) applied Grey Taguchi method to optimize cutting factors in turning of AISI304L steel on surface roughness, cutting forces and surface hardness. They observed with confirmation experiments, the optimal cutting parameters are obtained with this method. Kedar Mallik, et al.(19) fabricated Co-Cr-Mo alloy samples using laser sintering and applied grey Taguchi method to study the effect of heat treatment on electro chemical properties, microstructure. M. Ayand A. Etyemez(20) used Grey Taguchi method to study the effect of cutting parameter and tool path on milling of Al7075. Emel Kuram and Babur Ozcelik(21)conducted experimental analysis to optimize cutting parameters using grey based taguchi method for micro milling of Al7075. Surface roughness, cutting forces and Tool wear are analyzed to get optimum response values. M. Nurhaniza et al.(22)analyzed the effect of process on surface integrity using Taguchi orthogonal array for milling of Al-CFRP composites. From the results, they observed good surface integrity at high speed, low feed and low DOC within the selected limits. Biswajit Das et al.(23)experimented on milling of Al–4.5%Cu–TiC metal matrix composites using Grey based fuzzy logic to optimize the response parameters.

From the above discussion it was observed that most of the researchers focused on single objective optimization. Very few researchers addressed pocket milling. Influence of step over in combination with speed and feed is not reported in pocket generation. Selection of tool trajectory for multi response optimization in pocket milling can also be studied. Grey Taguchi method gives better combination for multi response optimization. Hence in the current study, Multi-response optimization using Taguchi based Grey relational analysis is applied to find the better combination of parameters for generating pockets using two types of tool paths.

2. Material and methodology:

2.1 Material: For the present study, austenitic stainless steel-AISI304 grade workpieces of size 80mmX70mmX10mm are selected. The chemical composition of the material by weight percentage is shown in table 1.

Table1: Chemical composition (Wt.%) of AISI304

Element	C	Cr	Ni	Mo	Mn	Si	N	Others	Iron

Composition (Wt. %)	0.057	18.07	8.07	-	1.72	0.29	0.08	0.075	Balance

2.2 **Methodology:** Taguchi orthogonal arrays (OA) give effective and efficient results with minimum number of experiments and reduce cost. Based on the number of factors and number of levels for each factor suitable OA is selected. "Lower the better", Signal to Noise ratio is used for surface roughness and "higher the better" for MRR.

In the present study, the machining parameters chosen are Speed, Feed, and stepover. The levels of the parameters are fixed based on literature review, machine specifications and material properties. Various levels of the machining parameters and responses are shown in table 2

Table 2: Machining parameters and their Levels selected for experimentation

Symbol	Parameters	Units	Level 1	Level 2	Level 3	Responses
S	Speed	RPM	2000	3000	4000	Surface Roughness (SR)(μm) Material Removal Rate (MRR) (gm/sec)
F	Feed	mm/ min	500	1000	1500	
SO	Step Over	%	20	30	40	

The combinations of experimental runs using L_9 orthogonal array for the specified parameters and level combinations are shown in table 3.

Table 3: Experimental runs in L_9 orthogonal array

Run No.	Speed (RPM)	Feed (mm/min)	Step Over (%)
1	2000	500	20
2	2000	1000	30
3	2000	1500	40
4	3000	500	30
5	3000	1000	40
6	3000	1500	20
7	4000	500	40
8	4000	1000	20
9	4000	1500	30

Number of researchers applied Grey Relational Analysis (GRA)(24–27) to get better combination of machining parameters in various fields of manufacturing to optimize the responses. In the present study, the experimental results are further processed with following steps to find the parameter levels for surface roughness and MRR using Grey relational Analysis (GRA).

1. Results obtained from the experiments are normalized using 'lower the better' for surface roughness using equations (1)

$$N_i(\,) = \frac{\omega_i(R_1) - min(\omega_i(R_1))}{max\,(\omega(R_1)) - min\,(\omega_i\,(R_1))} \text{ - - - - - - - - } (1)$$

and 'higher the better for MRR using equation (2).

IOP Conf. Series: Materials Science and Engineering 1112 (2021) 012006 doi:10.1088/1757-899X/1112/1/012006

$$N_i(R_2) = \frac{max\ (\omega_i(R_2)) - \omega_i(R_2))}{max\ (\omega(R_2)) - min\ (\omega_i(R_2))} - - - - - - - - (2)$$

Where, $N_i\,(R_1)$and $N_i\,(R_2)$ are the normalized responses, for surface roughness and MRR respectively.$\omega_i(R)$is the experimental value of the corresponding response, $max\ (\omega_i(R))$ and $min\ (\omega(R))$ are maximum and minimum values of the corresponding response.

2. Grey relational Coefficient (GRC) is calculated with the normalized data

$$(R) = \frac{\Delta_{min} + 0.5\Delta_{max}}{\Delta\omega(R) + 0.5\Delta_{max}} - - - - - - -(3)$$

Where $\Delta\omega(R)$=1-$N_i(R)$, for the corresponding response, Δ_{max}andΔ_{min} are maximum and minimumvalues of the $\Delta\omega_i(R)$.

3. The Grey relational Grade (GRG) is the mean of the grey relational coefficients obtained in the previous step.

$$GRG = \frac{1}{j}\sum_{R=1}^{j} \mathrm{P}(R) - - - - - - - - - - - (4)$$

Where, 'j' is the number of responses.

4. Combination of parameters having the highest GRG is considered as the optimal combination of parameters.

3. Simulation and Experimental set up:

3.1 Simulation: The Geometry of the pocket that is to be generated on the workpiece is modelled using Siemens NX 11.0 software. The tool trajectories, viz, follow periphery and zigzag are simulated on the pocket geometry separately as shown in fig. 1. Zigzag tool trajectory require finish cut also to generate the geometry accurately. NC part program for each tool path was also generated after simulation.

3.2 Experimental setup: Vertical Machining center of with Fanuc O controller is used to generate pockets with 10mm Carbide coated four flutes cutter. Experiments were conducted as per the run order generated by Minitab 11.0 software as shown in table 3. Each experiment was repeated for five times. A new cutter is used after 10 experiments for each tool path.

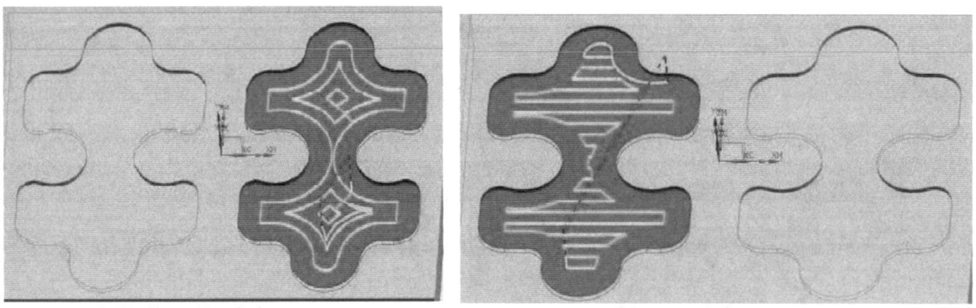

a) Follow Periphery. b) Zigzag.

Figure 1: selected Tool Trajectories for Pocket Milling.

Weight of the Work pieces are noted before and after machining using a precision balance with 0.01gm least count. Machining time for each experiment was noted during experimentation. Surface roughness (SR) is measured using Mitutoyo SJ201P surftest. MRR is calculated with equation (5)

IOP Conf. Series: Materials Science and Engineering 1112 (2021) 012006 doi:10.1088/1757-899X/1112/1/012006

$$MRR = \frac{_1 - W_2}{t_m}\, gm/sec ------(5)$$

Where, W_1, W_2 are weights of the workpiece before and after machining respectively and t_m is machining time in seconds

4. Results and discussion:

Experiments were conducted as per the order discussed in previous section and the surface roughness for each workpiece was measured. Material removal rate (MRR) was also calculated by weighing the work piece before and after machining and dividing their absolute difference by machining time. Signal to noise ratios were also calculated for each response in both tool trajectories as shown in table4.

Table 4: Response values for both tool trajectories

Expt. Run	Follow periphery				Zigzag			
	SR (μm)	S/N Ratio	MRR (gm/sec)	S/N Ratio	SR (μm)	S/N Ratio	MRR (gm/sec)	S/N Ratio
1	1.5846	-3.998	0.1706	-1.3779	1.9392	-5.753	0.10282	-19.759
2	1.5352	-3.723	0.272	2.6717	1.907	-5.607	0.15807	-16.024
3	2.175	-6.752	0.3367	4.5252	2.2328	-6.977	0.17397	-15.191
4	1.4478	-3.214	0.2565	2.163	1.793	-5.072	0.16298	-15.758
5	1.7142	-4.708	0.2135	0.5667	1.8226	-5.215	0.17812	-14.986
6	1.8656	-5.417	0.2909	3.2566	1.8226	-5.215	0.18119	-14.837
7	1.8484	-5.338	0.238	1.5154	1.931	-5.716	0.14096	-17.019
8	1.7361	-4.792	0.2353	1.4138	1.6374	-4.284	0.15701	-16.082
9	1.9344	-5.731	0.3642	5.2074	1.8872	-5.516	0.21511	-13.347

The means plots from ANOVA for follow periphery and Zigzag trajectories are shown in figure 2. From the plot it is observed that medium speed, low feed and medium step over gives minimum surface roughness value, whereas the means plot for zigzag trajectory indicates medium speed, medium feed and lower stepover gives minimum surface roughness value.

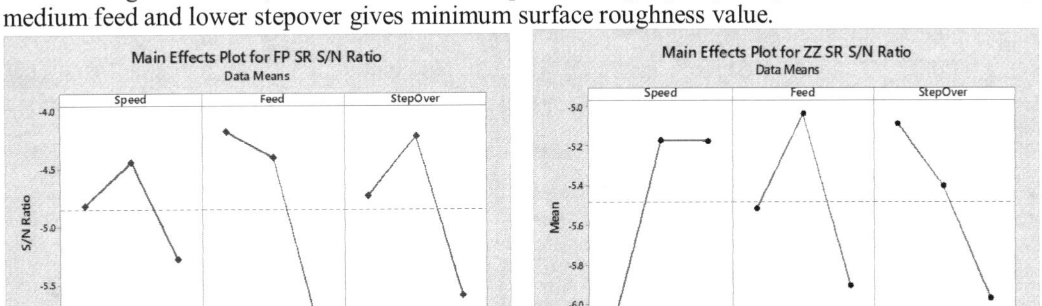

a) Follow Periphery. b) Zigzag.

Figure.2: Means plot for Surface Roughness.

The mean plots also indicate that feed has more influence on surface roughness, next comes the speed in follow periphery and speed is the most significant factor in zigzag trajectory, next comes the feed. The means plots for signal to Noise ratio from ANOVA for Material Removal Rate for both the trajectories is also shown in figure 4.

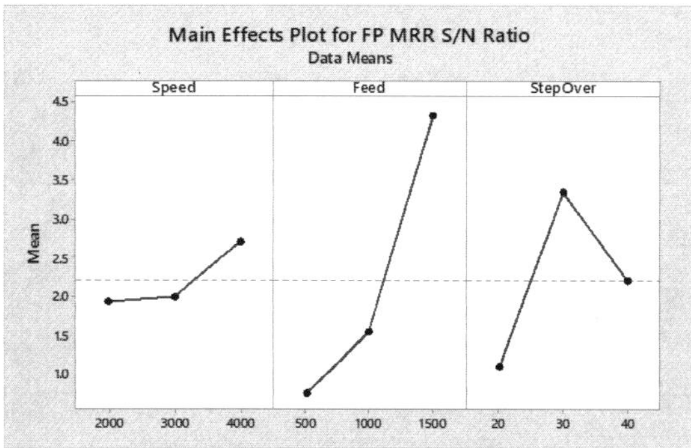

a) Follow Periphery.

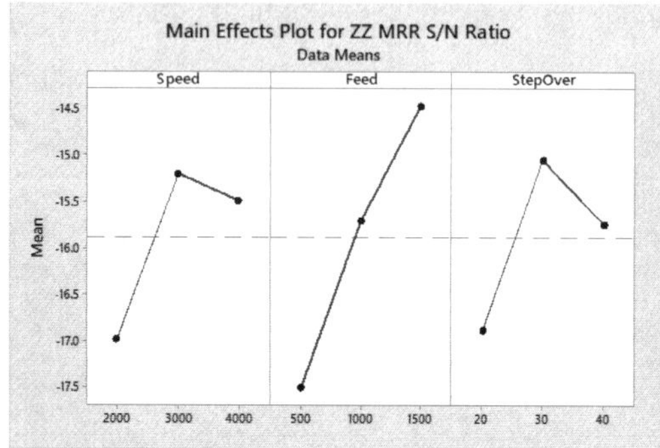

b) Zigzag.

Figure 3: Means plot for Material Removal rate.

From figure 4(a), It was observed that at higher speed, higher feed and medium step over gives maximum material removal rate and Feed is the most influencing in the parameters selected. Figure 4(b) indicates that for zigzag trajectory also higher speed, higher feed and medium step over gives maximum material removal rate, but speed is the most influencing parameter, then comes the feed.

Grey relational Grade for the experimental results are calculated for the two tool trajectories using the equations (1) to (4) and tabulated. Table 5 shows the GRG value and the ranks allotted for Follow periphery tool trajectory.

Table 5: Grey Relational values for Follow periphery.

Expt. No	SR	MRR	Normalized		Delta		Grey coefficient		GRG	Rank
			SR	MRR	SR	MRR	SR	MRR		
1	1.5846	0.1706	0.8119	0	0.1881	1	0.7266	0.3333	0.53	5
2	1.5352	0.272	0.8798	0.5238	0.1202	0.4762	0.8062	0.5122	0.6592	3
3	2.175	0.3367	0	0.858	1	0.1420	0.3333	0.7788	0.5561	4
4	1.4478	0.2565	1	0.4437	0	0.5563	1	0.4733	0.7367	1
5	1.7142	0.2135	0.6337	0.2216	0.3663	0.7784	0.5771	0.3911	0.4841	8
6	1.8656	0.2909	0.4255	0.6214	0.5745	0.3786	0.4653	0.5691	0.5172	6
7	1.8484	0.238	0.4492	0.3481	0.5509	0.6519	0.4758	0.4341	0.455	9
8	1.73614	0.2353	0.6035	0.3342	0.3965	0.6658	0.5577	0.4289	0.4933	7
9	1.9344	0.3642	0.3309	1	0.6691	0	0.4277	1	0.7139	2

From Table 5, it is observed that medium speed, low feed and medium stepover gives highest GRG value. This combination gives minimum surface roughness and near to optimum material removal rate. The grey relational grade is 0.7367. From the level means plot shown in figure 5, it is identified that low speed, high feed and medium stepover is the combination that gives highest Grey relational Grade.

Figure 4: GRG means plot for follow periphery.

Similarly, grey relational calculations for zigzag trajectory are shown in table 6. It is identified that higher speed, higher feed and medium step over gives optimal combination of parameters to get surface roughness closer to minimum and maximum material removal rate.

Table 6: Grey Relational values for Zigzag.

Expt. No	SR	MRR	Normalized		Delta		Grey coefficient		GRG	Rank
			SR	MRR	SR	MRR	SR	MRR		
1	1.9392	0.1028	0.4931	0	0.5069	1	0.4966	0.3333	0.415	9
2	1.907	0.1581	0.5472	0.492	0.4528	0.508	0.5248	0.496	0.5105	6
3	2.2328	0.174	0	0.6337	1	0.3663	0.3333	0.5772	0.4552	8
4	1.793	0.163	0.7387	0.5357	0.2613	0.4643	0.6568	0.5185	0.5876	5
5	1.8226	0.1781	0.6889	0.6706	0.3111	0.3294	0.6164	0.6028	0.6097	4
6	1.8226	0.1812	0.6889	0.698	0.3111	0.302	0.6164	0.6234	0.6199	3
7	1.931	0.141	0.5069	0.3397	0.4931	0.6603	0.5035	0.4309	0.4672	7
8	1.6374	0.1570	1	0.4826	0	0.5174	1	0.4914	0.7458	2
9	1.8872	0.2151	0.5805	1	0.4195	0	0.5438	1	0.7719	1

The level means plot drawn for grey relational data of zigzag trajectory shown in figure 6. From the figure also it was indicated that higher speed, higher feed and medium stepover gives optimum response values.

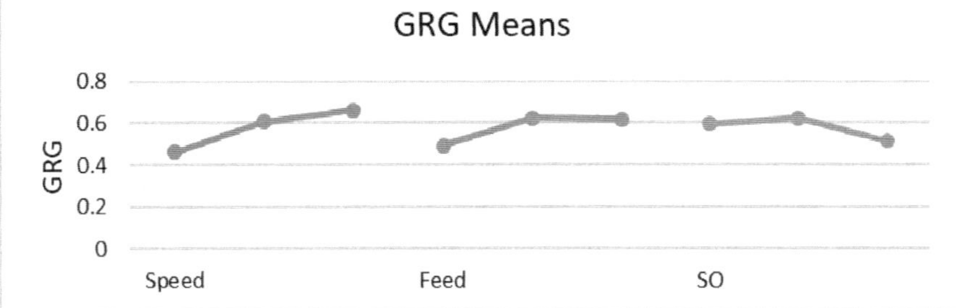

Figure 5: GRG means plot for Zigzag

Confirmation experiments are conducted for follow periphery to test the optimal combination of parameters. The test results confirm that low speed, high feed and medium stepover gives surface roughness and material removal rates closer to the optimal value. The Grey relational Coefficient obtained for the 2000RPM speed, 1500mm/min feed and 30% stepover is 0.8417. The surface roughness value obtained is 1.5832µm and the material Removal Rate is 0.3596gm/sec. There is 14.25% improvement in the GRG Value for the optimal combination of parameters.

Conclusions

Present study is concentrated on multi response optimization of machining parameters using Taguchi and Grey Relational Analysis. One linear and one nonlinear tool trajectories viz, zigzag and follow periphery are used to generate pockets. For each tool trajectory, Experimental results are further processed with Grey relational analysis. Following conclusions can be drawn from the study.

1. For follow periphery tool path, the combination of parameters from the mean plots is low speed, high feed and medium stepover. Confirmation test ensures the result with 14.25% increase in the GRG value.
2. For zigzag tool path, the optimal combination from the mean plot is higher speed, higher feed and medium stepover and experimental GRG value confirms the combination as the best combination for generating surface roughness closer to minimum and MRR closer to maximum values.
3. Taguchi method when combined with Grey relational analysis gives better results to find the optimal combination of parameters in multi-response optimization. Follow periphery gives better surface roughness and material removal rate when compared to zigzag within the limits of the parameters selected
4. ANOVA results shows that for follow periphery feed is the influencing factor and for that of zigzag, speed is most influencing factor.

References

1. Bieterman MB, Sandstrom DR 2003 *J Manuf Sci Eng Trans ASME* **125** 709–15.
2. Toh CK. 2004 *J Mater Process Technol.* **152** 346–56.
3. Gologlu C, Sakarya N. T 2008 *J Mater Process Technol.* **206** 7–15.
4. Romero PE, Dorado R, Díaz FA, Rubio EM 2013 *Procedia Eng [Internet].* **63**:523–31.
5. Perez H, Diez E, Perez J, Vizan A. 2013 *Procedia Eng [Internet]* **63** 573–81.
6. Held M, Spielberger C 2009 *CAD Comput Aided Des [Internet]* **41** 539–50.
7. Bouard M, Pateloup V, Armand P. 2011 *CAD Comput Aided Des [Internet].*;**43** 1099–109.
8. Medina-Sánchez G, Torres-Jimenez E, Lopez-Garcia R, Dorado-Vicente R. 2017. *Procedia Manuf.* **13** 59–66.
9. Ghani JA, Choudhury IA, Hassan HH. A 2004 *J Mater Process Technol.* **145** 84–92.
10. Mohammed T. Hayajneh MSTJB. 2007 *Jordan J Mech Ind Eng.* **1** 1–5.
11. Selvam MD, Karuppusami G, Dawood AKS. 2012 *An Int J (ESTIJ.)* **2** 2250–3498.
12. Jabbaripour B, Sadeghi MH, Faridvand S. 2010 *Jordanian International Mechanical Engineering Conference.*
13. Haq AN, Marimuthu P, Jeyapaul R. M. 2008 *Int J Adv Manuf Technol.* **37** 250–5.
14. Maiyar LM, Ramanujam R, Venkatesan K, Jerald J. 2013 *Procedia Eng [Internet].* **64**:1276–82.
15. Nayak SK, Patro JK, Dewangan S, Gangopadhyay S. 2014 *Procedia Mater Sci [Internet]* **6** (Icmpc):701–8.
16. Santhakumar J, Iqbal UM, Prakash M. 2018 *IOP Conf Ser Mater Sci Eng.*; **402**.
17. Rajyalakshmi M, Babu PS. 2018 *Journal of Production Research & Management* **8** 49–56.
18. Ay M. 2017 *Acta Phys Pol* **A**. **131** 349–53.
19. Mantrala KM, Das M, Balla VK, Rao CS, Kesava Rao VVS. 2015 *Front Mech Eng* **1** 1–7.
20. Ay M, Etyemez A. 2018 *O Acta Phys Pol A.* **134** 125–8.
21. Kuram E, Ozcelik B. 2013 *Meas J Int Meas Confed.* **46** 1849–64.
22. Nurhaniza M, Ariffin MKAM, Mustapha F, Baharudin BTHT. 2016 *Int J Manuf Eng.* **2016** 1–9.
23. Das B, Roy S, Rai RN, Saha SC. 2016 *Eng Sci Technol an Int J.* **19** 857–65.

24. Gajalakshmi K, Senthilkumar N, Prabu B. 2019 *Meas Control (United Kingdom).*;**52** 540–53.
25. Kibria G, Doloi B, Bhattacharyya B. 2013 *Opt Laser Technol [Internet]* **48** 16–27.
26. Pradhan MK. 2013 *Int J Adv Manuf Technol.* **67** 2051–62.
27. Mallik MK, Srinivasa Rao C, Kesava Rao VVS, Abdul Munaf SK. 2016 *ARPN J Eng Appl Sci.* **11** 12188–91.

Six Sigma Approach for Material Handling System for a Tobacco Threshing Plant

DR. MD.KHAJA MOINUDDIN FAROKHI*1, DR. P V CHALAPATHI[2], DR. K.V.L. SOMASEKHAR[3], M. VINITHA[4] P Naga Srinivas[5]

[1] Research Scholar, Department of Mechanical Engineering, KLEF Guntur, India
[2] Professor, Department of Mechanical Engineering, K L E F, Guntur, India
[3,4,5]Department of Mechanical Engineering, Vasireddy Venkatadri Institute of Technology, Guntur, A.P., India
*Correspondence email: farooqui.mm@rediffmail.com

Abstract
The prevailing global market which is competitive and fierce demands superior quality and high flexibility of production with shorter lead times. In order to stay competitive in the market, companies need to attain both customer satisfaction and cost reductions in production operations. Material Handling Systems (MHS) is the place to accomplish this goal, since they have direct impact on production. Material handling and its storage are features of all the companies although they are unexplored areas. Significant contribution made in the material handling area will go a long way in reducing the cost of production and improving profitability and rate of return. Therefore, the aim of this paper is to analyze how an in house MHS that could be efficient for the production it serves. With this intention, a case study has been conducted in A.S.KRISHNA&COMPANY PVT.LTD. Guntur. During the information is gathered through various sources, interviews, observations and measurements. Gathered data is further evaluated according to theoretical framework. By analyzing the findings from literature review and study, first problems and challenges related to present MHS's are identified. Thereafter, possible features that the system should possess are drawn out and a design is built out of selected features with a compatible plant layout.

Key Words
Six Sigma, Customer satisfaction, Cost reductions Global market, Material handling system

Introduction

Today's fierce competitive global markets, short product life cycles, and increased customer expectations have forced organizations to recognize the viral importance of investing and focusing on their logistics systems in terms of gaining competitive advantages. Material handling and storage features of all the companies although for many years they are unexplored areas. So to counter such problems proper research should be done. Logistics involves activities that ensure the necessary material is available at the right place and at the right time [1, 2].

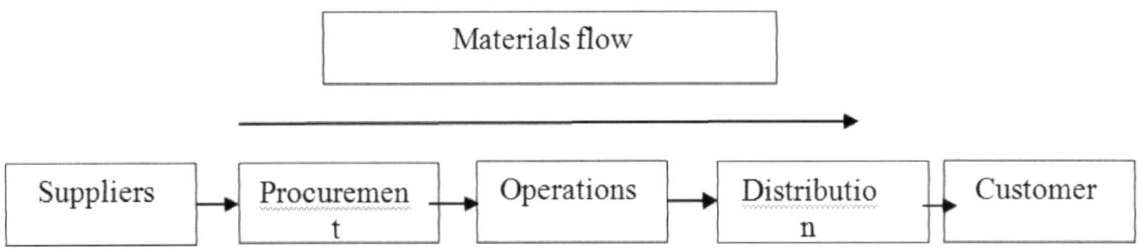

Figure: 1 Logistics Management Process

The primary objective of the store is to provide storage of materials. Some methods of storage are associated with particular method of handling materials and therefore task of handling or stock and that of storage must be taken together. Therefore provision of material handling is another essential objective of the store function. The plant should be designed in such a way that easy flow of items or materials held are ensured without hindrances and minimizing the waste of time. Today materials are the livelihood of any company and no company can operate

without them, they must be made available at the right time, right quality, right quantity, right place and the right price to promote profitability of the company [3, 4].
The company procures its tobacco products from reliable vendors. The timeframe can be still reduced by updating certain methods and management techniques being adopted in the threshing plant. The infrastructure alone cannot bring the best outcomes but the way they are being handled and proper sequence of operations can lead the company to greater profits in terms of money, man and material. The different types of tobacco leaves being processed are gathered from different corners of the A.P and surrounding areas which makes the farmers around the Guntur locality to sell their crops at good price with less transportation costs [5, 6].
Green Leaf Threshing Plant changes the raw tobacco into a processed tobacco leaves which had its extensive usage for manufacturing cigarettes and tobacco products. Blending, tipping, conditioning, classification and re-drying are the important unit operations in tobacco threshing. The quality specifications during threshing operations are stringent and difficult to adhere [7].

Table 1: Typical Plant Layout of a Conventional Threshing Plant

Quality Parameter of Lamina	**Approximate Test Interval (min)**	**Sample size(gms)**	**Procedure And Apparatus**
Particle Size	20	2500 to 3500	Lamina particle size tester (Degradation shaker)
Stem Content	20	2500 to 3500	Stem Tester
Moisture	5	50	Brabander (quick moisture analyzer) and Hearson oven
Quality Parameters of Stem			
Length	60	250	Stem Length Board
Diameter	60	250	Rotap Tester
Moisture	30	50	Brabander and Hearson oven

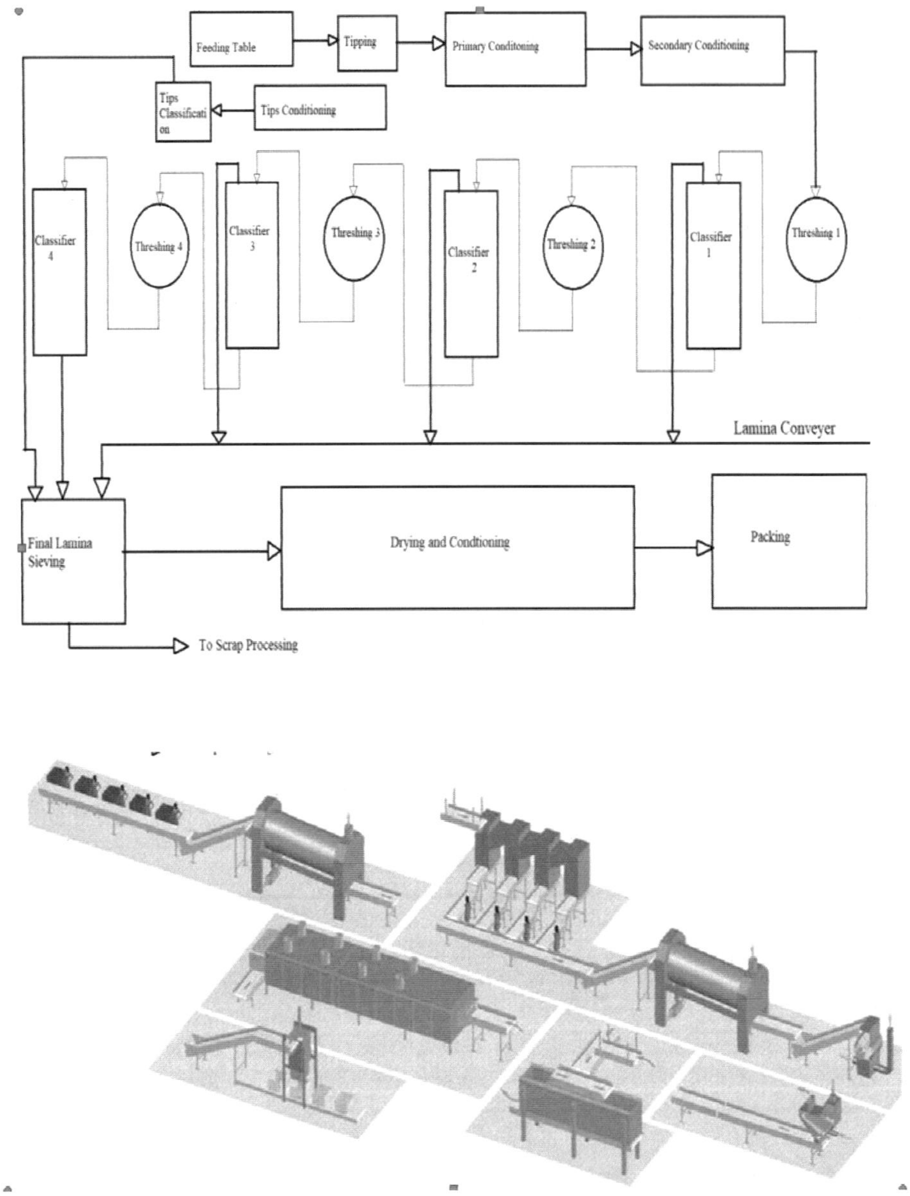

Figure 2: Flow process starting from the raw material (lamina) to the end material i.e., packing.

Experimentation

The present study was carried out on the MHS available in the company. To provide a deeper knowledge about the problems and challenges in the current in- house MHS and also to be able to suggest suitable solutions for the arisen issues a case study was conducted in this research. The DMAIC methodology is followed by a sequence of operations followed to evaluate the study. This study utilized descriptive survey which involves asking the same set of questions to a large number of individuals either by telephone, mail or in person. This design provides a precise picture of events and it also seeks to explain perceptions and behavior on the basis of data gathered at a particular time. Nevertheless results produced by this design can however be erratic because the questions which are normally asked seek to probe into private matters of the respondents and respondents may not be completely truthful about the response.

Table 2: Main classes and Attributes of Material Handling System

MHS Equipment Type	**Load Type**	**Load Capacity**	**Size**	**Nature**	**Speed of System**	**Accumulation Requirement**	**Distance**	**Frequency of Move**	**Flexibility of Path**	**Loading and Unloading Ability**
Robots	Discrete	Low – Medium	Medium	Solid	Low – Medium	no	short	Often	Low	Medium
AGVs	Discrete	Medium	medium	Solid - Fragile	Medium	no	Medium	Often	High	High
Rail Guided Vehicles	Discrete	High	Medium-large	Solid	High	no	Long	Low	Low	High
Gantry	Discrete	Low-medium	Medium	solid	Low-medium	no	Medium	Low	Low	Low-medium
Forklift	Discrete	High	Large	Solid	High	no	Long	High	High	Low
Conveyer	Continuous	Low-medium	Small-medium	Solid	Low-medium	yes	Short-medium	Low	Low	Low – Medium
Manual	Discrete	Low	medium	Solid	Low	no	short	high	High	Medium

Results and Discussion
To gain the complete knowledge of the production let us see the time attribute taking place in each step of the production.

Table 3: Table of the time attribute of the production processes inside the plant

Time Attributes

Process	Time Taken (seconds)
Unloading a bale on conveyer	45
Travelling Time on Conveyer 1	19.28
Conveyer to condenser	18.33
Inclined Conveyer to Condenser	12
Condenser operation	30
Non Tobacco material Removal	120 (2 mins)
Travelling in Pneumatic unit	600 (10 mins)
Inside Dehydrator	360 (6 mins)

Dehydrator to auto press	53
Auto press	600 (10 mins)

As the time attributes are calculated simultaneously the production data is also calculated in the plant. As a part of this case study the entire day's production is calculated. The details are like this. The production inside the plant starts at 8.40 am in the morning.

Table 4: Timings along with the processes at the starting of production in a day

S. No	Time	Process	Remarks
1	8.00 - 8.40	The machinery starts in the plant.	Trail run of machinery
2	9.01	Removal of the first bale and placing onto conveyer	Delayed to 9.01 from 8.47 because of labor problem
3	9.04	Outlet of steamed leaves from conditioner	Leaves get buffered in conditioner
4	9.24	Leaves coming into dehydrator from threshing plant	The leaves get threshed and segregated for 20 mins in pneumatic unit.
5	9.36	Leaves comes outside from the dehydrator for packing	14 mins of time needed to get equal moisture content.
6	9.42	The processed tobacco fills and gets rammed into box of required weight.	To maintain the density inside the plant the leaves get rammed with high pressure
7	9.5	The second comes to fill and the first box leaves the filling station.	The box gets strapped and leaves to the warehouse

The number of bales consumed per day: 500bales

The processed box comes from500 bales: 180 boxes

<u>Productivity</u>

On an average the input of raw tobacco is 50tons
The yield of processed tobacco is 60 to 70% of raw bales.
<u>Electricity Consumption</u>:

The power consumed for hour is 500 units
The power consumption for a day is nearly 4500 units. The cost of single unit is 9 rupees.
<u>Labor and their Wages:</u>
The daily wage workers inside the plant are about 80 members.
The office staff constitutes of 30 members
On the whole,The running cost of the plant per kg is 8Rs.
The running cost of the plant per day 4 Lacs.

<u>Financial Impact because of the Solution</u>

The solution lets the company to lessen the human fatigue at the different places of the production leading to save the production time and the cost of production incurred previously on the production happened in the plant.

Let us go through a case happening in the plant

Number of workers involved in loading and unloading of the bales and processed tobacco are 60 men.

The daily wage of a labor is 600 to 800 rupees depending up on their work. The money incurred on wages for the workers by the plant is:

Wage on an average 750Rs

Total money incurred per day on them 750*60=45,000
Rs. Total money incurred per day on power
4500*9=40,500R
s.
The power consumed by the machinery costs less than the wages for the workers. So, by using the above solution the loss for the company in this particular aspect can be removed by implementing the solution which cost less than the wage of labors.

Conclusion

The tobacco will go through several treatments for about 30 minutes in threshing process. Hence the processing loss is high. New methods and process optimization must be tried for all stages such as conditioning, threshing and classification to minimize the processing loss. In GLT operation of FCV tobacco, around 8 to12% of the tobacco is packed as scrap (by-products) and these products are sold for the production of bidi and other low quality products at a low price. In some developed countries, the tobacco by-products are converted into sheet tobacco or reconstituted tobacco and used along with the lamina. But due to high initial investment, this technology has not been used in India. Low cost methods to produce reconstituted tobacco must be developed to give value addition to the tobacco byproducts.

So our project which had been done on the Design and Analysis of Material Handling for the Tobacco threshing plant is mainly concentrated on the

- Reduction of the human effort
- Reducing the time being wasted in material movement in production area precisely at
  - Production lines
  - Raw material loading and unloading
  - Final Product Loading
- Reducing the facial issues incurred in material Management.
- Suggesting the best plant Layout.

References

[1] V.Cozzania, F.BarontinibT.McGrathc, M.NordlundcM.Smith, G.Zuberc," An experimental investigation into the operation of an electrically heated tobacco system" Thermochimica Acta ,Volume 684, pp.1-12

[2] Randi V.Wilfert Eckel Emmett P.Lampert, "Effect of tobacco etch virus on the seasonal growth of flue-cured tobacco", Crop Protection, Volume 12, No. 7, pp. 505-512

[3] Thamalampudi VenkataReddy, Samresh Dwivedi, Navin Kumar Sharma," Development of TILLING by sequencing platform towards enhanced leaf yield in tobacco", Industrial Crops and Products, Volume 40, pp.324-335

[4] AtsushiNagai, ToshikiMine, TakeshiYamamoto, HiroyukiWariishi, "Transfructosylation reaction in cured tobacco leaf (*Nicotiana tabacum*)", Journal of Bioscience and Bioengineering, Volume 116, No. 6, pp. 666-671

[5] WenhuaZiaYubao, ChenaYihong Panb, Yougang Zhanga, Yufeng Hec, QiangWanga" Pyrolysis, morphology and microwave absorption properties of tobacco stem materials" Science of The Total Environment, Volume 683, pp. 341-350

[6] SungbumJuna, SeokcheonLeeb, YuehwernYihb," Pickup and delivery problem with recharging

for material handling systems utilising autonomous mobile robots" European Journal of Operational Research, Volume 289, No 3, pp. 1153-1168

[7] Chris Siu KeiLeungHenry Ying KeiLau, "A hybrid multi-objective AIS-based algorithm applied to simulation-based optimization of material handling system", Applied Soft Computing, Volume 71, pp. 553-567

IOP Conf. Series: Materials Science and Engineering 1112 (2021) 012008 doi:10.1088/1757-899X/1112/1/012008

Hydrodynamic Performance of Positive Surface Textured Patterns under Mixed Lubrication

Venkateswara Babu Peddakondigalla*[1], Kiran Kumar Veeranapu[2], Ismail Syed[3], Vasavi Boggarapu[4]

[1,2]Department of Mechanical Engineering, Vasireddy Venkatadri Institute of Technology Guntur, India-522508
[3,4]Department of Mechanical Engineering, National Institute of Technology Warangal, India-506004
*Corresponding author E-mail address: venky5041@gmail.com

Abstract: It is proved that the optimal design of surface textures (positive or negative) improve the lubrication performance by generating extra hydrodynamic pressure at the interface of the sliding pairs. In the present work, the effect of various positive surface textures on hydrodynamic performance of parallel sliding contact under mixed lubrication is researched. The modified Reynolds equation and asperity contact model are solved simultaneously to assess the hydrodynamic pressure and minimum film thickness for different shape of textures. The results depict that elliptical shaped texture generate high hydrodynamic pressure as well as minimum film thickness at the conjunction than the other texture shapes.

Keywords: Hydrodynamic pressure, Minimum film thickness, Mixed lubrication, Surface texture

1. Introduction

Numerous research works have reported the effects of surface texturing on hydrodynamic and frictional performance under sliding lubrication conditions. In this context, Henry et al. [1] examined the influence of square-shaped dimple textures on load carrying and frictional performance of parallel thrust pad bearing. The results depicted a friction reduction of 30% compared to un-textured surfaces. In contrary, Zhang et al. [2] considered rectangular and circular shaped dimples on piston ring surfaces to assess the frictional performance at top and dead centers of an engine. The rectangular textured surface has shown lower average friction co-efficient than circular shape textures. Furthermore, Rahmani et al. [3] derived the optimal geometrical conditions of a parallel slider bearing with square shaped dimples to evaluate the tribological properties along with the load support, frictional force and friction coefficient. The study concluded the dimple length ratio (the ratio of dimple length to distance between two consecutive dimples) of 1 and 1.5-1.7 exhibit a superior tribological performance. The authors have further extended their work by considering rectangular, equilateral as well isosceles triangular shaped textures. The rectangular shaped textures have shown superior tribological performance than the other. The optimal area density of dimple shaped textures in reducing the friction at different lubrication regimes was reported by Gachot et al. [5]. The authors have found the optimal results at area densities of 10-30%, 7-12% and 5-7% under fluid-film, mixed and boundary lubrication conditions respectively.

IOP Conf. Series: Materials Science and Engineering 1112 (2021) 012008 doi:10.1088/1757-899X/1112/1/012008

The research works reported on positive textures (protrusions) displayed a remarkable influence in lowering the friction at the interface under lubricating conditions. In this regard, Sawant et al. [6] examined the spot textures (protrusions) on machining performance of high speed steel (HSS) tools. The protrusions on rake face of the tool were fabricated through plasma transferred arc powder deposition technique. The authors have proved the reduction in cutting and thrust forces during machining with the aid of spot textured cutting tools than un-textured tools. Furthermore, Obikawa et al. [7] introduced the square shaped protrusions on rake face of carbide cutting tools that yield a better lubricating conditions in orthogonal machining. More recently, Venkateswara Babu and his research group [8-11] carried out an extensive work on positive surface texturing, which exhibited a significant impact on tribological performance. The authors conducted sliding experiments by introducing square-shaped protrusions on the sliding surfaces, which reduced the friction under mixed lubrication conditions [8]. In continuation of their research work, the authors have developed mixed lubrication model to validate their experimentation results as cited in [9].

However, conducting experiments for various textured shapes involves high cost and time. Keeping in view, the present work considers a theoretical model of textured parallel sliding contact as mentioned in [9] to study the effect of positive texture geometry on hydrodynamic pressure generation and minimum film thickness.

2. Numerical Model

To evaluate the hydrodynamic performance of various texture shapes (protrusions), the parallel sliding contact is mathematically modeled by including textures with square, triangular, circular and elliptical shapes. The configuration of sliding contact with square-shaped textures is shown in Fig.1.

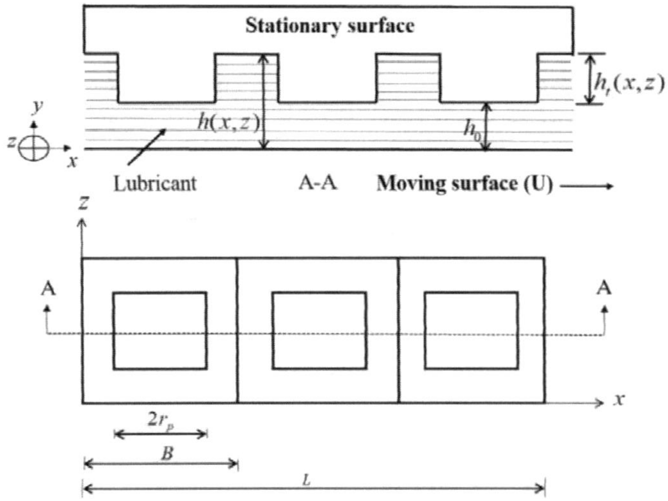

Fig.1 Square-shaped textured parallel sliding contact configuration

The symbols in Fig.1 are defined as follows;

B, L are axial width and length of the stationary surface respectively, h is average oil fil thickness, h_0 is minimum oil film thickness, h_t is height of texture, $2r_p$ is base length of texture, U is sliding velocity of moving surface and x, y, z is global coordinate system.

2.1 Film thickness equation

The oil film thickness (h) can be estimated by the following equation [9].

$$h(x,z)=\begin{cases} h_0 & \text{above texture} \\ h_0+h_t(x,z) & \text{elsewhere} \end{cases} \tag{1}$$

The oil film thickness profiles for various texture shapes are shown in Fig.2, corresponding to an area density (ratio of area of texture to the area of unit cell), S_p of 0.1 and h_t of 10 μm.

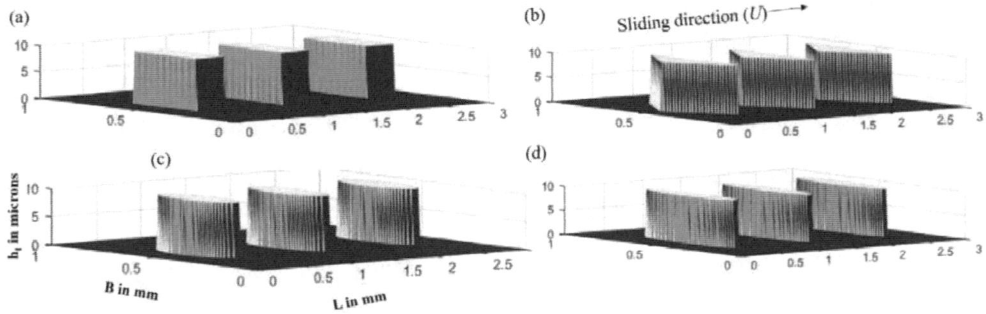

Fig.2 Oil film thickness profiles for texture shapes of (a) square, (b) triangular, (c) circular, (d) elliptical

2.2 Modified Reynolds equation

To evaluate the hydrodynamic pressure (p), the modified Reynolds equation, which includes the fluid flow factors is applied [12, 13].

$$\frac{\partial}{\partial x}\left(\phi_x\frac{h^3}{\mu}\frac{\partial p}{\partial x}\right)+\frac{\partial}{\partial z}\left(\phi_z\frac{h^3}{\mu}\frac{\partial p}{\partial z}\right)=6U\phi_c\frac{\partial h}{\partial x}+6U\sigma\frac{\partial\phi_s}{\partial x} \tag{2}$$

The symbols in Eq. (1) are defined as follows;
p is hydrodynamic pressure, μ is viscosity of the lubricant, σ is the composite roughness of interacting surfaces and the factors $\o_c$, $\o_s$, $\o_x$, $\o_z$ are factors of contact, shear and pressure flows respectively. These are calculated based on the relations given in [12-14].

The boundary conditions to estimate p in x- and z-directions are considered as follows;

$$p_{(x=0)}=p_{(x=L)}=0 \text{ and } p_{(z=0)}=p_{(z=B)} \tag{3}$$

Eq. (2) is solved by satisfying above boundary conditions using finite difference method to obtain hydrodynamic pressure, p.

2.3 Pressure balance criteria

In mixed lubrication, the external load (W) acting on the contact is responsible for generation of both hydrodynamic pressure (p) and asperity pressure (p_{asp}), which can support the total load. Therefore, total pressure is summation of hydrodynamic and asperity pressures. However, it is required to obtain greater values of p than p_{asp} as it enhances the total oil film load carrying capacity and decreases the frictional force. Furthermore, the higher minimum film thickness values are beneficial to minimize friction as well as wear. Therefore, the present work aims on obtaining the optimal values of p and minimum film thickness for various shapes of texture under mixed lubrication.

2.4 Hydrodynamic and asperity pressure equations

The oil film pressure is obtained by estimating Eq. (2). The asperity pressure (p_{asp}) can be obtained by solving Greenwood-Tripp model [15], which is given as;

$$p_{asp} = K'E'F_{2.5}(h/\sigma) \tag{4}$$

$$K' = \frac{16\sqrt{2}}{15}\pi(\eta\beta\sigma)^2\sqrt{\frac{\sigma}{\beta}} \tag{5}$$

$$E' = \frac{1}{\frac{1-v^2_1}{E_1}+\frac{1-v^2_2}{E_2}} \tag{6}$$

Herein, the parameters η, β, $F_{2.5}$ (h/σ) are related to asperity distribution of the surfaces, which depends on the ratio of film thickness, h/σ. All these parameters are adopted from the previous studies [9].

3. Results and discussion

It is well known that, the higher values of hydrodynamic pressure and minimum film thickness are beneficial to obtain lower friction and wear of the sliding pairs. In this context, the simulations are performed to obtain optimal values of hydrodynamic pressure and minimum film thickness for different texture patterns. To analyze the different shapes of textured patterns on the hydrodynamic performance, the parameters included in numerical simulations are given in Table 1.

Table 1 Parameters of numerical analysis [9]

Parameter	Value
Material parameters	
Elastic modulus of moving surface, E_1	215 GPa
Elastic modulus of stationary surface, E_2	200 GPa
Poisson's ratio of moving surface, v_1	0.3
Poisson's ratio of stationary surface, v_2	0.3
Composite roughness, σ	0.546 μm
Texture parameters	
Area density, S_p	0.1, 0.2, 0.3
Texture height, h_t	10 μm
Operating parameters	

Dynamic viscosity of fluid μ	0.121 Pa s
Axial length of the textured surface, L	3 mm
Unit cell size, B	1 mm
Sliding velocity, U	1.67 m/s
External load, W	0.5 N

The hydrodynamic pressure percentage of various shapes of surface texture at different area densities are shown in Fig.3. The higher area densities generated the more hydrodynamic pressures for all the shapes of texture patterns. However, elliptical patterns generate maximum hydrodynamic pressure percentage among all the texture patterns. Because, the hydrodynamic pressure developed in elliptical textures occupy more area under the curve as shown in Fig.4 (d), due to which hydrodynamic pressure is maximum when compared to other texture shapes (can be seen in Fig.4). It confirms that elliptical textures are superior in generating additional fluid pressure, thereby increasing the ability to support the total load. Moreover, among all the shapes, circular shape of texture exhibited low performance in generating extra fluid film pressures at all area densities.

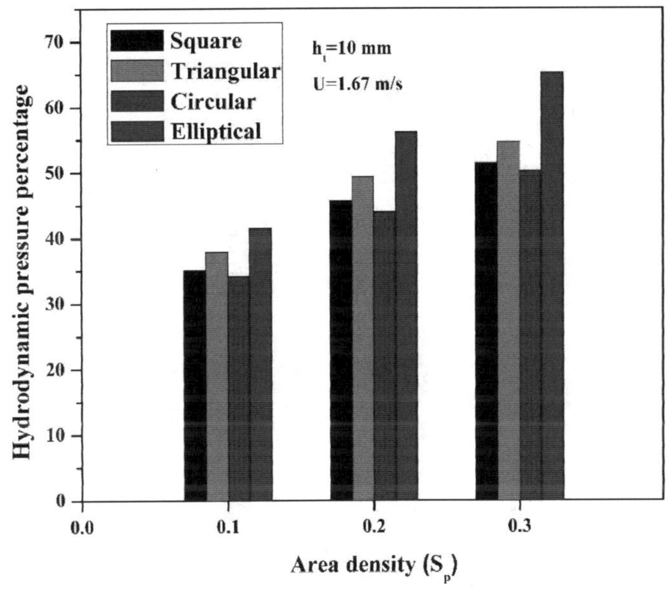

Fig.3 Hydrodynamic pressure percentage of various texture shapes

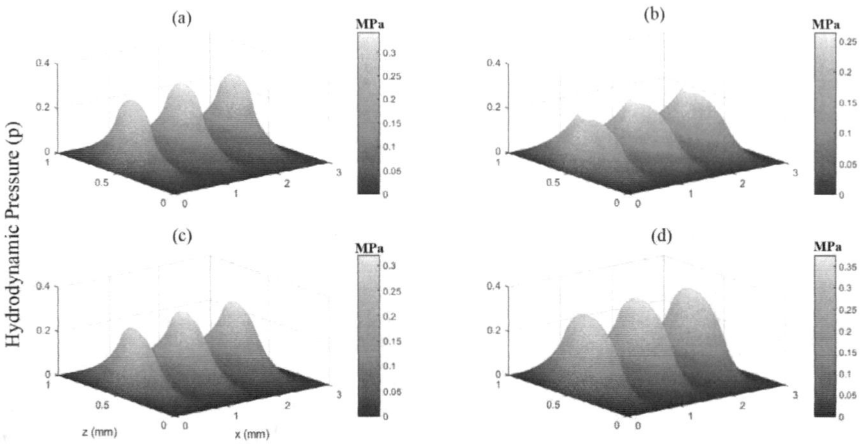

Fig.4 Hydrodynamic pressure distribution of various texture shapes of (a) square, (b) triangular, (c) circular, (d) elliptical at S_p=0.1 and h_t=10 μm

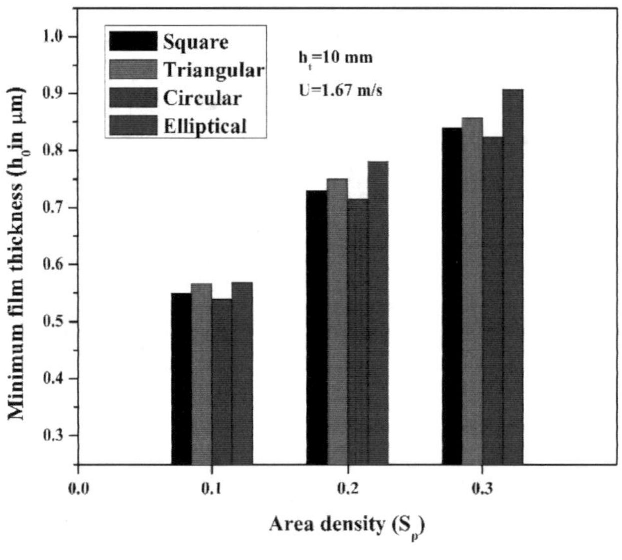

Fig.5 Minimum film thickness variation for different texture shapes

The minimum oil film thickness for various texture shapes is described in Fig.5. For all texture shapes, the area density has shown predominant effect on minimum film thickness. In particular, larger area densities exhibited greater minimum film thickness values. Furthermore, as for the shape of texture's concern, the elliptical patterns exhibited the superiority in generating higher minimum film thickness values. This is due to the domination of elliptical textures in generating more fluid film pressures (see Fig.4), which resulted in higher minimum film thickness values as an extra hydrodynamic lift at the junction separates the sliding surfaces.

4. Conclusion

The concluding remarks of the present study are as follows;

- The texture area density has shown a prominent impact on hydrodynamic pressure generation and minimum film thickness. In this context, the higher area densities are superior to lower area densities.
- As for texture shape's concern, elliptical shape of texture exhibited better performance by generating extra hydrodynamic pressure and minimum film thickness at the conjunction.

References

[1] Henry Y, Bouyer J and Fillon M 2015 An experimental analysis of the hydrodynamic contribution of textured thrust bearings during steady-state operation: A comparison with the untextured parallel surface configuration *Proc. Inst. Mech. Eng. J* **229** 362-375.

[2] Zhang Y, Zhang X, Wu T and Xie YB 2016. Effects of surface texturing on the tribological behavior of piston rings under lubricated conditions *Ind. Lubr. Tribol.* **68** 158-169.

[3] Rahmani R, Shirvani A and Shirvani H 2007 Optimization of partially textured parallel thrust bearings with square-shaped micro-dimples *Tribol. Trans.* **50** 401-406.

[4] Rahmani R, Mirzaee I, Shirvani A and Shirvani H 2010 An analytical approach for analysis and optimisation of slider bearings with infinite width parallel textures *Tribol. Int.* **43** 1551-1565.

[5] Gachot C, Rosenkranz A, Hsu SM and Costa HL 2017 A critical assessment of surface texturing for friction and wear improvement *Wear* **372** 21-41.

[6] Sawant MS, Jain NK and Palani IA 2018 Influence of dimple and spot-texturing of HSS cutting tool on machining of Ti-6Al-4V *J. Mater. Process. Technol.* **261** 1-11

[7] Obikawa T, Kamio A, Takaoka H and Osada A 2011 Micro-texture at the coated tool face for high performance cutting *Int. J. Mach. Tool Manuf.* **51** 966-972.

[8] Babu PV, Syed I and Ben BS 2019 Influence of positive texturing on friction and wear properties of piston ring-cylinder liner tribo pair under lubricated conditions *Ind. Lubr. Tribol.* **71** 515-524.

[9] Babu PV, Syed I and Ben BS 2020 Experimental and numerical studies of positive texture effect on friction reduction of sliding contact under mixed lubrication *Proc. Inst. Mech. Eng. J* 1350650120930911.

[10] Babu PV, Syed I and Ben BS 2020 Experimental investigation on effects of positive texturing on friction and wear reduction of piston ring/cylinder liner system *Mater. Today. Proc.* **24** 1112-1121.

[11] Babu PV, Syed I and Ben BS 2020 Optimization of Texture Geometry for Enhanced Tribological Performance in Piston Ring-Cylinder Liner Contact under Pure Hydrodynamic and Mixed Lubrication *Innovative Product Design and Intelligent Manufacturing Systems* Springer Singapore 799-808. (Editors: Deepak BBVL, Parhi DRK, Jena and Pankaj C)

[12] Patir N and Cheng HS 1978 An average flow model for determining effects of three-dimensional roughness on partial hydrodynamic lubrication *J. Lubr. Technol.* **100** 12-17.

[13] Patir N and Cheng HS 1979 Application of average flow model to lubrication between rough sliding surfaces *J. Lubr. Technol.* **101** 220-229.

[14] Meng F, Wang QJ, Hua D and Liu S 2010 A simple method to calculate contact factor used in average flow model *J. Tribol.* **132** 024505.

[15] Greenwood JA and Tripp JH 1970 The contact of two nominally flat rough surfaces *Proc. Inst. Mech. Eng. J* **185** 625-633.

Optimization Of Process Parameters In 3d Printing –Fused Deposition Modeling Using Taguchi Method

Dr M Sumalatha[1], J N Malleswara Rao[2]and B Supraja Reddy[3]

[1,3] Assistant Professor Mechanical Engineering Department, V. R. Siddhartha Engineering College, Vijayawada, Andhra Pradesh.
[2]Associate Professor, ME Dept, Dhanekula Institute of Engg & Tech, Ganguru, Vijayawada, Andhra Pradesh.
Corresponding Author: sumalathame@vrsiddhartha.ac.in,

ABSTRACT:
3D printing or Rapid Prototyping (RP) or Additive Manufacturing (AM) is an innovative manufacturingmethod that is mostly used for prototypes in various Industries such as Aerospace, Defense, Medical, Automotive, etc. One of the worthy and widely used 3 D printing techniques is Fused Deposition Modeling (FDM), which involves adding melted material layer by layer. FDM has several benefits over other manufacturing methods. While setting the 3 D printing options, we have to take several parameters into account, such as speed, layer thickness and infill density, etc.The present experimental work illustrates the performance of ABS built parts fabricated by AION 500make FDM machine. The quality of parts produced by FDM process mainly depends on the selection of process parameters. The present research work consists of process parameters such as layer thickness, infill density and speed of deposition, etc. This study aims to find effects of process parameters on different performance parameters i.e., mechanical considerations (impact strength), build time, surface roughness in a systematic manner with less number of experimental runs. Taguchi design of experiment (DOE) approach has been used to save cost and time of experimentation. Statistical significance of process parameters is analyzed using analysis of variance (ANOVA). The significant parameters for optimum results and optimal parameter setting has been suggested using S/N ratio.

Keywords: - 3D Printing, FDM, Build & Support Materials, ABS Plastic, Impact Strength, Surface roughness, MINITAB-18, ANOVA and S/N Ratio.

1.INTRODUCTION:

Prototyping is one of the important steps to finalize the product design. Manual prototyping is out dated method. In the recent times, Rapid Prototyping is used. This process was introduced in 1980's. Rapid Prototyping is a group of techniques used to fabricate a small scale model of physical part,quickly by using the three-dimensional Computer Aided Design (CAD). It is the latest, advanced technology, which saves time, especially for most complicated products. Rapid Prototyping is also known as Additive Manufacturing. These processes make three-dimensional parts directly from CAD models by adding materials layer by layer.

Fundamentally, all the manufacturing techniques are of two types: i. Additive and ii. Subtractivemanufacturing. Subtractive type is a material removal technique from solid piece until the desired shape and size is obtained as per the drawing of the Component. Additive type RP is opposite to subtractive type. In this Rapid Prototyping, the material is added layer by layer up to the desired shape and size of the Component. Additive fabrication has the capability to create any shape and geometry. RP technology requires less time for classical prototyping in the sense that the creating – testing – redesigning – rebuilding and repeating the process.

RP or 3 D Printing process makes the model by accumulating layers in required shape on an x-y plane. The z-axis is generated by stacking one layer over the other, thereby the product will be

IOP Conf. Series: Materials Science and Engineering 1112 (2021) 012009 doi:10.1088/1757-899X/1112/1/012009

fabricated. RP processes can be divided into different groups according to their layer formation of materials.To make the physical model, different processes, such as StereoLithography, Fused Deposition Modeling, Selective Laser Sintering, Laminated Object Manufacturing, and 3D Printing etc., are used.

The various Computer Aided Design packages use a number of different algorithms to represent solid objects. To establish consistency, the .STL file format is being used as the standard of the rapid prototyping industry. Therefore, it is required to convert the CAD file into .STL file format.

FDM consists of three steps: i. Pre-Processing, ii. Production, and iii. Post Processing. In the pre-processing, a CAD model is constructed, and later, it is converted into .STL file format for FDM process. The layers are built until completion of the model. In post processing procedure supports are removed by washing or stripping away. The surface of the model is then finished and cleaned using different techniques. FDM Technology produces parts layer-by-layer by heating thermoplastic material to a semi- liquid state and extruding it according to computer-controlled paths.

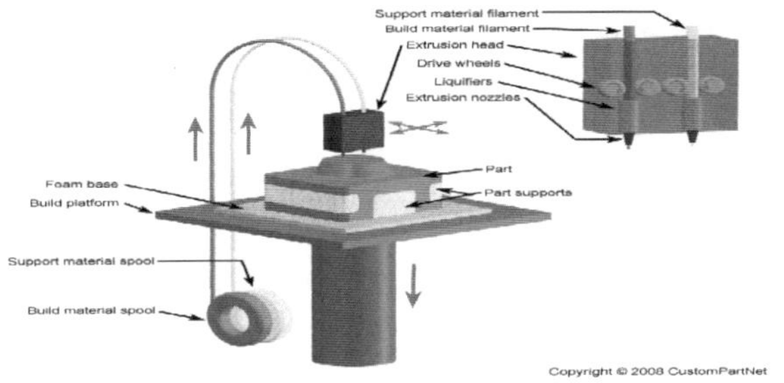

Figure 1: Fused Deposition Modeling Process

FDM uses two materials: i. Build or modeling material and ii. Support material. Modeling material develops the finished piece, and Support materialacts as scaffolding. Material filaments are fed from the 3D printer's material bays to the printer head, which moves in X and Y coordinates, depositing material to complete each layer. The base moves down along the Z axis and the next layer starts. For building complex shaped parts, FDM uses support material apart from the build material which supports the overhanging structures during the process. This helps to maintainstructural integrity of component until it is strengthened before the support material is removed by breaking or by dissolving using appropriate liquid solvents.

2.LITERATURE REVIEW:

CanyMendonsa, et.,al.[1]The objective of this paper is to study the influence of process parameters like Print speed (A), Layer thickness(B),Infill density(C) on the build time and optimization of these parameters to get Fused Deposition Modeling part in lower lead time using Taguchi and ANOVA approaches.The results show that the build time depend on infill densityandthe layer thickness.Positively decreasing the layer thickness and negatively reducing the infill density can reduce the build time for a given print. In addition, it is concluded that Layer thickness is the most significant factor for Build time in FDM components.

Xinhua Liu, et.,al.[2]: In this paper, tensile strength, impact strength, and flexural strength are considered as three evaluation indexes to characterize the mechanical properties of a FDM part.The orthogonal test of five factors and three levels was designed. The Taguchi method is used to optimize and study the influence of various process parameters on the three performance indexes.

J. Santhakumar, et.,al. [3]: In this paper, the improvement of Impact Strength is achieved by Optimizing the FDM Process Parameters using Polycarbonate material. This study includes four important process parameters like layer thickness, build orientation, raster angle and raster width whose influence on Impact Strength is studied. Using Taguchi Design of Experiments methodology experiments were conducted out. This work finds out the optimum process parameters required to obtain maximum impact strength on Polycarbonate Material. Analysis of Variance test (ANOVA) was carried to find the most effecting process parameters on Impact strength. Also, it is concluded that Layer thickness is the most vital factor for Impact strength for a Polycarbonate Material.

M. Tharunkumar,et.,al.[4]:In this paper, the effect of FDM parameters on Impact Strength is experimentally investigated. Experiments were conducted according to DOE with three different process parameters such as direction of rotation, build orientation angle and model interior. The process parameters were optimized by using Taguchi method, S/N ratio is evaluated and the effect of the process parameters on Impact strength was estimated by the analysis of variance. In addition, it is concluded that Angle of Rotation is the most vital factor for Impact Strength of ABS Plus FDM parts.

R. Anitha, et.,al. [5]: In this paper, it is studied the effect of various process parameters, i.e. layer thickness, road width and speed of deposition of fused deposition modeling parts. The FDM parts which are fabricated depend upon the process parameters. Taguchi approach was used for design of experiments. L18 orthogonal array was selected for experiments. In addition, it is concluded that the layer thickness is the most crucial factor for Minimum Surface Roughness of FDM parts.

Raju Bangalore Singe Gowda,et.,al.[6]: In this paper,the influence of the physical build parameters over the part quality are studied. An L9 orthogonal array was designed with the minimum number of experimental runs with desired parameter settings and by analysis tools such as ANOVA (analysis of variance).

M.Alhubail,et.,al. [7]: This paper presents the results of the experimental work on the effect of the Fused Deposition Modelling (FDM) process, variable parameters such as layer thickness, air gap, raster width, contour width and raster orientation on the quality characteristics such as surface roughness (SR) and tensile strength (TS). Taguchi design (L32) was used to obtain the experimentation runs. The results showed that higher tensile strength could be obtained by changing layer thickness and raster width at a low level and the air gap at -0.01 mm.

P. Chennakesava and Y. Shivraj Narayan [8]: In this review paper, an attempt has been made to know the insights of one such additive manufacturing process i.e., Fused Deposition Modeling (FDM).

AshayKohad,et.,al. [10]:This study includes identification of various process parameters involved in FDM, which affects quality of Parts fabricated. This study aims to find effects of process parameters on different performance parameters based on previous research done. This study also aims to compare different techniques for optimization of process parameters in Fused Deposition Modeling.

3. MATERIALS AND METHODS:

For extrusion and deposition purposes, most existing FDM machines use thermoplastic materials in a filament type. Acrylonitrile Butadiene styrene (ABS) and Polylactide (PLA) thermoplastics are predominantly used in the process.

S.no	**Name of material**	**Normal boiling point**	**Temperature used in FDM**
1	ABS	105^0 C	230^0 C
2	PLA	65^0 C	180^0 C

Table 1: Properties of materials

3.1 Process Parameters:

1.*Layer thickness:* - The thickness of the layer deposited by the nozzle depends on the nozzle type used.

2. *Speed of deposition:* - It is the rate at which the nozzle deposits the raster.

3.*Fill Density:*-This percentage number determines the amount of plastic infill there will be in an object. Usually a setting of around 0.30 is used.

3.2 Performance Parameters:

Performance parameters are those for which the optimization is going to perform.

1. *Impact strength.*
2. *Built Time:* Fabrication of part using FDM is quite time-consuming process so there must be need to minimize the total time required to complete part fabricated using FDM.
3. *Surface Roughness:* To getfunctioning part fabricated from FDM, the component must possess good surface roughness as per the application where that part is going to perform.

Advanced Fusion Plastic Modeling (AFPM) technology was the India's first industrial grade professional 3D printer . The AION 500 is a high-performance 3D printer at an affordable price that provides a professional-level build volume. This AION 500 offers a large build size, ultra fast print speeds, and more precision. It is designed for versatility.

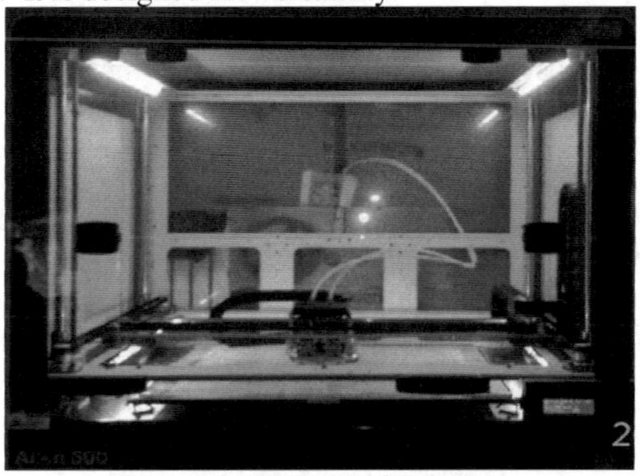

Figure2: AION 500 Machine

The standard specimen for ***ASTM D256*** is 64 x 12.7 x 4.0 mm. The depth under the notch of the specimen is 10.2 mm.

The parameters which can be adjusted in KISSlicer software for optimizing the performance parameters.

Layer Thickness: - This is the vertical resolution setting in millimeters, usually set to 0.2, but can be set as less as 0.05, as per the requirement. Layer thickness directly affects printing time, as thinner layers will cause more layers, which will take more time or vice versa with thicker layers.

Infill density: - It is necessary to set the density of printed material within the boundary of the loops. 100% will print an entirely solid piece but it is mostly required. Having a sparser infill will decrease printing time, reduce warping due to heating, conserves material, lightens the object, and may actually increase rigidity. This setting needs to be adjusted according to the size of the part.

Infill Style:- It is required to set the lattice shape of the sparse infill structure like straight, round, or hexagonal.

Speed of deposition:- It is the rate at which the nozzle deposits the material.

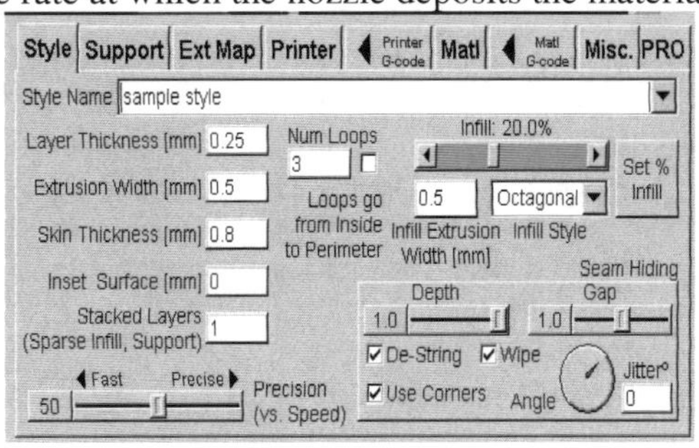

Figure3:KISSlicer Software settings bar

3.3 Design Of Experiments:

Taguchi method is used to reduce the variation of processes through robust design of experiments. The objective of the method is to produce high quality product at minimum cost to the manufacturer. Genichi Taguchi developed the Taguchi method. For the design of experiments, a method was built to investigate how various parameters influence the mean and variance of a process output attribute that determines how well the process operates.. The experimental design is proposed by Taguchi method which involves orthogonal arrays to organize the process parameters affecting the process and the levels at which they should be varied.The Taguchi method measures pairs of combinations instead of trying to test all possible combinations, such as factorial design.. This facilitates the collection of the necessary data to determine which factors most affect the product quality with optimum or less number of experiments. This saves time and resources. The Taguchi method is widely used when there are an intermediate number of variables (3 to 50), few interactions between variables, and when only a few variables contribute significantly.

In this experimental work, an attempt has been made to demonstrate the application of Taguchi's Method to improve the impact strength and surface roughness characteristics of an ***ASTM D256*** test specimens that were produced on FDM (***Fused deposition modeling***) by optimizing the process parameters (control factors) like speed, layer thickness and infill density.

When the process parameters are more, a large number of experiments have to be conducted. In order to solve this problem, the Taguchi method uses a special design of orthogonal arrays to study the entire parameter space with only a small number of experiments. Three process parameters (controlling factors) are i. Speed, ii. Layer thickness and iii.Infill density for experiments. Each parameter has three levels, denoted by 1, 2 and 3 respectively.

If three parameters and three levels are used for each parameter, according to the Taguchi method, the L9 Orthogonal array should be used for experimentation. Orthogonal arrays involve special standard experimental design that requires only a small number of experimental trials to find the main factors effects on output.

Based on this orthogonal array (OA) is to be selected which has at least nine rows i.e., 9 experimental runs.

In this work, L9 is sufficient. It would require 27 experiments to optimize the parameters. L9 orthogonal array suggests nine experiments which are necessary to optimize the parameters. The experiments are selected as per the level combinations depending on the orthogonal array. It is important that all experiments be conducted. The performance parameter (output) are noted for each experimental run for analysis.

Process parameters

1. Speed of deposition
2. Layer thickness
3. Infill density

Parameters	**Level 1**	**Level 2**	**Level 3**
Speed, mm/s	40	55	70
Layer thickness, mm	0.4	0.3	0.2
Infill density, %	50	33.33	25

Table2: Process parameters with their levels

The fabrication of parts is carried out accordingly by changing the parameters suggested by Taguchi method.

Run no.	**Level A**	**Level B**	**Level C**
1	1	1	1
2	1	2	2
3	1	3	3

4	2	1	2
5	2	2	3
6	2	3	1
7	3	1	3
8	3	2	1
9	3	3	2

Table3: Taguchi Orthogonal Array Design - L9(3*3)

1. **EXPERIMENTAL PROCEDURE:** Impact Test specimen has been modeled according to **ASTM D256.**

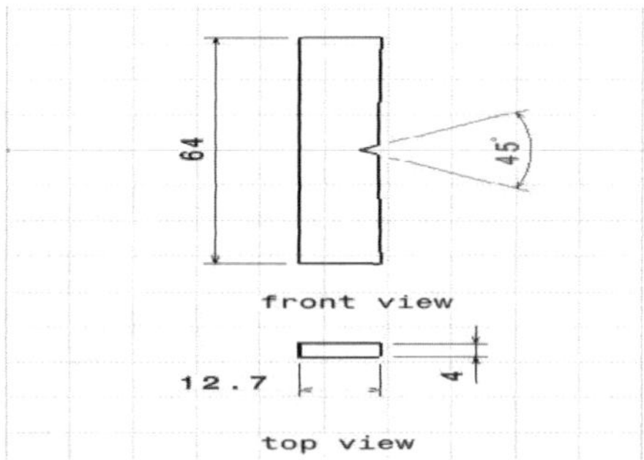

Figure4: Dimensions of standard test Specimens

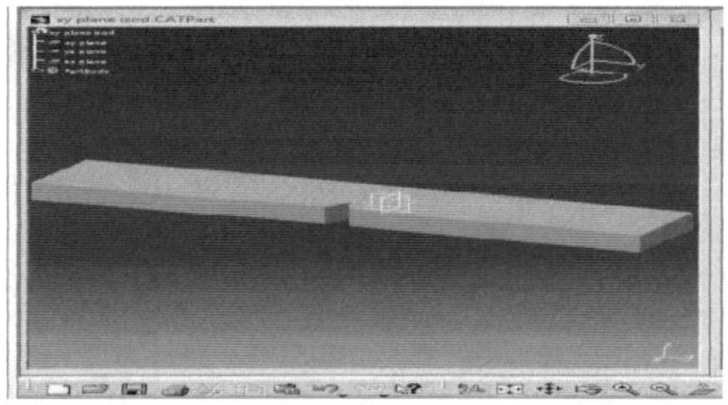

Figure5: ASTM D 256 standard IZOD notched CATIA V5 mode

For manufacturing the Specimens in FDM process, we need to convert the CATIA file into Stereolithography file format (STL). This file is supported by many other software packages. It is mostly used for 3D printing, rapid prototyping and computer - aided manufacturing. After drawing the Izod Impact Test Specimen (ASTM D256) in CATIA, we save the design in STL format by changing the file extension of Test specimen from .cat to .stl (Stereolithography) file format.

Impact Strength Parts Fabricated by FDM Process

Figure6: Fabricated Components by FDM

5.TESTING:

5.1 Impact Test:

Izod Impact Testing (Notched Izod) ASTM D256 is a common test to understand notch sensitivity in plastics.

Notched Izod Impact is a single point test that measures the material resistance to impact from a swinging pendulum. Izod impact is defined as the kinetic energy needed to initiate fracture and continues the fracture until the specimen is broken. Izod specimens are notched to prevent deformation of the specimen upon impact. This test can be used as a rapid and easy quality control check to determine if a material meets specific impact properties or to compare materials for general toughness.

ASTM impact energy is expressed in J/m. Impact strength is calculated by dividing impact energy in J by the thickness of the specimen. The test result is typically the average of three specimens. The higher the resulting numbers the tougher the material.

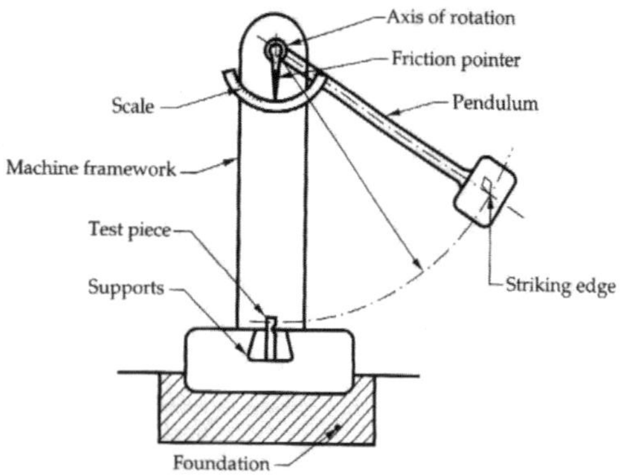

Figure7:Izod Impact Testing Equipment

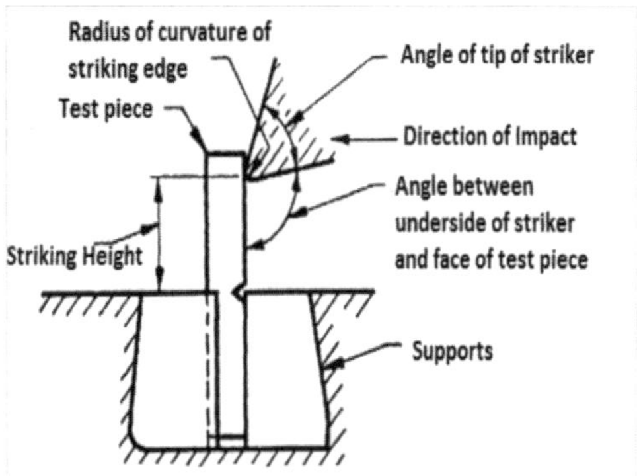

Figure8:Izod Impact testing specimen

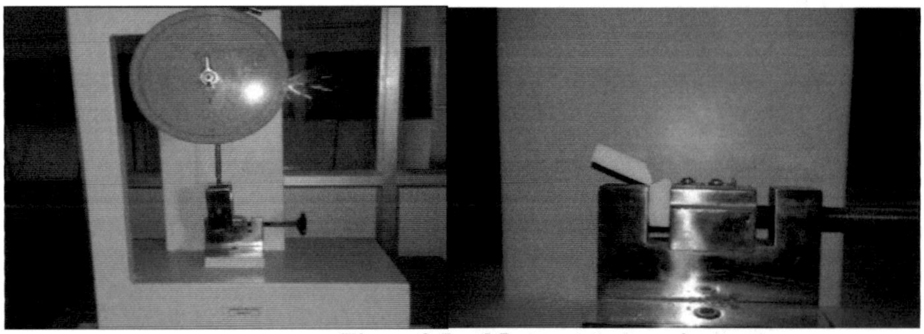

Figure9:Izod Impact test equipment

Izod impact testing is an ASTM standard method of determining the impact resistance of materials. A pivoting arm is raised to a specific height and then released. The arm swings down hitting the sample, breaking the specimen. The energy absorbed by the sample is calculated from the height the arm swings to after hitting the sample. A notched sample is generally used to determine impact energy and notch sensitivity.

5.2 Surface Roughness:

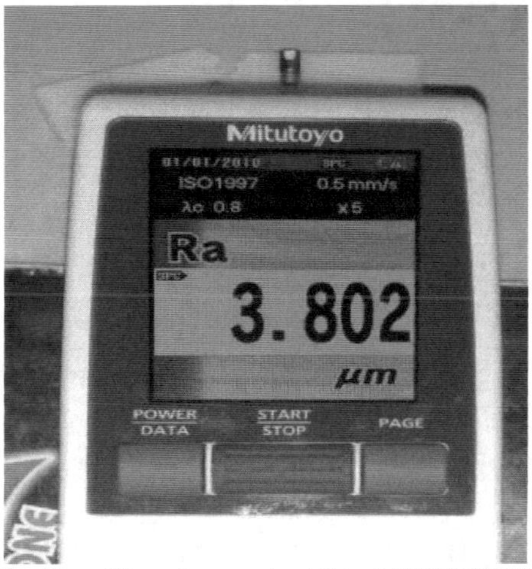

Figure10: Surface Roughness test for ASTM D256 samples

The surface roughness of **ASTM D256** samples is tested by using Mitutoyo, Japan made surface roughness tester. The values of R_a (µm) in X & Y directions are noted down. R_a (µm) is the average roughness of the sample.

5.3 Build Time:

After opening the STL file in KISSlicer software, we adjust the process parameters i.e., Layer thickness, Infill density and Speed. After adjusting the process parameters we need to slice the file and note down the time required for printing of Specimens and Volume of material required for printing of specimens.

6.EXPERIMENTAL RESULTS:

6.1 Experimental Impact Strength of ABS Parts:

Run No	IS 1	IS 2	IS 3	Mean value
1	142.857	142.857	176.47	154.063
2	135.135	210.526	131.579	159.080
3	147.059	176.47	147.059	156.863
4	236.842	153.846	153.846	181.511
5	138.88	138.88	138.88	138.88
6	142.857	147.059	117.65	135.856
7	147.059	147.059	147.059	147.059
8	131.35	162.162	108.108	131.135
9	171.429	171.429	171.429	171.429

Table4: Impact strength values

*6.2 Surface Roughness:*The surface roughness of **ASTM D256** samples are given below:

Run order	R_a in X direction	R_a in Y direction	R_a Mean
1	6.57	5.91	6.24
2	3.47	3.995	3.735
3	14.335	10.468	12.402
4	11.383	12.303	11.843
5	7.6905	9.1405	8.416

6	6.958	4.6665	5.812
7	12.510	13.041	12.776
8	5.0415	5.5325	5.287
9	4.2073	3.8113	4.009

Table 5: Surface Roughness values for ASTM D-256 samples

*6.3 Build Time:*The time required for printing of Specimens:

Run Order	Speed	Layer Thickness	Infill Density	Build Time (min.)
1	40	0.4	50	9.830
2	40	0.3	33.33	11.227
3	40	0.2	25	16.167
4	55	0.4	33.33	10.027
5	55	0.3	25	12.277
6	55	0.2	50	19.460
7	70	0.4	25	11.253
8	70	0.3	50	14.893
9	70	0.2	33.33	20.933

Table 6:Build Time

7.STATISTICAL ANALYSIS:

A statistical model is developed by correlation of input parameters such as Speed of deposition, Layer thickness and Infill density based on analysis of data.

7.1Impact Strength:

Taguchi optimization plan utilizes data obtained from the experiment and analysis with the help of S/N ratio gives the optimum possibilities for the number of trails selected according to the plan.
Taguchi Analysis:-
Impact strength versus Speed, Thickness, Density

Level	Speed of deposition	Layer Thickness	Infill Density
1	156.7	160.9	140.4
2	152.1	143.0	170.7
3	149.9	154.7	147.6
Delta	6.8	17.8	30.3
Rank	3	2	1

Table 7: Response Table for Means

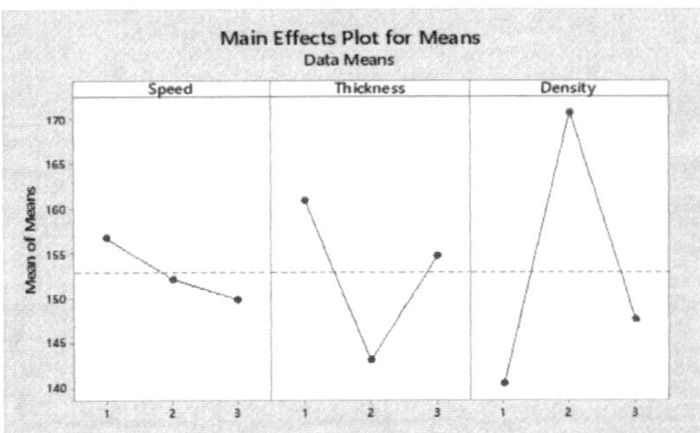

Figure 11: Main Effects Plot for Means

Signal to Noise Ratio Calculations

The signal to noise ratio is obtained from Minitab software, which presents the percentage contribution of each factors taken. The response output of statistical analysis is maximum impact strength for the combination of process parameters from the design of experiments. Hence, better formula is chosen for calculation.

S/N Ratio formula for larger the better

$$S/N = -10*\log(\Sigma(1/Y^2)/n)$$

For these calculations, we have three Impact strength values i.e., Y1, Y2 and Y3.

Then **"n"** value is **"3"**.

S/N Ratio values:

Run. No	Impact strength	S/N ratio
1	154.063	43.7540
2	159.080	44.0323
3	156.863	43.9104
4	181.511	45.1781
5	138.890	42.8534
6	135.856	42.6616
7	147.060	43.3499
8	131.135	42.3544
9	171.430	44.6817

Table 8: S/N Ratio values

Level	Speed of deposition	Layer Thickness	Infill Density
1	43.90	44.09	42.92
2	43.56	43.08	44.63
3	43.46	43.75	43.37
Delta	0.44	1.01	1.71
Rank	3	2	1

Table 9: Response Table for Signal to Noise Ratios

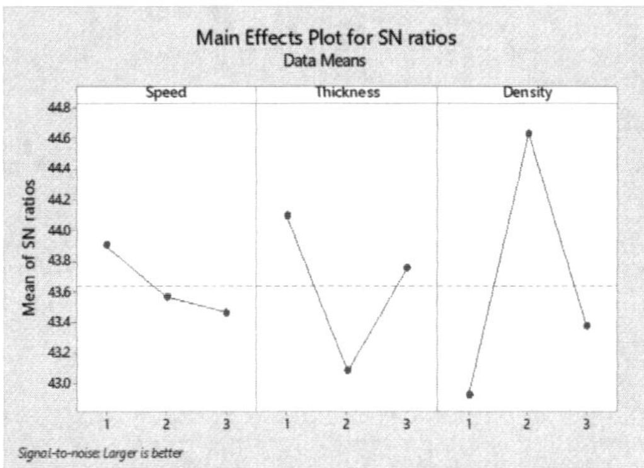

Figure 12:Main Effects Plot for SN ratios

From the analysis of experiments, done highest S/N ratio (45.1781) is obtained for experimentnumber 4 having the 181.511 j/m impact strength.

From the S/N ratio graphs, the best combination for Speed of deposition is 40 mm/s, Layer thickness is 0.4 mm and Infill Density is 33.33 %.

7.2Surface Roughness:

Taguchi Analysis:

Surface Roughness versus Speed,Thickness and Density.

Level	Speed of deposition	Layer thickness	Infill Density
1	7.459	10.286	5.780
2	8.690	5.813	6.529
3	7.357	7.408	11.198
Delta	1.333	4.474	5.418
Rank	3	2	1

Table 10: Response Table for Means

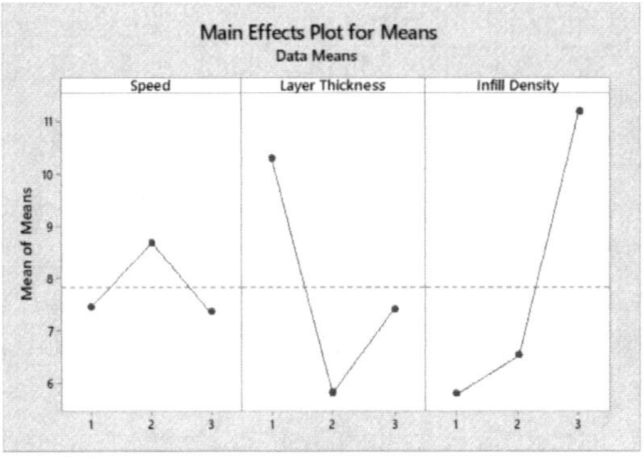

Figure 13: Main Effect plots for Means

S/N Ratio Values:

Run order	R_a Mean	S/N Ratio
1	6.24	-15.9037
2	3.735	-11.4458
3	12.402	-21.8698
4	11.843	-21.4692
5	8.416	-18.5021
6	5.812	-15.2865
7	12.776	-22.1279
8	5.287	-14.4642
9	4.009	-12.0607

Table 11: S/N Ratio for SR

Level	Speed of Deposition	Layer thickness	Infill Density
1	-16.41	-19.83	-15.22
2	-18.42	-14.80	-14.99
3	-16.22	-16.41	-20.83
Delta	2.20	5.03	5.84
Rank	3	2	1

Table 12: Response Table for S/N Ratios

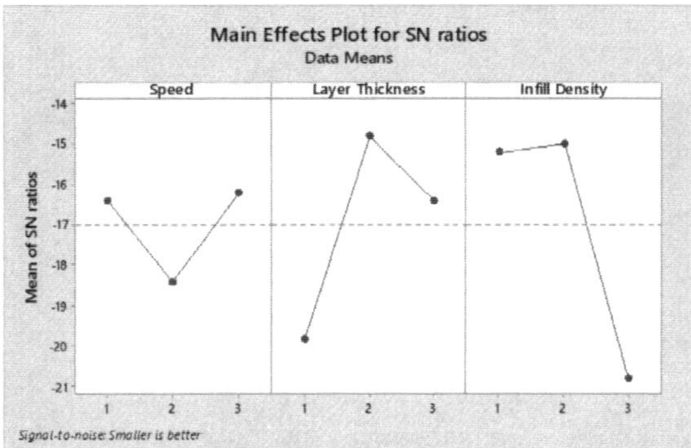

Figure 14: Main effects plot for Signal to Noise Ratios

From the analysis of experiments conducted, highest S/N ratio (-11.4458) is resulted for experimentation number 2 having the 3.735 µm surface roughness.

From the S/N ratio graphs, the best combination of Speed of deposition of 70 mm/s, Layer thickness of 0.3 mm and Infill Density of 33.33% for Minimum Surface Roughness.

7.3Build Time:

Taguchi Analysis

Build Time versus Speed, Layer thickness &Infill Density

Level	Speed of deposition	Layer thickness	Infill Density
1	12.41	10.37	14.73
2	13.92	12.80	14.06
3	15.69	18.85	13.23
Delta	3.28	8.48	1.50
Rank	2	1	3

Table 13: Response Table for Means

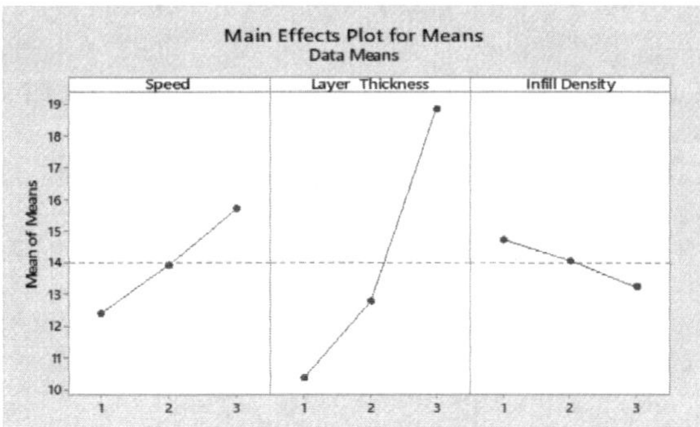

Figure 15: Main effects plot for Means

S/N Ratios for Build time

Run Order	Build Time (min.)	S/N Ratio
1	9.830	-19.8511
2	11.227	-21.0053
3	16.167	-24.1726
4	10.027	-20.0234
5	12.277	-21.7818
6	19.460	-25.7829

7	11.253	-21.0254
8	14.893	-23.4596
9	20.933	-26.4166

Table 14: S/N Ratios for Build time

Level	Speed of deposition	Layer thickness	Infill Density
1	-21.68	-20.30	-23.03
2	-22.53	-22.08	-22.48
3	-23.63	-25.46	-22.33
Delta	1.96	5.16	0.70
Rank	2	1	3

Table 15: Response Table for S/N Ratios

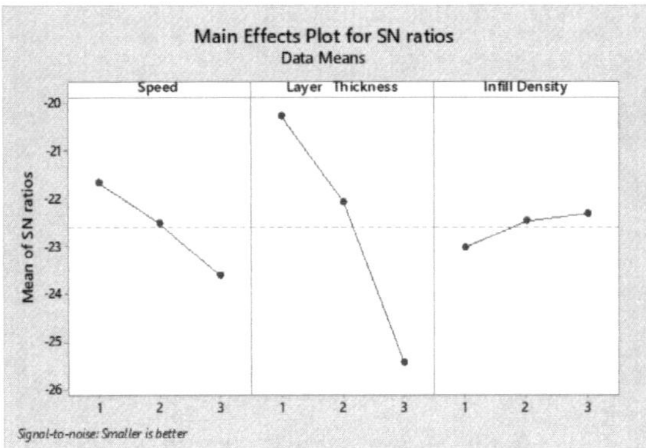

Figure16: Main effects plot for S/N Ratios

From the analysis of experiments done, highest S/N ratio (-19.8511) is obtained for experimentation number 1 having the 9.830 min. Build time.

From the S/N ratio graphs, the best combination of Speed of deposition of 40 mm/s, Layer thickness of 0.4 mm and Infill Density of 25% for Minimum Build Time.

8.CONCLUSIONS:

1. From the analysis of experiments, done highest S/N ratio (45.1781) is obtained for experimentnumber 4 having the 181.511 j/m impact strength.
2. From the S/N ratio graphs, the best combination of Speed of deposition is 40 mm/s, Layer thickness is 0.4 mm and Infill Density is 33.33 % for Highest Impact strength.
3. From the analysis of experiments done, highest S/N ratio (-11.4458) is obtained for experiment number 2 having the 3.735 μm surface roughness.
4. From the S/N ratio graphs, the best combination of Speed of deposition of 70 mm/s, Layer thickness of 0.3 mm and Infill Density of 33.33% for Minimum Surface Roughness.
5. From the analysis of experiments done, highest S/N ratio (-19.8511) is resulted. For experiment number 1 having the 9.830 min. Build time.
6. From the S/N ratio graphs, the best combination of Speed of deposition of 40 mm/s, Layer thickness of 0.4 mm and Infill Density of 25% for Minimum Build Time.

REFERENCES:

[1] CanyM, Naveen K V, PrathikU andVyasDS,2015 *Int J. Sc. Res.* 4

[2] Liu X, Zhangl M, Li S, Sil L, Junquan P and Hul Y, 2016 *Int J AdvMfg Tech.*

[3] Kumar S J, Maggirwar R, Srinivas G, Karthekeyan S and Naveen K, 2016 *Ind J ScTech*. 9.

[4] R. Anitha, S. Arunachalam, and P. Radhakrishnan, "Critical Parameters Influencing the Quality of Prototype inFused Deposition modelling" *Journal of Materials Processing Technology*, Volume 118, Issues 1–3, 3 December 2001, Pages 385-388 (Elsevier).

[5] M. Tharunkumar, M. Sumalatha, J. NagaMalleswaraRao, "Experimental Investigation Of Impact Strength For Abs Plus F. D. M. Parts Using Taguchi", *International Research Journal of Engineering and Technology (IRJET)* Volume: 04 Issue: 09 | Sep -2017.

[6] Raju Bangalore Singe Gowda,ChandraSekharUdayagiri, and DrakshayaniDoulatNarendra, "Studies on the Process Parameters of Rapid Prototyping Technique (Stereolithogra- phy) for the Betterment of Part Quality",*Hindawi Publishing Corporation International Journal of Manufacturing Engineering*Volume 2014, Article ID 804705.

[7] M Alhubail1, D Alenezi and B Aldousiri, "Taguchi-based Optimization of Process Parameters of Fused Deposition Modelling for Improved Part Quality", *International Journal of Engineering Research & Technology (IJERT)* Vol. 2 Issue 12, December – 2013.

[8] P. Chennakesava and Y. Shivraj Narayan "Fused Deposition Modeling – Insights", *International Conference on Advances in Design and Manufacturing (ICAD&M'14)* ISBN 978-93-84743-12-3 © 2014 Bonfring.

[9] Omar A. Mohamed Syed H. Masood,Jahar L. Bhowmik "Optimization of fused deposition modeling process parameters: A review of current research and future prospects", *Advances in Manuf.* DOI 10.1007/s40436-014-0097-7. Received: 13 June 2014/Accepted: 30 December 2014 Shanghai University and Springer-Verlag Berlin Heidelberg 2015.

[10] AshayKohad, RajendraDalu, "Optimization of Process Parameters in Fused Deposition Modeling: A Review", *International Journal of Innovative Research in Science, Engineering and Technology* An ISO 3297: 2007 Certified Organization Volume 6, Special Issue 1, January 2017.

[11] Ojasdandgaval, Pranitabichkar, "Rapid Prototyping Technology - Study Of Fused Deposition Modeling Technique", *International Journal of Mechanical And Production Engineering*, ISSN: 2320-2092, Volume- 4

IOP Conf. Series: Materials Science and Engineering 1112 (2021) 012010 doi:10.1088/1757-899X/1112/1/012010

Multi-objective optimization using TOPSIS in turning of Al 6351 alloy

S R Rao[1*], S A K Jeelani[2], and V Swamulu[3]

[1]Department of Mechanical Engineering, Tirumala Engineering College, Narasaropet, Andhra Pradesh, India
[2]Department of Mechanical Engineering, VVIT, Guntur, Andhra Pradesh, India
[3]Department of Mechanical Engineering, Narasaraopeta Institute of Technology, Narasaropet, Andhra Pradesh, India
[*]Corresponding author:ramaraosadineni1@gmail.com

Abstract. Surface finish and machining rate are the important machining criteria in turning process and plays a vital role in manufacturing environment. This work focused on the parametric optimization of cutting velocity, feed and depth of cut in turning of AA 6351 alloy with carbide tool for maximization of machining rate and minimization of surface roughness simultaneously using technique for order Preference by similarity to ideal solution (TOPSIS) method. Contribution of every machining parameter on multi objectives is spotted by employing analysis of variance (ANOVA). Finally results are validated with a confirmation experiment. Significant improvement within the responses is observed using the TOPSIS method.

1. Introduction

The quality of surface and material removal rate are the predominant parameters in any manufacturing industry to assess the production rate of machine tools and the quality of the machined components. Hence, obtaining the required surface finish is essential for the functional requirement of designed parts. Surface quality affects on the several properties like fatigue strength, wear rate, friction coefficient, lubrication and corrosion resistance of the machined parts [1]. But the machining rate (MR) and surface roughness (SR) are the two conflicting machining criteria. When we try two increase the one the other affected worse. So, products must be produced with the required surface quality with optimum cost. This can be achievable when simultaneously maximize the MR and minimize the SR.

The influence of speed and feed on SR and MR were observed by Ranganath et al.[2] using Taguchi experimental design in turning of A 6061 alloy and observed that good surface finish at high cutting speed. MR and time of machining (TM) were optimized in machining (turning) of Al 7075 aluminium alloy with carbide tool by taking depth of cut, speed and feed as process parameters and observed that speed is the most important parameter that effects on MR and TM [3]. In [4] Al6061 alloy, silicon carbide (10%wt) and graphite (3%wt) hybrid metal matrix composites were turned on a CNC lathe and MRR and SR were optimized by using response surface methodology (RSM) with depth of cut, speed, feed and tool material as input parameters. Surya et al.[5] did the multi response (MRR, SR and tool wear rate) optimization using grey relational analysis (GRA) and ANOVA in machining of Al 7075 aluminium alloy with speed, depth of cut and feed as process parameters.

The affect of depth of cut, speed and feed on MR and SR were investigated by Goyal et al. [6] in machining of AISI 1020 low-carbon steel using WNMG332RP carbide insert tip and observed that signal to noise (S/N) ratio decreases on increasing cutting speed, while S/N ratio increases with an increase in depth of cut and feed. Parametric optimization was carried out in CNC turning process [7] using Taguchi method with grey analysis for optimizing the different performance measures such as MR, total machining cost and SR in turning of LM 25 Al alloy with speed, feed, depth of cut and flow rate of cutting fluid as input parameters. Rahul and Vijaykumar [8] obtained the empirical model to predict MR in terms of cutting velocity, feed and depth of cut using regression models and also used genetic algorithms for optimal combination of machining parameters to optimize MR in turning of aluminium alloy. Affect of depth of cut, nose radius, feed and cutting speed on MR and surface finish was investigated by Ahmeda and Rajesh [9] in machining of Al 6351 T6 and the influence of the input parameters were analyzed using ANOVA. ANOVA results showed that feed the most influencing machining parameter on SR followed by speed and for MR the significant parameters are speed and the feed followed by the wt% of reinforced particles in turning of TiB_2 particles reinforced Al 6063 metal matrix composites using K20 carbide insert [10].

TOPSIS method was used in [11] for the simultaneous optimization of SR and MRR in abrasive water jet machining of Inconel 625 alloy. Balasubramaniyan and Selvaraj [12] did the multi objective optimization in turning of EN25 steel with carbide tools using combined Analytic Hierarchy Process (AHP) and TOPSIS methods and obtained the optimal input parameters. TOPSIS analysis was used for optimization of SR and MR simultaneously in turning of GFRP composites with speed, feed, and depth of cut as machining parameters [13].

TOPSIS method was used in various multiresponse optimization problems like optimal selection of stir casting parameters to obtain maximum tensile strength of Al–15%Mg2Si composite [14], multiobjective optimization of mechanical properties of aluminum composites with different reinforcing particles [15], to select optimal machining parameters in turning of Titanium [16], optimization of the machining parameters in EDM of Al–24%SiC metal matrix composite[17], etc.

From the above literature it is notified that less work was done on multiresponse optimization in turning using TOPSIS method. In this work MRR and SR are simultaneously optimized by using TOPSIS method in turning of Al 6351 alloy.

2. Experimentation

In the present work Al alloy 6351 was machined on CNC lathe with a carbide tool. Experiments were designed based on Taguchi's L27 orthogonal array. Experiments were conducted by varying the machining parameters. Machining rate is measured using Eq. (1) and the surface roughness values were measured by using Talysurf tester (Mituto211). The considered input parameters with their levels are shown in Table 1.

$$\mathbf{MR = V * f * d \; mm^3/min} \qquad (1)$$

Table 1. Levels of considered input parameters

Input parameters	Cutting velocity (V)	Feed (f)	Depth of cut (d)
Level 1	15	0.06	0.45
Level 2	20	0.09	0.60
Level 3	25	0.12	0.75

3. Methodology

Experiments are designed using Taguchi L_{27} full factorial orthogonal array. TOPSIS is one of the multiple attribute decision-making (MADM) methods used for solving many-objective problems. In this method feasible solutions are obtained by constructing the ideal solutions and negative ideal solutions to the multi-objective problems and then finding the closeness to the ideal solutions and being far away from the negative ideal solutions [18]. The steps in the TOPSIS are:

Step1: Formulate the experimental decision matrix (M) using multiple objectives.

$$M_{ij} = \begin{bmatrix} R_{11} & R_{1j} & R_{1m} \\ R_{i1} & R_{ij} & R_{im} \\ R_{n1} & R_{nj} & R_{nm} \end{bmatrix} \quad (2)$$

Where i is the number of experimental trails (= 1,2,3,4, . . ,n) and in this case, it is 27. j represent the number of responses (j = 1,2, …m)and in this study, it is equal to two (MRR &SR).

Step2: Normalize the responses to eliminate the difference in measuring units and bring them on a same scale in the range of 0 and 1 using Eq. (3)

$$N_{ij} = \frac{R_{ij}}{\sqrt{\sum_{i=1}^{n} R_{ij}^2}} \quad (3)$$

Step3: Find the weighted normalized matrix by multiplying the weight of each response (w_j) as given in Eq. (4).

$$W_{ij} = w_j \times N_{ij} \quad (4)$$

Step4: Then, calculate the ideal positive (P^+) and ideal negative (P^-) alternative of the response using the Eqs. (5) & (6).

$$P^+ = \mathrm{Max}(W_{ij}) \text{ for larger the better response or } \mathrm{Min}(W_{ij}) \text{ for smaller the better} \quad (5)$$

$$P^- = \mathrm{Max}(W_{ij}) \text{ for smaller the better response or } \mathrm{Min}(W_{ij}) \text{ for larger the better} \quad (6)$$

Step5: Now determine the separation measures (S^+ *and* S^-) from positive ideal (P^+) and negative ideal (P^-) to obtain the separation of every substitute to the ideal substitute and evaluated with Euclidean distance by using the Eqs. (7) and (8).

$$S^+ = \sqrt{\sum_{i=1}^{n} \left(W_{ij} - P^+\right)^2} \quad (7)$$

$$S^- = \sqrt{\sum_{i=1}^{n} \left(W_{ij} - P^-\right)^2} \quad (8)$$

Step6: The multi response index (MRI) or closeness coefficients of each substitute to the ideal solution are determined using the Eq. (9). This MRI is used as a single objective function in the Taguchi method.

$$MRI = \frac{S^-}{S^+ + S^-} \quad (9)$$

4. Results and discussions

Experiments were carried out by changing the depth of cut, cutting speed and feed as per the Taguchi design and the responses MR and SR were measured for each experiment. The experimental matrix and the measured responses are shown in Table 2.

Table 2. Experimental results

Exp. No.	V	f	d (mm)	MR(mm^3/s)	SR(μm)
1	15	0.06	0.45	6.750	0.402
2	15	0.06	0.60	9.000	0.430
3	15	0.06	0.75	11.250	0.455
4	15	0.09	0.45	10.125	0.471
5	15	0.09	0.60	13.500	0.485
6	15	0.09	0.75	16.875	0.509
7	15	0.12	0.45	13.500	0.532
8	15	0.12	0.60	18.000	0.550
9	15	0.12	0.75	22.500	0.577
10	20	0.06	0.45	9.000	0.374
11	20	0.06	0.60	12.000	0.397
12	20	0.06	0.75	15.000	0.425
13	20	0.09	0.45	13.500	0.430
14	20	0.09	0.60	18.000	0.441
15	20	0.09	0.75	22.500	0.454
16	20	0.12	0.45	18.000	0.485
17	20	0.12	0.60	24.000	0.500
18	20	0.12	0.75	30.000	0.521
19	25	0.06	0.45	11.250	0.386
20	25	0.06	0.60	15.000	0.401
21	25	0.06	0.75	18.75	0.435
22	25	0.09	0.45	16.875	0.405
23	25	0.09	0.60	22.500	0.432
24	25	0.09	0.75	28.125	0.461
25	25	0.12	0.45	22.500	0.433
26	25	0.12	0.60	30.000	0.440
27	25	0.12	0.75	37.500	0.465

4.1 TOPSIS results

The normalized and weighted normalized values are determined using Eqs. (3) and (4) respectively and are given in Table 3. In this work equal importance is given for MR and SR. Therefore the weight of each response is taken as 0.5.

After that the ideal and negative-ideal solutions of each response was computed using Eqs. (5) and (6) and presented in Table 4. In the presents work MRR is treated as the "higher the better" quality characteristic and SR is considered as the "lower the better" one.

Table 3 Normalized and weighted normalized responses

Exp.	Normalized values		Weighted Normalized	
No.	MR	SR	MR	SR
1	0.066852	0.168908	0.033426	0.084454
2	0.089135	0.180672	0.044568	0.090336
3	0.111419	0.191176	0.05571	0.095588
4	0.100277	0.197899	0.050139	0.09895
5	0.133703	0.203782	0.066852	0.101891
6	0.167129	0.213866	0.083564	0.106933
7	0.133703	0.223529	0.066852	0.111765
8	0.178271	0.231092	0.089135	0.115546
9	0.222838	0.242437	0.111419	0.121218
10	0.089135	0.157143	0.044568	0.078571
11	0.118847	0.166807	0.059424	0.083403
12	0.148559	0.178571	0.074279	0.089286
13	0.133703	0.180672	0.066852	0.090336
14	0.178271	0.185294	0.089135	0.092647
15	0.222838	0.190756	0.111419	0.095378
16	0.178271	0.203782	0.089135	0.101891
17	0.237694	0.210084	0.118847	0.105042
18	0.297118	0.218908	0.148559	0.109454
19	0.111419	0.162185	0.05571	0.081092
20	0.148559	0.168487	0.074279	0.084244
21	0.185699	0.182773	0.092849	0.091387
22	0.167129	0.170168	0.083564	0.085084
23	0.222838	0.181513	0.111419	0.090756
24	0.278548	0.193697	0.139274	0.096849
25	0.222838	0.181933	0.111419	0.090966
26	0.297118	0.184874	0.148559	0.092437
27	0.371397	0.195378	0.185699	0.097689

IOP Conf. Series: Materials Science and Engineering 1112 (2021) 012010 doi:10.1088/1757-899X/1112/1/012010

Table 4. Ideal positive and ideal negative solutions

Response	Ideal positive	Ideal negative
MRR	0.185699	0.033426
SR	0.078571	0.121218

Now separation measures are obtained using the Eqs. (7) and (8). Then closeness coefficient or multi response index is obtained using Eq. (9). Finally S/N ratios of MPI are calculated using Eq. (10). These values are shown in Table 5.

Table 5. Separation measures, MPI and S/N ratio of each experiment

Exp. No.	Separation measures S^+	S^-	MPI	S/N ratio
1	0.152387	0.036764	0.194364	-14.2277
2	0.141621	0.03283	0.188192	-14.508
3	0.131099	0.033962	0.205757	-13.7329
4	0.137084	0.027842	0.168817	-15.4517
5	0.121114	0.038611	0.241735	-12.3332
6	0.105999	0.052134	0.329683	-9.6381
7	0.123396	0.034737	0.219668	-13.1647
8	0.103401	0.055997	0.351305	-9.0863
9	0.085652	0.077993	0.476599	-6.4369
10	0.141131	0.044078	0.23799	-12.4688
11	0.126368	0.045889	0.2664	-11.4893
12	0.111934	0.051852	0.316587	-9.9901
13	0.119428	0.045508	0.275911	-11.1846
14	0.097584	0.062609	0.390833	-8.1602
15	0.076157	0.082162	0.518964	-5.6973
16	0.09934	0.058967	0.372485	-8.5778
17	0.071902	0.086939	0.547335	-5.2349
18	0.048302	0.115732	0.705535	-3.0296
19	0.130014	0.045898	0.260915	-11.67
20	0.111564	0.055101	0.330609	-9.6137
21	0.09373	0.066491	0.414996	-7.6391
22	0.102342	0.061802	0.376512	-8.4844
23	0.075273	0.083731	0.526598	-5.5704
24	0.049893	0.108617	0.685236	-3.2832
25	0.075307	0.083655	0.526257	-5.576
26	0.039644	0.118676	0.749595	-2.5035
27	0.019118	0.15408	0.889617	-1.0159

The main effect graph for S/N ratio is shown in Fig. 1. From the Fig. 1 it is observed that third level of cutting speed, feed and depth of cut gives the maximum S/N ratio, i.e. V3, f3 and d3 gives the optimum multi machining criteria. From the delta values it is observed that cutting speed is the main parameters that affects on multi-objective optimization followed by feed and depth of cut.

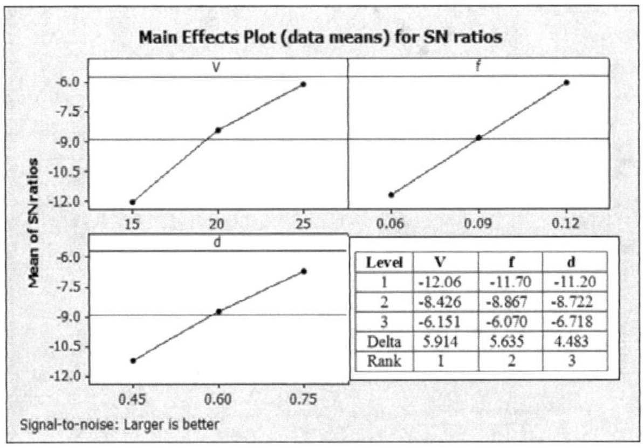

Figure 1. Main effect graph for S/N ratios

4.2 Results of ANOVA

ANOVA results for S/N ratios are given in Table 6. From this it is observed that all the three process parameters have the significant influence on the multiple machining criteria since their P values are less than 0.05. From the ANOVA results it is confirmed that cutting velocity is the most important parameter followed by feed and depth of cut.

Table 6. ANOVA results for S/N ratios.

Source	DF	SS	MS	F	P value	Contribution (%)
V	2	160.160	80.080	78.99	0.000	38.68
f	2	142.886	71.443	70.47	0.000	34.51
d	2	90.756	45.378	44.76	0.000	21.91
Error	20	20.277	1.014			4.9
Total	26	414.079				

4.3 Confirmation experiment

A confirmation test was conducted to confirm the optimal input parameter setting expected by the TOPSIS method. The machining criteria measured from confirmation test conducted at the optimal parametric combination i.e., cutting velocity=25 m/min, depth of cut=0.75 mm and feed=0.12 mm/rev are compared with the responses measured at the initial parameter setting (cutting speed=15 m/min, feed=0.06 mm/rev and depth of cut=0.45 mm). The results of confirmatory test are given in Table 7. From this test it is noticed that a satisfactory results within the improvement of responses.

Table 7. Confirmation test results

Responses	Initial parameter setting		TOPSIS parameter setting		Improvement	
	Response value	S/N ratio	Response value	S/N ratio	Response value	%
MRR	6.750	16.5861	37.5	27.0437	30.75	82.00
SR	0.402	-7.91548	0.465	-4.77648	0.063	13.55
Parameter setting	$V_1f_1d_1$		$V_3f_3d_3$		$V_3f_3d_3$	

5. Conclusions

In this work Al 6315 alloy was turned on CNC machine by varying the cutting velocity, feed and depth of cut and the responses machining rate and surface roughness are measured and multi-objective optimization was done using TOPSIS. The conclusions of this work are as follows.

- From the results of ANOVA, it is noticed that the major process parameter affecting on the multi responses is cutting velocity with a contribution of 38.68% followed by 34.51% contribution of feed and the contribution of depth of cut is 21.91%.
- Predicted optimal combination by the TOPSIS method for maximization of MR and minimization SR is cutting velocity=25 m/min, feed=0.12 mm/rev and depth of cut=0.75 mm.
- The responses are improved significantly with the help of TOPSIS method.

References

[1] Naidu K M and Rao S R 2013 ANN based surface roughness prediction in turning of AA 6351 *Int. J. Eng. Res. Appl.* **3(4)** 1455-1459.

[2] Ranganath M S, Vipin R and Mishra S 2014 Optimization of surface roughness and material removal rate on conventional dry turning of Aluminium (6061) *Int. J. Adv. Res. Innov.* **1** 62-71.

[3] Alagarsamy S V, Raveendran P, Sagayaraj S A V and Vendan S T 2016 Optimization of machining parameters for turning of aluminium alloy 7075 using Taguchi method *Int. J. Eng. Technol.* **3(1)** 316-322.

[4] Kumar A, Singh H and Garg H K 2016 Optimization of material removal rate and surface roughness while turning of hybrid Aluminium metal matrix composite on CNC lathe using response surface methodology *Int. J. Emerg. Technol.* **7(1)** 117-1 25.

[5] Surya M S, Vepa K S and Malleswari K 2019 Optimization of machining parameters using ANOVA and grey relational analysis while turning Aluminium 7075, *Int. J. Recent Technol. Eng.* **8(2)** 5682-5686.

[6] Goyal S, Singh V and Yadav K P 2016 Experimental study of turning operation and optimization of MRR and surface roughness using Taguchi method, *Int. J. Innov. Res. Adv. Eng.* **3(3)** 44-50.

[7] Dhanalakshmi S and Rameshbabu T 2020 Multi-aspects optimization of process parameters in CNC turning of LM 25 alloy using the Taguchi-grey approach, *J. Met.* **10(8)** 1-17.

[8] Rahul V and Vijaykumari J S 2015 Optimization of material removal rate of AlMgSiCu in turning operation using genetic algorithm, *WSEAS Trans. Appl. Theor. Mech.* **10** 95-101.

[9] Shofique A and Rajesh A 2017 Optimization of turning parameters of Aluminum 6351 T6 using Taguchi decision making technique, *Int. J. Netw. Sci.* **1** 27–38.

[10] Emmanual L and Karthikeyan T 2018 Investigation of surface roughness and material removal rate on machining of TiB_2 reinforced Aluminum 6063 composites: A Taguchi's approach, *Mech. Mech. Eng.* **22(4)** 1319–1327.

[11] Kishore G V 2015 Selection of optimum parameters for MRR & SR simultaneously in AWJM of Inconel-625 alloy using TOPSIS method, *Int. J. Mech. Eng. Technol.* **6(9)** 1-9.

[12] Singaravel B and Selvaraj T 2015 Optimization of machining parameters in turning operation using combined TOPSIS and AHP method, *Technical Gazette,* **22(6)** 1475–1480.

[13] Kumar P A and Chandra R B 2014 Multiresponse optimization of process parameters in turning of GFRP using TOPSIS method, *Int. Sch. Res. Notices.* **2014** 1-10.

[14] Khorshidi R, Hassani A, Rauof A H and Emamy M 2013 Selection of an optimal refinement condition to achieve maximum tensile properties of Al–15%Mg2Si composite based on TOPSIS method, *Mater. Des.* **46** 442-450.

[15] Akbari M, Asadi P, Zolghadr P and Khalkhali A 2017 Multicriteria optimization of mechanical properties of aluminum composites reinforced with different reinforcing particles type *P. I. Mech. Eng. E-J. Pro.* **232(3)** 323-337.

[16] Khan A and Maity K 2017 Application of MCDM-based TOPSIS method for the selection of optimal process parameter in turning of pure titanium *Benchmark. Int. J.* **24(7)** 2009-2021.

[17] Kumar B R, Routara B C and Kumar P A 2015 An approach for optimization the process parameter by using TOPSIS method of Al–24%SiC metal matrix composite during EDM *Mater. Today Proc.* **2(4-5)** 3116-3124.

[18] Zhongyou X 2012 Study on the application of TOPSIS method to the introduction of foreign players in CBA games *Phys. Procedia.* **33** 2034 – 2039.

Effect of Number of Passes on Surface Properties of Burnished Aluminium Alloy

Koona Bhavani[1], V.S.N. VenkataRamana[2], Rahul[3*], K. Sri Ram Vikas[4], Ch. Kishore Reddy[5]

[1,2]Department of Mechanical Engineering,GITAM Institute of Technology, GITAM (Deemed to be University), Visakhapatnam,Andhra Pradesh, 530045, India.
[3]Department of Mechanical Engineering,Chaitanya Bharathi Institute of Technology(A), Gandipet, Hyderabad,500075, Telangana, India.
[4,5]Department of Mechanical Engineering, Prasad V. Potluri Siddhartha Institute of Technology, Vijayawada, 520007, Andhra Pradesh, India.
*Corresponding author: luhar6100@gmail.com, rahul_mech@cbit.ac.in

Abstract: Basically, roller burnishing process is one of the surface finishing processes. In this process, a hard and highly polished roller will be pressed against a rotating specimen,thiscauses plastic flow on specimen surface from the peaks of surface irregularities into the valleys, which results in a reduction in the surface roughness considerably. This is a type of cold working whichalso results in improvementof micro-hardness, refinement of microstructure on the surface of the specimen under consideration. This work aims to understand the versatility of roller burnishing processand to understand the effect of the number of burnishing passes to enhance surface properties. Surfaceroughness, electrical conductivityand microhardness were studied in the present work. Experimental work was done on aluminium AA7075- T6 alloy. The parameters under which the burnishing was done were burnishing force, spindle feed and speed and are kept constant in the present work.

Keywords: Roller burnishing, microhardness, surface roughness, AA7075-T6.

1. Introduction

Finishing processes are becoming predominant in the production of machines and instrument components, the burnishing process has fascinated great interest among the engineers and researchers due to the versatility involved in the process.To increase productivity, and to reduce machining times and to sustain competitiveness in the market of machining production systems, it is becoming crucial to improve existing technologies. Burnishing process is one surface finishing technique applied widely to improve surface properties. Many researchers are working in this area and developed analytical models, for studying surface properties. Lars Hiegemannet.al.,[1] conductedvarious experimentson ball burnishing and developed an analytical model which was used to predict the roughness after ball burnishing for a thermally sprayed coating. Kable et.al., [2] examined surface roughness, microhardness on medium carbon steel by changing different parameters like speed, feed andnumber of passes on the drilled hole, by using Taguchi analysis, a significant improvement in hardness was observed i.e., from 377Hv to 528Hv and surface roughness was reduced from 2.44 μm to 0.13 μm. Recently, a newlydesigned ball burnishing tool made of HSS ball of 8mm diameter have been employed on AA6061 using a conventional lathe and process parameters have been optimized using Taguchi method [3]. Many researchers have worked on roller burnishing using lathe on various ferrous and non-ferrous materials [4-9]. In reference [10] roller burnishing is carried on Al_2O_3/A356 composite specimens with varying process parameters, it is clearly evident from this reference that multi passing improves surface roughness and also by this process microhardness and subsurface hardness improves. Even in mild steel roller burnishing increases hardness, Malleswara Rao et.al, [11] has investigated surface hardness by varying process parameters. After the third pass hardness slightly decreased. Surface finish has a positive and longevity effect on the functioning of the machined parts. When compared to permanent metallic alloys, the biodegradable magnesium-calcium (MgCa) alloy proved to be an attractive orthopaedic biomaterial [12]. A correlation was developed between mechanical, structural properties and corrosion resistance of 18-9 stainless steel after burnishing [13]. To improve the ductility of materials burnishing force and number of burnishing tool passes were observed to be the important parameters [14].Many researchers have worked on roller burnishing on different types of materials in different conditions. But few researchers have worked on the effect of the number of passes on surface properties of aluminium alloys. Keeping this in view the present work concentrated on the study of the effect of the increase in passes in the modification of surface properties of AA7075 alloy.

2. Materials and Methods

In the present investigation burnishing operation was performed on AA7075-T6 rod of length 610mm and diameter 25mm(Fig. 1). This AA7075-T6 is an aerospace-grade alloy which possesses excellent strengthand it is comparable with various steels, additionally, it is good at fatigue strength, but the machinability of this alloy is average. This alloy is poor at corrosion resistance because of anodic dissolution of η precipitates. Its relatively high cost limits its usage to aerospace applications. Table 1 gives the chemical composition of the base alloyAA7075-T6 is typically used in Aircraft fittings.

Table 1. Chemical composition of AA7075-T6

Element	Si	Fe	Cu	Mn	Mg	Cr	Zn	Ti	Al
Wt.%	0.45	0.5	1.2-2	0.3	2.1-2.9	0.18-0.28	5.1-6.1	0.2	87.1-91.4

Fig. 1 Burnishing process setup.

The burnished and unburnished samples were tested for surface roughness on Mitutoyo surface roughness testingequipment. The hardness of the surface of both burnished and unburnished samples was measured by micro Vickers Hardness equipment as per (ASTM E 384). The microstructure of the burnished and un-burnished surfaces was observed for any refinement in the grainstructure, using an optical microscope (LEICACTR6000).The electrical conductivity of burnished and unburnished specimens wasmeasured using theKelvin Double Bridge.

3. Results and Discussions

In the present work, base metal (AA7075-T6 alloy) properties were tested and readings were taken before and after burnishing and later various tests were conducted on the burnished surfaces.

From the optical microstructure of base metal(Fig. 2) $MgZn_2$ particles were observed throughout the α matrix [15]. After burnishing it is clearly evident from the optical micrographs (Fig. 3 (a) – 3 (f)) that the size of the

$MgZn_2$ particles has considerably reduced and these were uniformly dispersed in α matrix. This is attributed to the fact that the force applied duringthe burnishing process is sufficient to break the $MgZn_2$ particles and can also help in uniform dispersion of the same.

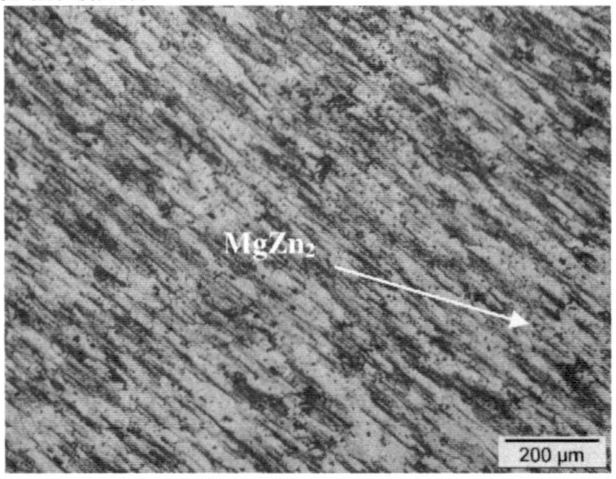

Fig. 2 Optical microstructure of base metal AA7075-T6 showing $MgZn_2$ Particles.

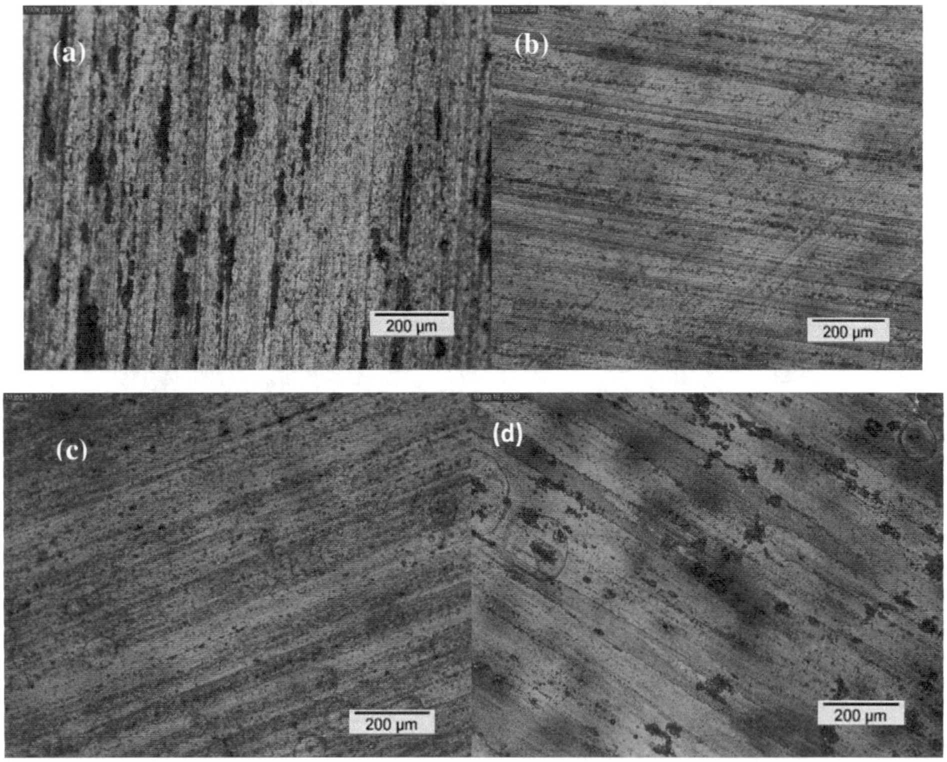

Fig. 3 Optical microstructures of AA7075-T6 alloys specimen after burnishing
(a) 1st pass (b) 2nd pass (c) 3rd pass (d) 4th pass (e) 5th pass (f) 6th pass.

3.1 Surface roughness

The surface roughness of the specimens was examined before and after the burnishing process using Mitutoyo surf test (SJ-210)and surface roughness were measured five times and an average was takenand represented in Table 2 andthe same values are shown in a graph in (Fig. 4).

Table 2: Surface roughness values of AA7075T6 prior and after burnishing with differentpasses at Load= 12Kgf speed = 220 rpm, feed = 0.14mm/rev

Number of passes	Surface roughness, Ra (µm)
Unburnished(Bases metal)	0.30
1	0.47
2	0.27
3	0.3
4	0.22
5	0.2
6	0.31

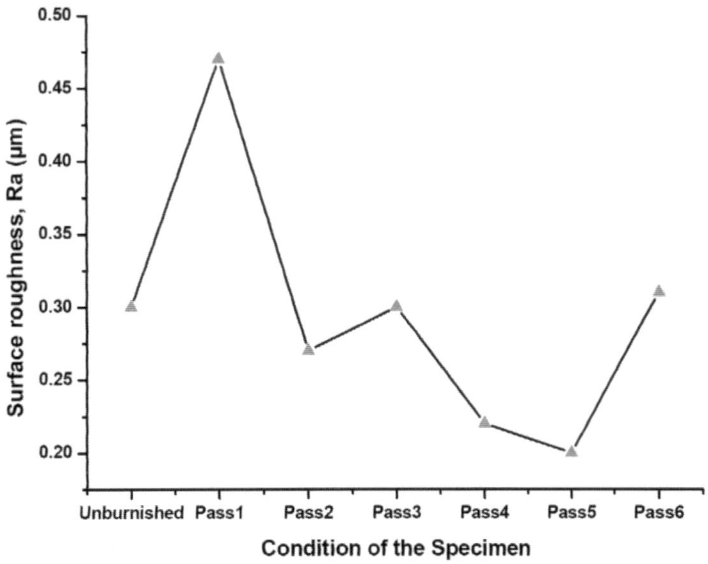

Fig.4 Graph showing surface roughness values of AA7075T6 before and after burnishing after differentpasses.

It is observed that the specimen is showing minimum roughness value after 5th pass of burnishing when compared with the unburnished specimen. Additionally, it is also observed that at initial passes the surface roughness is low compared to that after final passes. This trend can beattributed to the fact that as the number of passes is increasing the surface on the metal is becoming flaky, similar behaviour has been observed by D. G. Mahto et. Al., [16] and Gharbi, F. et. Al [17]. Hence, surface roughness is higher at the final pass.

3.2 Electrical Conductivity

The electrical conductivity on the surface of the specimens prior and after the burnishing process was measured and these values are presented in Table 3, andFig.5 shows the graph of electrical conductivity values of the AA7075-T6 specimens before and after burnishing.

Table 3: Electrical conductivity of AA7075T6 Prior and after burnishing for various number of passes

Number of passes	Electrical Conductivity(S/m)
Unburnished (Bases metal)	531.09
1	614.78
2	461.09
3	474.37
4	395.31
5	367.92
6	434.82

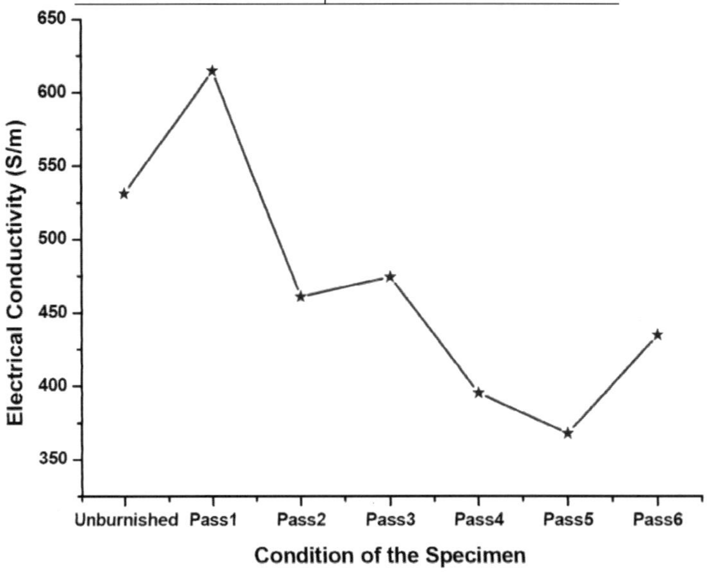

Fig.5 Graph of electrical conductivity values of AA7075-T6 Prior and after burnishing after differentpasses.

From Table 3, it is clearly understood that the electrical conductivity is decreasing, eventhough it has increased in the first pass. This behaviour is due to sensitivity towards stress. It is widely known that when aluminium alloys are subjected to cold work, electrical conductivity decreases, this can be attributed to an increase in dislocation population [18]. But in the final pass slightly electrical conductivity has been increased this can be attributed to the transformation of precipitates [19]. Other factors which affect electrical conductivity include precipitate size and distribution, electron scattering, lattice faults etc.,[20].

3.3 Microhardness

The microhardness test was conducted on the specimen as per (ASTM E 384) standards, for unburnished and burnished specimens 300 gf load is applied with dwell 15 seconds. Five readings were taken and average hardness values were shown in Table 4.

Table 4 Microhardness values before and after burnishing at various number of passes

Number of passes	Microhardness,VHN
Unburnished (Bases metal)	258
1	264
2	272
3	298
4	323
5	334
6	345

Hardness has been significantly enhanced with increase in passes. Previous studies also show that the increase in the number of passes enhances surface hardness[21-22]. Hardness enhancement can be attributed to the work-hardening of the surface layer [23].

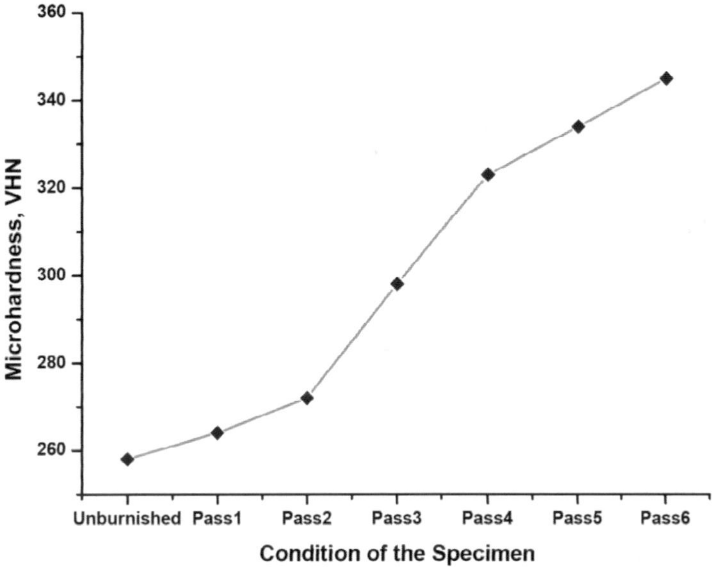

Fig.6 Graph showing microhardness values of AA7075T6 before and after burnishing at differentpasses.

4.Conclusions:

In this present investigation, burnishing operation was done on AA7075-T6 with different passes and various tests were performed from which the following conclusions can be made.

1. The optical microstructure reveals $MgZn_2$ particles and their distribution. Clusters of these particles were observed in base metal and the same particles were disintegrated after burnishing process and uniformly distributed in the matrix.
2. The surface finish is improved after the burnishing process. The best surface finish was obtained at 5th passof burnishing. After 5th pass surface roughness slightly increased due to surface flakiness on metal.
3. Electrical conductivity is decreasing, eventhough it has increased in the first pass. This behaviour is due to sensitivity towards stress.
4. Considerable improvement in hardness was noticedby increasing in the number of passes this is due to work hardening behaviour.

References

[1] Hiegemann, Lars, et al. "Analytical prediction of roughness after ball burnishing of thermally coated surfaces." Procedia Engineering 81 (2014): 1921-1926.
[2] Kamble, P. S., and V. S. Jadhav. "Experimental study of Roller burnishing process on plain carrier of planetary type gear box." Parameters 1.2 (2012): 3.
[3] Patel, P. N., N. B. Patel, and T. M. Patel. "Parametric Optimization of Ball Burnishing Process Parameter for Hardness of Aluminum Alloy 6061." IOSR Journal of Engineering (IOSRJEN) 4.08 (2014): 21-26.
[4] Loh, N. H., S. C. Tam, and S. Miyazawa. "Ball burnishing of tool steel." Precision Engineering 15.2 (1993): 100-105.
[5] Loh, N. H., S. C. Tam, and S. Miyazawa. "Investigations on the surface roughness produced by ball burnishing." International Journal of Machine Tools and Manufacture 31.1 (1991): 75-81.
[6] Kotiveerachari, B., and R. L. Murty. "Study of some aspects of burnishing." International Journal of Production Research 23.3 (1985): 499-521.
[7] Pande, S. S., and S. M. Patel. "Investigations on vibratory burnishing process." International Journal of Machine Tool Design and Research 24.3 (1984): 195-206.
[8] Neema, M. L., and P. C. Pandey. "Effects of Asperity Deformation on the surface of machined surfaces." 9th AIMTDR Conference, IIT Kanpur. 1980.
[9] Dabeer, P. S., and Dr GK Purohit. "Investigation on Surface finish by Ball Burnishing." Proceeding of National conference on Design for Product Life Cycle, Birla Institute of Technology and Science, Pillani, Rajasthan. 2006.
[10] Ibrahim, A. A. "An Investigation for Surface Integrity of Al2O3/A356 Composites Using Roller Burnishing Process on a Milling Machine." Applied Mechanics and Materials. Vol. 330. Trans Tech Publications Ltd, 2013.
[11] Rao, Malleswara JN, ChennaKesava A. Reddy, and Rama PV Rao. "The effect of roller burnishing on surface hardness and surface roughness on mild steel specimens." International Journal of Applied Engineering Research 1.4 (2010): 777.
[12] Salahshoor, Meisam, and Yuebin Guo. "Biodegradable orthopedic magnesium-calcium (MgCa) alloys, processing, and corrosion performance." Materials 5.1 (2012): 135-155.
[13] Pałka, K., A. Weroński, and K. Zalewski. "Mechanical properties and corrosion resistance of burnished X5CrNi18-9 stainless steel." Journal of Achievements in Materials and Manufacturing Engineering 16.1-2 (2006): 57-62.
[14] Ghodake, A., R. Rakhade, and A. Maheshwari. "Effect of Burnishing Process on Behavior of Engineering Materials-A Review." Journal of Mechanical and Civil Engineering 5.5 (2013): 9-20.
[15] Kumar, P. Vijaya, G. Madhusudhan Reddy, and K. Srinivasa Rao. "Microstructure, mechanical and corrosion behavior of high strength AA7075 aluminium alloy friction stir welds–Effect of post weld heat treatment." Defence Technology 11.4 (2015): 362-369.
[16] Mahto, D. G., and Ashutosh Verma. "Roller Burnishing-A Literature Review of Developments and Trends in Approach to Industrial Application." Asian Journal of Engineering and Technology 1.3 (2013).
[17] Gharbi, F., et al. "Effect of ball burnishing process on the surface quality and microstructure properties of AISI 1010 steel plates." Journal of materials engineering and performance 20.6 (2011): 903-910.
[18] Morozov, Maxim, Gui Yun Tian, and Philip J. Withers. "Elastic and plastic strain effects on eddy current response of aluminium alloys." Nondestructive Testing and Evaluation 28.4 (2013): 300-312.
[19] Şimşek, Ijlal. "Investigation of the effect of second phase precipitates on the corrosion and electrical conductivity of 7075 aluminum alloys." Anti-Corrosion Methods and Materials (2019).
[20] Grimvall, Göran. Thermophysical properties of materials. Elsevier, 1999.
[21] Procesu, Uporaba Grey-TaguchijeveMetodePri. "Use of grey based Taguchi method in ball burnishing process for the optimization of surface roughness and microhardness of AA 7075 aluminum alloy." Materiali in tehnologije 44.3 (2010): 129-135.
[22] Hassan, Adel Mahmood, and AimanSharef Al-Bsharat. "Influence of burnishing process on surface roughness, hardness, and microstructure of some non-ferrous metals." Wear 199.1 (1996): 1-8.
[23] Hassan, Adel Mahmood. "The effects of ball-and roller-burnishing on the surface roughness and hardness of some non-ferrous metals." Journal of materials processing technology 72.3 (1997): 385-391.

Effect of Cerium Oxide Nanoparticles Additive Blended in Palm Oil Biodiesel as Alternative Fuel Used in Diesel Engine

Dr. T Srinivasa Rao[1], Dr. H Suresh Babu Rao[2]*, S A K Jilani[3] and Avinash Mutluri[4]

[1,3,4]Department of Mechanical Engineering, Vasireddy Venkatadri Institute of Technology, Nambur, Peddakakani Mandal, Guntur District, 522508, Andhra Pradesh, India.

[2]Department of Mechanical Engineering, Chaitanya Bharathi Institute of Technology, Proddatur, India.

* Corresponding author email: hsbr110479@gmail.com

Abstract

In the present study, the effects of adding of cerium oxide (CeO_2) nanoparticle additive as a fuel additive to a palm oil biodiesel-diesel fuel blend on engine performance, and exhaust emissions were studied in a single cylinder four stroke diesel engine under different torque conditions. The test fuels used were fossil diesel fuels, B20 blend biodiesel (as 20% biodiesel and 80% diesel) with cerium oxide (CeO_2) nanoparticle additive (30ppm, 60ppm, 90ppm and 120ppm). The experimental results demonstrated that B20 blend biodiesel at 90ppm of CeO_2 nanoparticles additive found significant reduction emissions nitrogen oxide (NOX), carbon dioxide (CO), unburnt hydrocarbons (UBHC) and smoke. However, brake specific fuel consumption (BSFC) is decreased with significant from increased brake thermal efficiency (BTE) by doping of CeO_2 from 30ppm to 90ppm. B20 biodiesel blend with 90ppm of CeO_2 nanoparticles additive as optimum blend out of the test blends.

Key words: Biodiesel; Calophyllum inophyllum; Cerium oxide nanoparticle; Performance; Exhaust emissions.

1. Introduction

A diesel engine has higher brake thermal efficiency than a petrol engine. Also, they have an advantage of using lesser fossil fuel. Petroleum diesel plays a significant role in the transportation sector, the agricultural sector and industrial growth. However, the total petroleum reserves are depleting day by day due to indiscriminate extraction and lavish consumption [1, 45, 47]. Therefore, the researchers are suggested that biodiesels are a viable alternative to diesel fuel in the diesel engine. Biodiesel can be prepared from various renewable feedstocks like vegetable oils - palm oil, sunflower oil, peanut oil etc., are edible oils and jatropha, calophyllum inophyllum, karanja, cotton seed, waste plastic oil, etc., are non-edible oils [3, 6, 8, 10, 44, 46]. The biodiesel properties are like rich oxygen content, higher kinematic viscosity, reduced smoke emission and diluted level of pollutants from engine exhaust (CO, HC and particulate matter). The biodiesel's emission characteristics are better than petro-diesel fuel except for NO_X emission [13]. To control NO_X emissions from the C.I engines, various researchers have been described strategies like engine modification like EGR (Exhaust Gas Recirculation), increasing/decreasing injection timing, injection pressure and adulteration of fuel [25, 27]. Among these fuel alterations with nanoparticles in biodiesel results in reducing the level of pollutants at the engine exhaust and eventually, increases engine performance. Nanoparticle blended with fuel shown better thermal properties, which is mainly due to the higher surface area to volume ratio of the nanoparticles also resulting in increased oxidation of HC and acting as an oxygen buffer against NO_X [19].

Many researchers used different additives to biodiesel and found reduced emissions and increased performance [41, 42, 43]. Acetone blended with Palm biodiesel given a significant reduction in HC, NO_X, CO and smoke emissions when compared to palm biodiesel [27, 29]. V Sajith et al., 2010 conducted experiments with cerium oxide as nanoparticle additive in Jatropha biodiesel (20ppm, 40ppm & 80ppm) had shown increased brake thermal efficiency by 1.5% and there is a significant reduction of NO by 30% and hydrocarbon emissions by 40%. Ajin et al., 2011 experimentally observed that the CeO_2 as a nano-additive in diesel improved the BTE by 5% and a reduction of HC and NOX emission by 45% and 30% respectively. From the various metal oxides copper, iron, cerium and cobalt have used as fuel additives. Among all nanoparticle's additives, cerium oxide has exhibited a high catalytic activity due to its high surface-to-volume ratio, which improves the brake thermal efficiency and reduces the emission [15]. Selvan et al., 2009 experimentally investigated the effect of cerium oxide nanoparticle with diesel and biodiesel as fuel in diesel engine and observed the brake thermal efficiency was increased and a drastic reduction in HC and NO_X emission. The performance and emission characteristics of the CI engines improves due to the doping of alumina nanoparticles in the biodiesel emulsion fuels, which leads to ignition delay and reduced peak pressure.

About palm oil

Oil palm, was growing wild and was a native to West Africa and nowadays it is an agricultural crop. Annually, Oil palm can produce 10 to 34 tons per hectare of oil palm fresh fruit bunches. Palm oil biodiesel has most widely recognized as a potential alternative fuel [2, 13].

Figure 1 shows the various biodiesel feedstock oil yield. Palm oil has shown a high productivity when it is compared to other oils. The palm oil productivity is about 5945 litres per hectare and which is about 14 times higher than soybean oil.

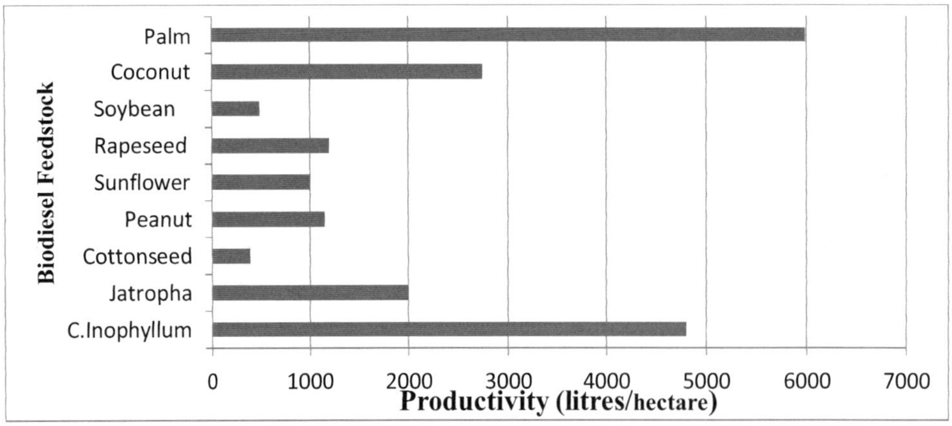

Figure 1. Various biodiesel feedstock production oil yield [16].

2. Materials and Methods

Preparation of palm oil biodiesel

Different researchers have developed methods for biodiesel preparation from various feedstocks [14, 15, 16]. The following is the process for converting palm oil to biodiesel.

Palm oil is collected and is preheated to 60°C separately. A 0.5g of NaOH and methanol (9:1 methanol to oil ratio) are mixed properly to form Sodium methoxide. The preheated oil is slowly poured in Sodium methoxide, keeping the stirrer on. The mixture is allowed stir in a magnetic stirrer for 1hour at 400 rpm maintaining 55°C reflux. Then the mixture is poured in separating funnel and left it undisturbed for 1hour in a separating funnel and thus formed glycerin is removed which are formed at the bottom layer. The top layered oil is required one and it is separated to the next step.

Water washing has done to wash the soaps out of the biodiesel with warm water, 3 to 4 times. Add a small amount of dilute acetic acid in biodiesel before adding the water. Thus, the pH of the solution should be closer to neutral and takes out any catalyst suspended in the biodiesel. The biodiesel is water washed gently with warm water at 50°C temperature to remove impurities for three to four times. Then the washed oil is heated above 100°C to remove water particles and excess methanol. Moreover, thus obtained product is the required palm oil biodiesel.

Characterization of Cerium Oxide nanoparticle

There are various methods to synthesis CeO_2 precipitation, hydrothermal, sol-gel method, microemulsion method and other methods [23-30]. Among the various preparation methods, microwave-assisted solution method is an inexpensive and simple method to prepare the nanoparticles. The synthesis of CeO_2 nanoparticles was done by method - Solvothermal technique [23].

The synthesis of CeO_2 nanoparticles has carried out by using Solvothermal technique. X-ray diffraction method (XRD) characteristic peaks corresponding to the (111), (200), (220), (311), (222), (400), (331) and (420) planes are located at 2θ = 28.6°, 33.1°, 47.5°, 56.4°, 59.1°, 69.4°, 76.7° and 79.1° respectively.

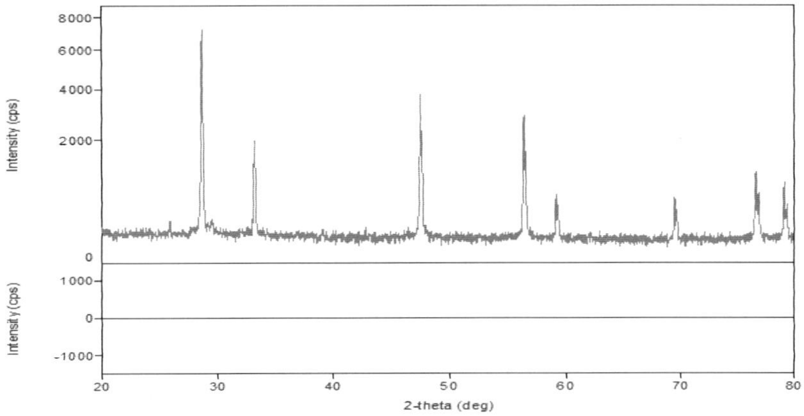

Figure 2. XRD of synthesized cerium oxide nanoparticles.

3. Experimental Setup

The experimental investigation was carried out and the properties of modified fuels are compared with diesel fuel. The properties were studied as the calorific value, viscosity, flash point and fire point. The standard ASTM procedures were used during experimentation. The performance tests were conducted on research engine (single cylinder four stroke diesel engine) using test fuels and also, the emission characteristics are studied by using exhaust gas emission analyzer.

A single cylinder, four stroke and water-cooled diesel engine was used to conduct the performance, emission and combustion test. An electrical generator used for loading the engine. AVL 444 N Di-Gas analyzer and AVL 437C Smoke meter were used for emission studies. The following Table 1 is the specifications of the engine:

Table 1: Specifications of the diesel engine.

Type	Single cylinder, four stroke, water cooled diesel engine
Stroke	110mm
Bore	88mm
Rated output	5.2 kW
Rated speed	1500 rpm
Compression ratio	17.5
Loading device	Electric generator

4. Results & Discussions

The findings of diesel engine performance, combustion, and emission characteristics for the diesel, biodiesel blend B20 and CeO_2 nanoparticle as additive to B20 (30ppm, 60ppm, 90ppm and 120ppm) test fuels are discussed in the subsequent sections.

Properties of fuels used

The properties of the above modified fuels along with diesel fuel are shown in Table 2. The various physicochemical properties of the above modified fuels were carried out and the thus obtained results are compared with the pure diesel (D100). The main objective of the investigation was to find the variations in the properties of the fuels also, the effect of the level of inclusion of CeO_2 nanoparticles added to the B20 biodiesel blend. The tested fuels were shown an increasing trend for the flash and fire points with the increase in dosing level, which indicates a continual decrease in the volatility of the test fuels with increases in the quantity of the CeO_2 nanoparticle additive [5, 8]. The CeO_2 nanoparticles additive to the modified fuels shown that the fluid layer resistance has increased and hence the kinematic viscosity [9] also.

Table 2 Properties of the modified fuels and Diesel.

	D100	B20	B20+30ppm	B20+60ppm	B20+90ppm	B20+120ppm
Kinematic Viscosity @40°C in cSt	2.5	3.29	3.32	3.38	3.53	3.98
Flash point ^{0}C	45	52	56	64	72	74
Fire point ^{0}C	56	59	63	72	79	81
Density @15°C in kg/m3	834	852	857	860	863	866
Calorific value in kJ/kg	42,538	41,094	41,162	41,214	41,319	41,325

Performance and emissions of test fuels on diesel engine

Brake Thermal Efficiency (BTE)

The biodiesel's BTE is less than diesel fuel [3] due to its lower calorific value. Figure 3 shows variations in the BTE with load for the palm oil biodiesel blends doped with CeO_2 nanoparticles and diesel fuel [41]. The result depicted that the BTE of the diesel engine continuously improved by the increased dosing levels of CeO_2 nanoparticles additive in B20 biodiesel blend. The enhancement in BTE for the CeO_2 nanoparticles additive in B20 biodiesel blend doped fuels from 30ppm to 120ppm is mainly due to the occurrence of better fuel combustion and the nanoparticles doping promoted the combustion in the engine [6, 43]. Above 90ppm dosing level of the CeO_2 nanoparticle increased, due to more agglomeration of particles, the catalytic action was lowed and lessens the BTE at 120ppm. The results depicted that the BTE of blend B20+90ppm of CeO_2 has highest efficiency in the modified blends of biodiesel but less than pure diesel i.e., D100 fuel at all loads.

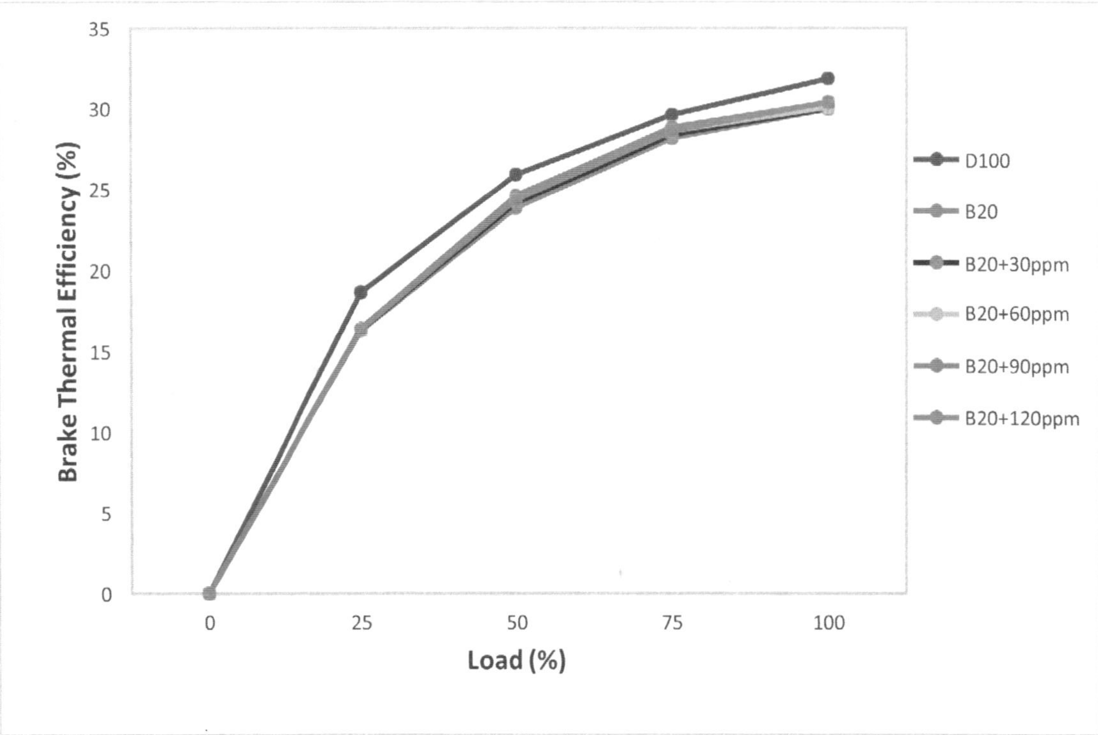

Figure 3. Variation of brake thermal efficiency (BTE) for modified fuels and diesel.

Brake Specific Fuel Consumption (BSFC)

Due to lower calorific value of the biodiesel's, BSFC of the biodiesel is higher when compared to diesel fuel [3]. Figure 4 shows that the variations in BSFC with load for the palm oil biodiesel blends with CeO_2 nanoparticles doped in biodiesel and diesel fuel. Due to increase in dose of CeO_2 nanoparticles in biodiesel from 30ppm to 120ppm, favoured to enhance the combustion. Therefore, the BSFC decreased with the increased in the dosing levels of the particles, except for B20+120ppm, due to more agglomeration and lowered catalytic action. B20+90ppm blend shown the lower BSFC out of other biodiesel blends except for diesel.

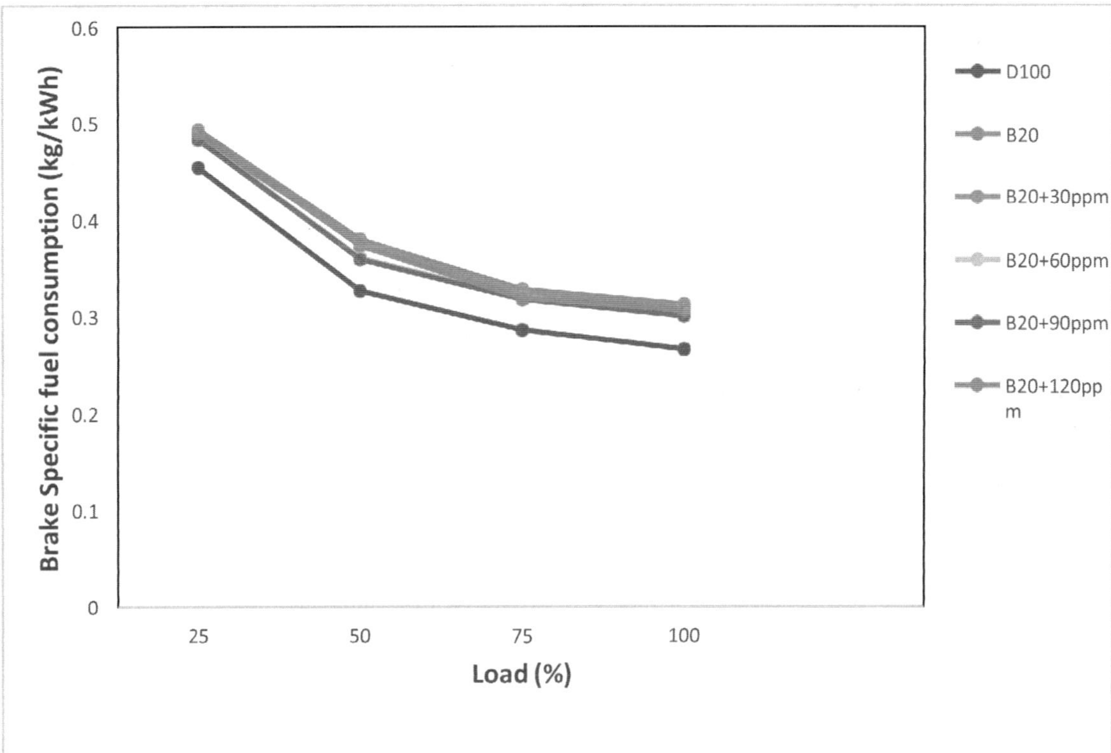

Figure 4. Variation of brake specific fuel consumption (BSFC) for the modified fuels and diesel.

Unburned hydrocarbon (UBHC) emission

UBHC emission of biodiesel is lower than diesel fuel due to lower calorific value and incomplete combustion of fuel. Figure 5 illustrates variation in UBHC emission with load for the palm oil biodiesel blends doped with CeO_2 nanoparticles and diesel fuel. Due to CeO_2 nanoparticles additive doped in biodiesel blends from 30ppm to 120ppm increases the oxidation of hydrocarbon. Therefore, a complete combustion is promoted in the engine. For B20+120ppm, due to more agglomeration and lower catalytic action, the UBHC of B20+90ppm blend is lower than other modified fuels and diesel fuel. The chemical reactions during combustion are:

$$(2x+y)CeO_2 + C_xH_y \rightarrow \left[\frac{(2x+y)}{2}\right] Ce_2O_3 + \frac{x}{2}CO_2 + \frac{y}{2}H_2O \quad (1)$$

$$4CeO_2 + C_{Soo} \rightarrow 2Ce_2O_3 + CO_2 \quad (2)$$

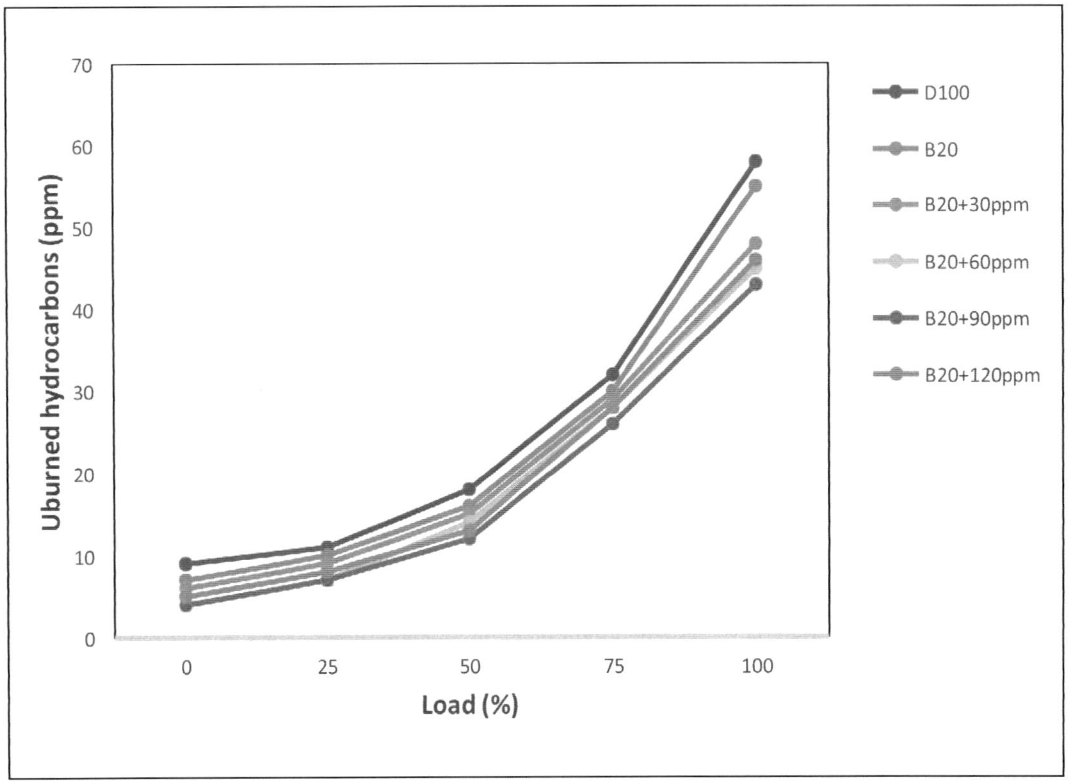

Figure 5. Variation of unburned hydrocarbon (UBHC) for the modified fuels and diesel.

Carbon monoxide (CO) emission

Due to low calorific value and incomplete combustion, carbon monoxide emission of the biodiesel is less than diesel. Figure 6 shows the variation in CO emission with load for the palm oil biodiesel blends with CeO_2 nanoparticles doped and diesel fuel. The additive CeO_2 nanoparticles doped in biodiesels promoted better combustion, CO emissions from the diesel engine depends on the fuel properties, due to the ability of nanoparticles to convert CO to CO_2. Therefore, CO of B20+90ppm biodiesel blend is lower than other modified fuels (except B20+120ppm biodiesel blends) and diesel fuel.

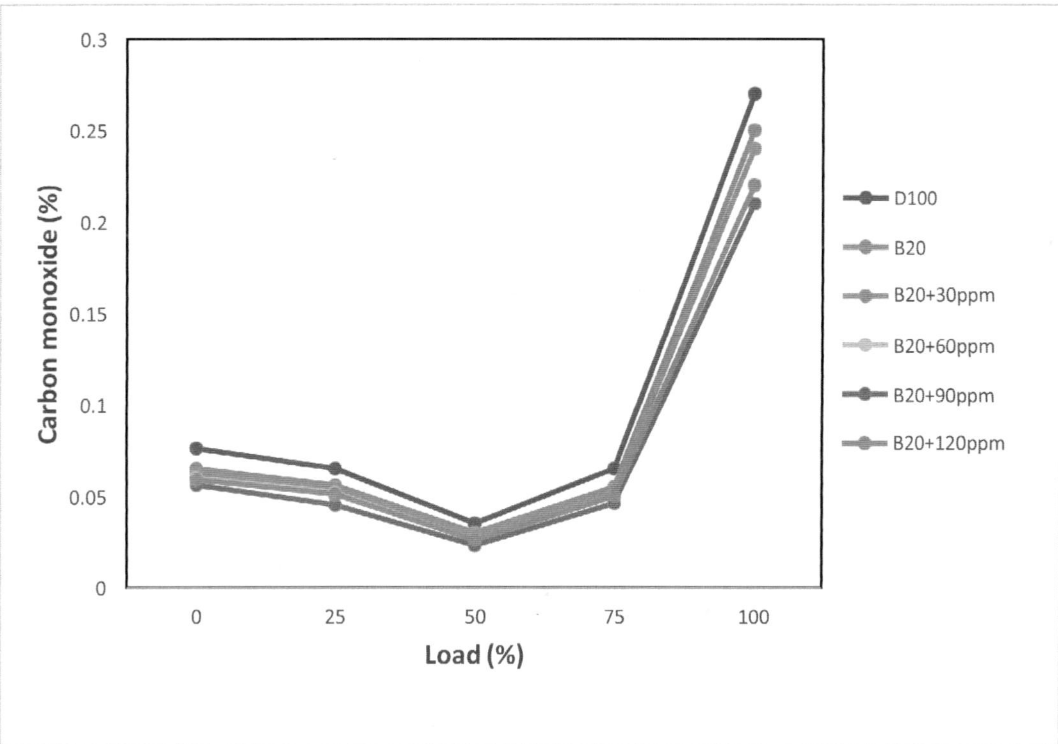

Figure 6. Variation of carbon monoxide (CO) for the modified fuels and diesel.

Nitrogen oxides (NOX) emission

Figure 7 illustrates the variation in NO_X emission with load for the palm oil biodiesel blends with CeO_2 nanoparticles doped and diesel fuel. The CeO_2 nanoparticles additive doped in biodiesel blends from 30ppm to 120ppm. The NO_X emission has increased, in order to reduce it, CeO_2 nanoparticles additive added to B20 biodiesel blend. Due to high thermal stability, oxidation of hydrocarbon forms Ce_2O_3 is formed equation (2) and again reduction of NO occur due to deoxidizing to CeO_2. The chemical reaction during combustion is:

$$Ce_2O_3 + NO \rightarrow 2CeO_2 + \frac{1}{2}N_2 \quad (3)$$

Therefore, due to the CeO_2 nanoparticles additive to biodiesel blends, lower NOx emission occurs for the modified fuels doped up to 90ppm, and the NO_X emission B20+90ppm biodiesel blend are lower than other modified fuels except for B20+120ppm biodiesel blend.

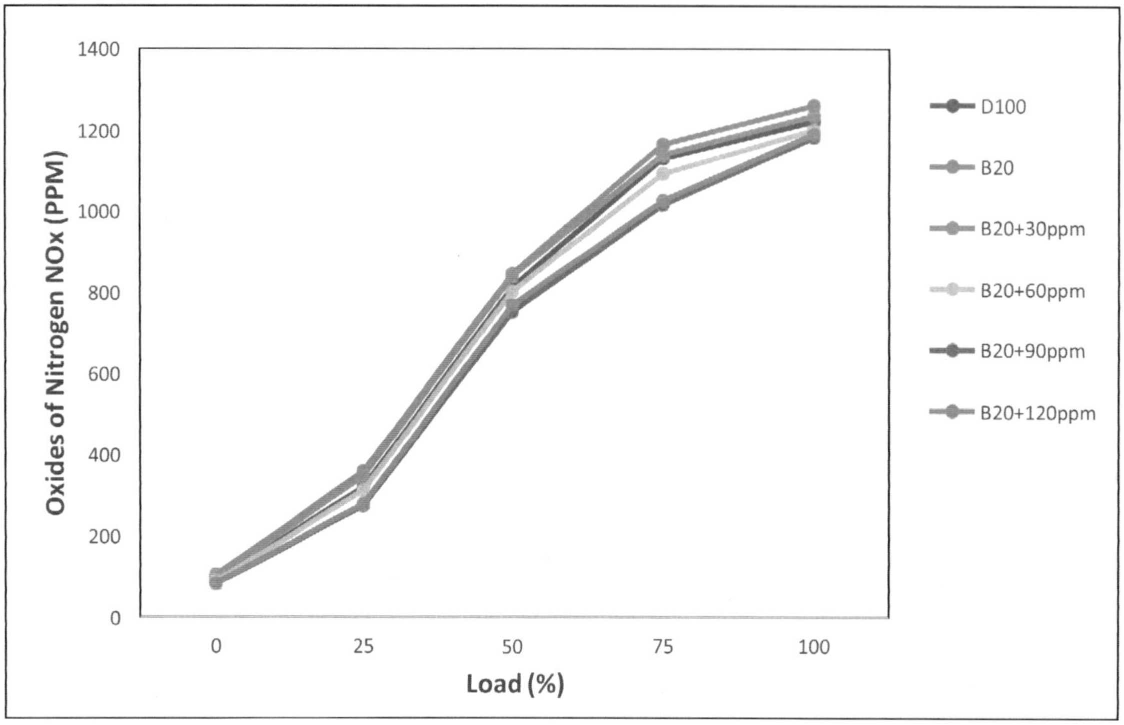

Figure 7. Variation of Oxides of Nitrogen (NO_X) for the modified fuels and diesel.

Smoke Emissions

Figure 8 demonstrates the variation in smoke emission with load for the palm oil biodiesel blends with CeO_2 nanoparticles doped and diesel fuel. The CeO_2 nanoparticles additive doped in biodiesel blends from 30ppm to 120ppm, the decreased in smoke emission are detected due to higher surface area to volume ratio and improved combustion. Therefore, with increase in dosing of the nanoparticles, the smoke emission was reduced except for B20+120ppm. Hence, smoke emission of B20+90ppm blend is lower than other modified fuels and diesel fuel.

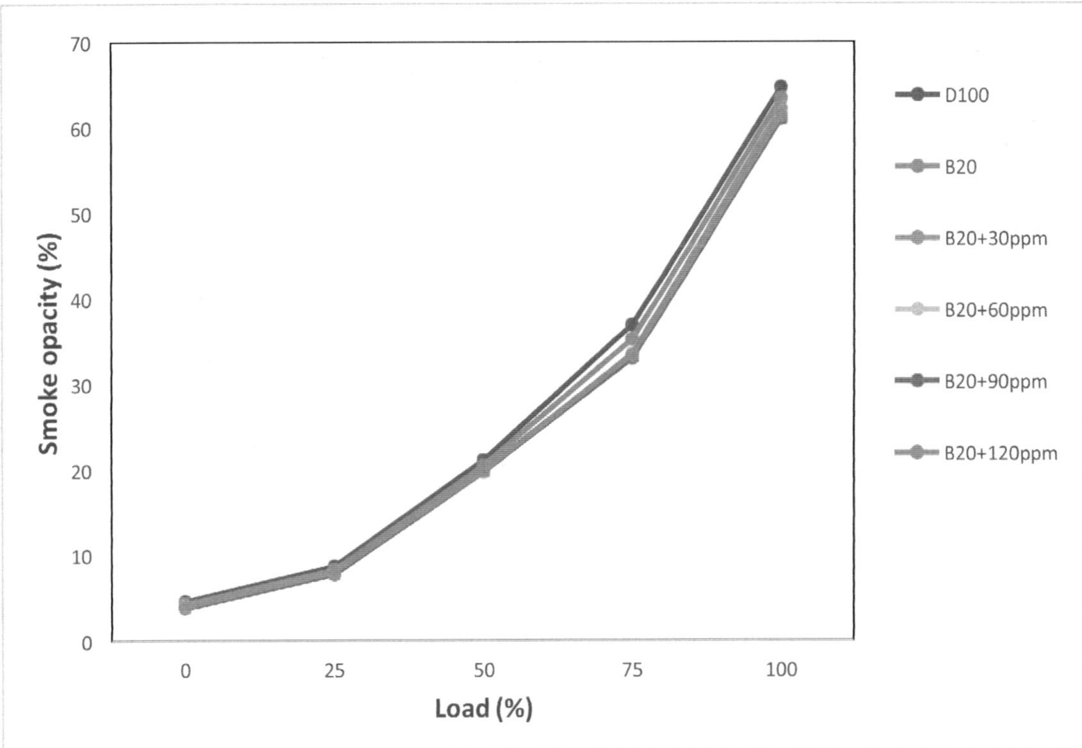

Figure 8. Variation of smoke for the modified fuels and diesel.

5.Conclusions

The outcome of the experimental investigation work was observed and the following effects on performance and emission of engine by CeO_2 nanoparticles doping in B20 palm oil biodiesel blends at 30ppm to 120ppm:

- A maximum increment of 1.23% in BTE and a maximum of 2.56% decrement in BSFC by CeO_2 nanoparticles doping in B20 biodiesel blend palm oil was found in B20+90ppm biodiesel blend at maximum load.
- A maximum reduction in unburned hydrocarbon and carbon monoxide emissions were observed by CeO_2 nanoparticles doping in palm oil B20 biodiesel blend is 19.7% and 15.4% respectively.
- Smoke and NOx emissions were reduced by CeO_2 nanoparticles doping in palm oil B20 biodiesel blend with maximum reduction of 4.6% and 5.9% respectively in biodiesel blend B20+90ppm.

Therefore, from the experimental investigation work concluded that B20+90ppm of palm oil biodiesel blend as best biodiesel blend of modified fuels and can be considered as an alternative fuel for the diesel engine.

References

[1] Agarwal A K, 2003 Biofuels (alcohols and biodiesel) applications as fuels for internal combustion engines*Progress in Energy & Combustion Science,* **33** 231–277.

[2] Mohamad Azri Sukiran, Soh Kheang Loh and Nasrin Abu Bakar 2018Conversion of pre-treated oil palm empty fruit bunches into bio-oil and bio-char via fast pyrolysis*Journal of Oil Palm Research* Vol **30** (1).121– 129.
[3] Ambarish Datta and Bijan Kumar Mandal 2012 Biodiesel Production and its Emissions and Performance: A Review*International journal of Scientific & Engineering Research*, vol. **3**, Issue 6.
[4] Sadhik Basha J, Anand R B 2013 The influence of nano additive blended biodiesel fuels on working characteristics of a diesel engine*Journal of Brazilian society of Mechanical Sciences and Engineering*,**35** 257-264.
[5] Ganesan.S, Dineshbabu Munuswamy, Prabhu Appavu, Arunkumar.T and Yuvarajan Devarajan 2019 Effect of EGR & Nanoparticles on performance and emission characteristics of a Diesel engine fuelled with palm biodiesel and diesel blends*Journal of Oil Palm Research,* Vol. **31** (1) 130 – 137.
[6] Sajith V, Sobhan C B, Peterson G B 2010 Experimental investigations on the effects of cerium oxide nanoparticle fuel additives on biodiesel",*Advances Mechanical Engineering*1- 6.
[7] Ajin C. Sajeevan and Sajith V 2013 Diesel engine emission reduction using catalytic nanoparticles: An experimental investigation", *Journal of Engineering.* 589382.
[8] Dutta P, Pal S, Seehra M S, Shi Y, Eyring E M and Erust R D 2006 Concentation of Ce^{3+} and oxygen vacancies in cerium oxide nanoparticles *Chemistry of Materials,* vol. **18**, no. 21, 5144 – 5146
[9] Selvan V A M, Anand R B and Udayakumar M 2009 Effects of cerium oxide nanoparticle addition in diesel and diesel-biodiesel-ethanol blends on the performance and emission characteristics of a CI engine", *Journal of Engineering and Applied Science,* vol.**4**, no. 7.
[10] Sadhik Basha J and Anand R B 2011 Role of nanoadditive blended biodiesel emulsion fuel on the working characteristics of a diesel engine*Journal of Renewable and Sustainable Energy* **3**, 023106.
[11] Ashraful A M 2014 Production and comparison of fuel properties, engine performance and emission characteristics of biodiesel from various non-edible vegetable oils: A review", *Energy conversion and management,* **80** 202-228.
[12] Suresh Babu Rao H, Dr. Venkateswara Rao T and Dr. Hema Chandra Reddy K, 2013 Palm oil and Calophyllum inophyllum oil are Potential feed stocks for future biodiesel in compression ignition engines: A Review*International Journal of Mechanical Engineering & Technology (IJMET).* vol.**4**, Issue. 5, 301-312.
[13] Girdhar Joshi, Jitendra K. Pandey, Sravendra Rana, Devendra S. Rawat 2017 Challenges and opportunities for the application of biofuel, Renewable and Sustainable Energy Reviews **79** 850–866.
[14] Sahoo PK, Das LM 2009 Process optimization for biodiesel production from Jatropha, Karanja and Polanga oils *Fuel* **88**(9):1588–94.
[15] Sahoo PK, Das LM, Babu MKG, Naik SN 2007 Biodiesel development from high acid value polanga seed oil and performance evaluation in a CI engine*Fuel* **86** (3):448–54.
[16] Ong H C, Silitonga A C, Masjuki H H, Mahlia, Chong, Boosroh 2013 Production and comparative fuel properties of biodiesel from non-edible oils: Jatropha curcas, Sterculia foetida and Ceiba pentandra *Energy Conversion and Management 73* 245-255.
[17] Ganapathy T, Gakkhar R P, Murugesan K 2011 Influence of injection timing on performance, combustion and emission characteristics of Jatropha biodiesel engine, *Appl. Energy* **88** (12) 4376–4386.
[18] Ashok B, Nanthagopal K, Thundil Karuppa Raj R, Pradeep Bhasker J, Sakthi Vignesh D 2017 Influence of injection timing and exhaust gas recirculation of a Calphyllum inophyllum methyl ester fuelled CI engine, *Fuel Processing Technology* **167** 18-30.
[19] Murugesan A, Subramaniam D, Vijayakumar C, A. Avinash, N. Nedunchezhian 2012 Analysis on performance, emission and combustion characteristics of diesel engine fueled with methyl–ethyl esters*J. Renewable and Sustainable Energy*. **4** (6) 063116.
[20] Rashedul H.K., MasjukiH.H,. KalamM.A, AshrafulA.M, Ashrafur RahmanS.M., ShahirS.A. 2014 The effect of additives on properties, performance and emission of biodiesel fuelled compression ignition engine, *Energy Conversion and Management* **88** 348-364.
[21] Bhupendra Singh Chauhan, Naveen Kumar, Haeng Muk Cho, Hee Chang Lim 2013 A study on the performance and emission of a diesel engine fueled with Karanja biodiesel and its blends, *Energy* **56** 1-7.

[22] Palash S M, Masjuki H H, Kalam M A, Masum B M, Sanjid A, Abedin M J 2013 State of the art of NOx mitigation technologies and their effect on the performance and emission characteristics of biodiesel-fueled Compression Ignition engines, *Energy Convers. Manag.* **76** 400–420.

[23] Kumar E, Selvarajan P, Muthuraj D 2013 Synthesis and Characterization of CeO_2 nanocrystal by Solvothermal route, *Materials Research* 16(2): 269 – 276.

[24] Chen PL and Chen IW 1993 Reactive Cerium (IV) Oxide Powders by the Homogeneous Precipitation Method. *Journal of American Ceramic Society* **76**:1577-1583.

[25] Djuricic B and Pickering S 1999 Nanostructured cerium oxide: preparation and properties of weakly-agglomerated powders. *Journal of the European Ceramic Society.*

[26] Wu NC, Shi EW, Zheng YQ and Li WJ.2002 Effect of pH of Medium on Hydrothermal Synthesis of Nanocrystalline Cerium (IV) Oxide Powders. *Journal of American Ceramic Society.* **85**:2462-2468.

[27] Hirano M and Kato E 1996 The hydrothermal synthesis of ultrafine cerium(iv) oxide powders. *Journal of Materials Science Letters* **15**:1249-1250.

[28] Li LP, Lin XM and Li GS, Inomata H 2001 Solid solubility and transport properties of $Ce_{1-x}Nd_{dx}O_{2-\delta}$nanocrystalline solid solutions by a sol-gel route. *Journal of Material Research* **16**:3207- 3213.

[29] Masui T, Fujiwara K, Machida KI, Adachi GY, Sakata T and Mori H. 1997 Characterization of Cerium (IV) Oxide Ultrafine Particles Prepared Using Reversed Micelles. *Chemistry of Materials.*

[30] Sumalin Phoka, Paveena Laokul, Ekaphan Swatsitang, Vinich Promarak, Supapan Seraphin, Santi Maensiria 2009 Synthesis, structural and optical properties of CeO_2 nanoparticles synthesized by a simple polyvinyl pyrrolidone (PVP) solution route, *Materials Chemistry and Physics* ***115*** 423–428.

[31] Can O, Oztürk E, Solmaz H, Aksoy F, Çinar C, Yücesu H S 2016 Combined effects of soybean biodiesel fuel addition and EGR application on the combustion and exhaust emissions in a diesel engine, *Appl. Therm. Eng.* ***95*** 115–124.

[32] Qi D, Leick M, Liu Y, Chia-fon F L 2011 Effect of EGR and injection timing on combustion and emission characteristics of split injection strategy DI-diesel engine fueled with biodiesel, *Fuel* **90** *(5)*1884–1891.

[33] Arul Prakasajothi Mahalingam, Dinesh Babu Munuswamy, Yuvarajan Devarajan and Santhanakrishnan Radhakrishnan 2018 Investigation on the emission reduction technique in Acetone-Biodiesel aspirated diesel engine". *Journal of Oil Palm Research Vol* **30** *(2)* 345 – 349.

[34] Yung Chee Liang, Harrison Lau Lik Nang and Choo Yuen May 2013 Physico-chemical properties of biodiesel produced from Jatropha curcas oil and palm oil *Journal of Oil Palm Research Vol* **25** *(2)*159 – 164.

[35] Ulf Neuling and Martin Kaltschmitt 2017 Review of biofuel production – feedstock, processes and markets *Journal of Oil Palm Research Vol* **29** *(2)* 137 – 167.

[36] Adeeb Hayyan, Shahidah Nusailah Rashid, Maan Hayyan, M Y Zulkifliy, Mohd Ali Hashim and Nurul Alfaa Osman 2017 Synthesis of novel eutectic catalyst for the esterification of crude palm oil mixed with sludge palm oil*Journal of Oil Palm Research Vol* **29** *(3)* 373 – 379.

[37] Denis J Murphy 2014 The future of oil palm as a major global crop: opportunities and challenges *Journal of Oil Palm Research Vol* **26** *(1)*1– 24.

[38] Mirzajanzadeh M, Meisam Tabatabaei, Mehdi Ardjmand, Alimorad Rashidi, Barat Ghobadian, Mohammad Barkhi, Mohammad Pazouki, 2015 A novel soluble nano-catalysts in diesel–biodiesel fuel blends to improve diesel engines performance and reduce exhaust emissions, *Fuel* **139** 374–382.

[39] Saifuddin M, Goh P E, MoneruzzamanW. S. H, K. M. and Fatima A, 2014 Biodiesel Production from Waste Cooking Palm Oil and Environmental Impact Analysis, *Bulgarian Journal of Agricultural Science,* **20** *(No 1)* 186-192.

[40] Sanjid A, MasjukiH. H.. KalamM. A, Ashrafur RahmanS. M. AbedinM.J. PalashS. M 2013 Impact of palm, mustard, waste cooking oil and Calophyllum inophyllum biofuels on performance and emission of CI engine *Renewable and Sustainable Energy Reviews* **27** 664–682.

[41] Devarajan Y, Jayabal R K, Ragupathy D and Venu H 2016 Emissions analysis on second generation biodiesel. *Frontiers Environ. Sci. Eng.,* **11**(1),3.

[42] Venkata Ramanan M and Yuvarajan D 2016 Emission analysis on the influence of magnetite nanofluid on methyl ester in diesel engine. *Atoms. Pollution Res.,* 7(3); 477-481.

[43] Shaafi T, Sairam K, Gopinath A, Kumaresan G, Velraj R 2015 Effect of dispersion of various nanoadditives on the performance and emission characteristics of a CI engine fuelled with diesel, biodiesel and blends—A review, *Renewable and Sustainable Energy Reviews* **49** 563–573.

[44] Radha Krishna Gopidesi, Goli Ravi Sankar, Appana Durga Pavan Kumar, Alladi Sukhesh Kumar and Bondalapati Srimal 2019 Evaluating the Performance and Emission Characteristics of Ci Engine with Waste Plastic Oil, *International Journal of Mechanical and Production Engineering Research and Development (IJMPERD),* **9** *(No 3)* 1015-1020.

[45] Avinash Mutluri, Radha Krishna Gopidesi and Srinivas Viswanath Valeti 2020 A Research on the Performance, Emission and Combustion Parameters of the Hydrogen and Biogas Dual Fuel Engine, *Incas Bulletin*, **12** (No 3) 129 – 136.

[46] Radha Krishna Gopidesi and S.R. Premkartikkumar 2019 Performance, emission and combustion analysis of diesel engine fuelled with emulsified biodiesel, *Progress in Industrial Ecology – An International Journal,* **13** (No. 3) 292–301.

[47] Radha Krishna Gopidesi, Srinivas Viswanth Valeti, Nagarjuna Kumma, Avinash Mutluri and Borigorla Venu 2019 Evaluation Dual Fuel Engine Fuelled With Hydrogen and Biogas as Secondary Fuel, *International Journal of Recent Technology and Engineering (IJRTE),* **8** (No 2) 1902-1905.

MODAL ANALYSIS OF ELECTRICAL MOTOR CASING

Srinivasan.T[1] , Sharavanan. S[2],Naveen.E[3], Sethuvelappan[4] and Ramanan. P[5]

[1]Professor, Department of Mechanical Engineering, Koneru Lakshmaiah Education Foundation ,Vaddeswaram,Guntur, Andra Pradesh,India
2Asst Professor, Department of Mechanical Engineering, Vels institute of science,Technology and Advance Studeis,Pallavaram, Chennai
[3] Asst Professor, Department of Mechanical Engineering, Sri Sairam Engineering College, Chennai
[4]UG Scholar, Department of Mechanical Engineering, St. Joseph's College of Engineering, OMR, Chennai
[5]Engineer, Synce Engineering Service, Chennai

Corresponding Author: seenusreeja@gmail.com

Abstract. Getting analytical and mathematical solutions for issues involving difficult geometries, loading and material properties, it is usually unacceptable. Analyticalsolutionsthat are given by a location in a body. This analytical solution usually needs standard or partial differential equations that are not obtained. Hence there is the need to rely on numerical strategies, like finite component strategies for acceptable solutions. Mostsensibleissues involve sophisticated domains (both material and material constitution), hundreds and non-linearities that forbid the event of analytical solutions exploitation numerical strategies. A numerical method, with the arrival of computer, is often used for the investigation and analysis of the results of varied parameters of the systemon analyzed. It is price effective and saves timeand material resources compared to the multitude of physical experiments required to realize a similar level of the understanding.The ability of numerical strategies and electroniccomputation, make incorporation of all relevant options during a mathematical model of a physical method not attainable without fear concerning its solutions by precise means.Those whoarefasttouse a computer programrather than think about the problemto be analyzed may find it difficult to interpret the input file to the computer program, a decent understanding oftheunderlyingtheory of the matter still as numerical methodology is needed.

Key Words: Electic motor casing, FEA, Model Analysis, Modeling,Ansys

1. Introduction

ANSYS is a general purpose finite element modeling package for numerical determination of a large sort of mechanical problems which issues include: static/dynamic structural analysis (both linear and non-linear), heat transfer and fluid problems, moreover as acoustic and electro-magnetic problems. ANSYS is general finite element analysis (FEA) software package [13]. Finite element analysis may be a numerical method for deconstructing a complex system into terribly tiny items (of user-designated size) called as elements. The software package implements equations that govern the behaviour of those parts and solves all, making a comprehensive clarification of the action of the system acts as a full. These results then are given in tabulated or graphical forms. This type of study is generally used for the planning and optimization of a system too advanced to for investigation by hand [11,16]. Systems that will work into this class are too advanced because of their pure mathematics, scale, or governing equations. ANSYS has the customary FEA teaching tool inside the engineering and science colleges. ANSYS provides economical methods to explore the performance of products or processes in very virtual surroundings [22]. This kind of development is termed virtual prototyping. With virtual prototyping techniques, users reiterate varied situations for optimizing the product long before production is started. This permits a discount within the level of risk, and within the price of ineffective styles.

The varied nature of ANSYS provides a method to confirm that users are ready to see the impact of a style on the total behavior of a product, be it electromagnetic, thermal, and mechanical, etc. In general, a finite element solution is also broken into the subsequent three stages [17-19]. This can be a general guideline that may be used for fitting any finite element analysis. investigated the results of flexural and compressive properties of kenaf combined with silicon dioxide nano particles in epoxy composites. The filler material utilized in this work was silicon dioxide Nano particles were used for boosting the standard of kenaf reinforced composite. [1] The composites were using the vacuum infusion method within which kenaf mats were set in. The silicon dioxide nano particles square measure was distributed into epoxy glue employing a homogenizer at 3000 rpm for 10 minutes before being infused into fibres. A big mechanical property with the inclusion of silicon dioxide nano particles was seen. Composites with 2% vol. silica recorded worth of 43.8 MPa and 3.05 GPa for flexural strength and flexural modulus and 40 MPa and 1.15 GPa for compressive strength and compressive modulus severally. studied the geo polymer composites reinforced with natural fibres and their mechanical properties. Here, the geo polymer composite was based on fly ash and reinforced with short natural fibres such as coconut, sisal, raffia and cotton. They concluded that the composite with coconut, sisal had mechanical properties but a composite with raffia had only poor properties. SEM images, showed the natural polymers as less coherent than the artificial ones. It is the best alternative for Portland cement.[2]investigate the reasons that explained the use of flax and hemp as the designer's first choice. They found higher specific stiffness than glass fibre composites in tension and plate bending, higher vibration damping capacity in those composites. They examined the technical aspects such as mechanical properties and non-technical aspects such as feel, touch, colour and texture of the fibres and their influence on the daily purpose items and the designer's choice of selecting these fibres.[3]conducted experiments to understand the tribological characteristics of oil palm and kenaf fibre with epoxy glue as medium and comparisons. Pin on disk check was administered to see the wear behavior. A pin with sample of diameter 10 mm was out through with an appropriate machining method. A dry sliding check was performed using the pin on disc tribometer. Check parameters like weight proportion of fibre, load, sliding speed and temperature range were taken as operational parameters. The result showed the rise in temperature resulted in slashed coefficient of friction and magnified wear rate for each of the composite. The wear rates for each of the fibres exhibited a higher performance through increase in the fibre composition. SEM was done and therefore the result showed the composites having blow holes, small cracks, fibre retreat and a few regions deboning were discovered.[4]administered various mechanical tests in various natural fibres and found that the hybrid composites had better mechanical properties and alternate material for various automobile and defense applications was also seen.[5] investigated the impact of kenaf with different yarn sizes and their mechanical properties. The kenaf fibre was fancied via pultrusion process. Mechanical tests like tensile, compression and flexural were administered and comparisons were made with different forms of kenaf yarn tex. Pultruded composites showed better mechanical properties created with smaller tex range in comparison to a large tex range, attributable to it, that helped providing higher wetting of fibre throughout production of composites and consequently helped extension of its properties. SEM analysis was done and therefore the result showed poor wetting of kenaf fibre recorded lower properties of kenaf yarn larger tex range than the smaller tex range.[6]investigated the low velocity impact properties of flax composites which showed the longevity and safety of the composites. Two composite matrix flax-epoxy and flax-MAPP were tested and compared. The energy absorbed by the flax-MAPP was seen as 50% higher than the flax-epoxy. The flax-epoxy resulted in a palpable decrease in properties after impact. Also the use of ductile thermoplastic matrix results in decreased impact area.[7]fabricated natural fibre and aluminium sheet hybrid composites. The composites were fabricated using the vacuum assisted resin transfer molding (VARTM) process. This hybrid composite had a high electromagnetic shielding. The introduction of aluminium sheets had no effect on the mechanical properties of the hybrid composites. The results obtained from the internal bond tests showed the interfacial bonds as strong.[8] investigated the effect of mechanical properties on hemp fibres and unidirectional fibre/epoxy composites. These effects were studied when the hemp fibres were oxidized of lignin by laccase. The oxidized fibres were taken for a study of mechanical properties such as tensile stress, thermal resistance, etc. From the results, they concluded that there was an increase in tensile stress, stiffness and thermal resistance of the laccase treated fibres. There was no cross linking by laccase seen. This increase in properties was due to the catalyzed polymerization of lignin in hemp fibres.[9] investigated the geometric optimization and compression design of natural fibre composite structural channel sections. It was carried out by the inclusion of geometric stiffeners which restricted the development of local buckling, creating less slender channel sections with greater compression strength. Compression strengths were compared with steel and timber wall stud strengths. They concluded that the developed composite was suitable for residential building applications. The combined plain channel and stiffened channel experimental data covers a broad range of section slenderness values.[10]

2. Modeling

Throughout the history of our industrial society, many innovations are seen proprietary and whole new technologies have evolved. May be the only development that has compact production of a lot of quickly and considerably than any previous technology is that the information processing system. Computers are being employed more and more for each style with particularization of engineering elements within the drawing workplace. Computer Aided Design (CAD) is outlined because the application of computers and graphics package to assist or enhance the products design from conceptualization to documentation. CAD is most commonly associated with the use of an interactive computer graphics system, referred to as CAD system. CAD design systems are powerful tools and are within the mechanical design and geometric modeling of product and elements.There are many reasons for employing a CAD system to support the engineering style function:

- To increase the productivity and quality.
- To uniform design standards and have a manufacturing knowledge base.
- To reduce inaccuracies caused by hand-copying of drawings and variation between drawings.

Computer Aided Manufacturing (CAM) stands out due to the effective use of engineering in producing designing and management. CAM is most closely related to functions in production engineering, like methods and production designing, machining, scheduling, management, internal control, and Numerical Control (NC) part programming. CAD and CAM are usually combined CAD/CAM systems.

This arrangement permits the transfer of data from the planning stage into the stage of coming up with the manufacture of a product, without the requirement to reenter the data on part geometry manually. The info developed throughout CAD is stored; then it is processed any, by CAM into mandatory knowledge and directions for operative and dominant production machinery, material handling instrumentation, and automatic testing and review for product quality.

3. Modeling of Electrical Motor Casing

The modeling of the electrical motor is done in 3D and it is shown in figure 1.

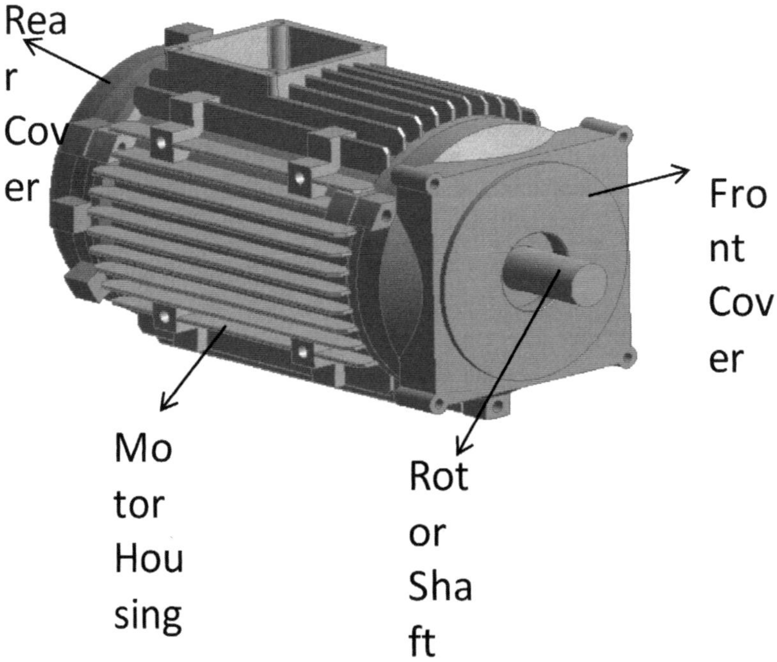

Figure 1. 3D modelling of an electric motor

4. Methodology

Methodology is that the systematic, theoretical analysis of the ways applied to a field of study. It includes the theoretical analysis of the body of ways and principles related to a branch of information. Typically, it encompasses ideas like paradigm, theoretical model, phases and quantitative or qualitative techniques.A methodology does not taken off to produce solutions - it is therefore, not constant as a technique. Instead, methodology offers the theoretical underpinning for understanding that method, set of ways, or supposed "best practices" may be applied to specific case, for instance, to shrewdaselected result. The methodology is that the general analysis strategy that outlines the means during which analysis is to be undertaken and among different things, identifies the ways to be employed in it. These ways, delineated within the methodology, outline the means that or modes of information assortmentor sometimes, however a selected result is to be calculated. Once correct to a study of methodology, such processes represent a constructive generic framework and will so be weakened into sub-processes, combined or their sequence modified.

5. FEA Analysis

In FEA analysis we can do many analyses for the fabricated composite material. Here the below mentioned three cases are done in FEA analysis.

5.1. Modal Analysis (To find the stiffness of the material)

a) Case A - Hemp
b) Case B - Flax
c) Case C - Hemp + Flax

6. Boundary Condition

The boundary condition of fixed support, rotational velocity, displacement and fixed mass is shown in figure

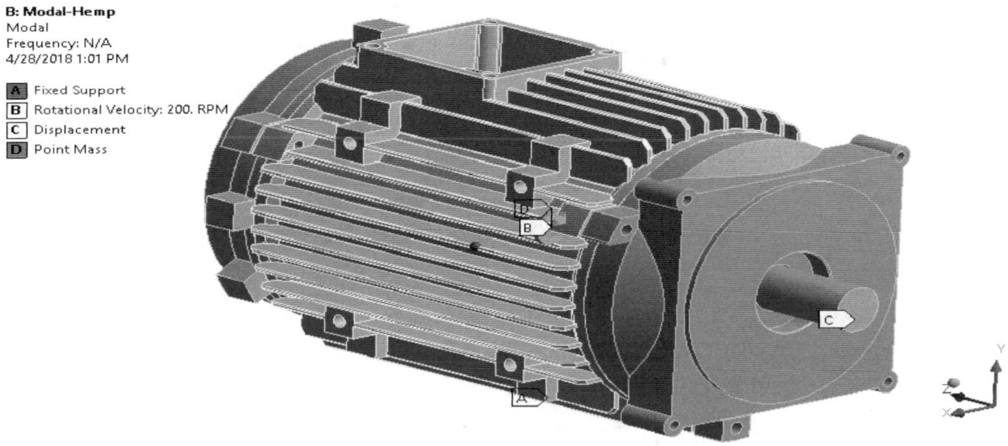

Figure 2. Boundary condition

For Modal Analysis **motor housing base** is constrained as a **fixed support**, i.e., there is no translation movement in X, Y and Z directions. To replace rotor coil, a point mass load is added to the assembly. Rotor shaft Z - translation has been constrained.

Also no separation added (Face-Face) between contact faces. In this pair, one part is acting as contact body and other one is target body.Motor housing base is constrained as a fixed support, i.e., there is no translation movement in X, Y and Z directions. To replace rotor coil, a point mass load is added to the assembly. Rotor shaft Z - translation has been constrained by applying rotational velocity from 200 to 1200 rpm. Figure 3, 4,5 and 6 shows the bonded contact for the electrical motor.

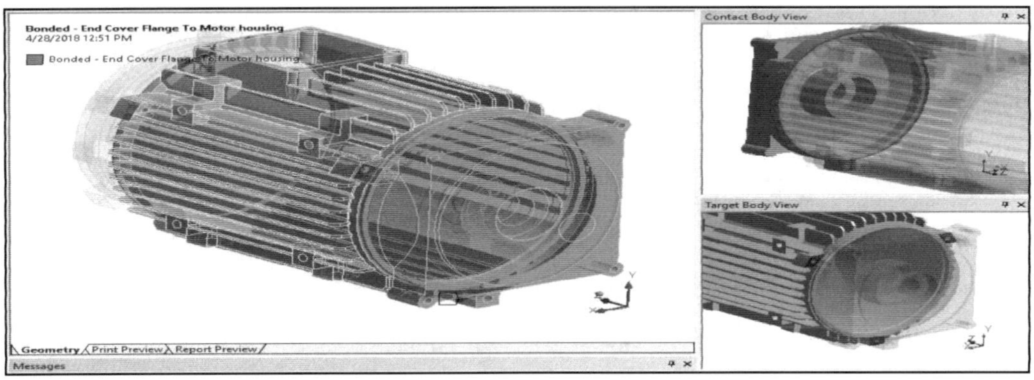

Figure 3. End cover flange to motor housing

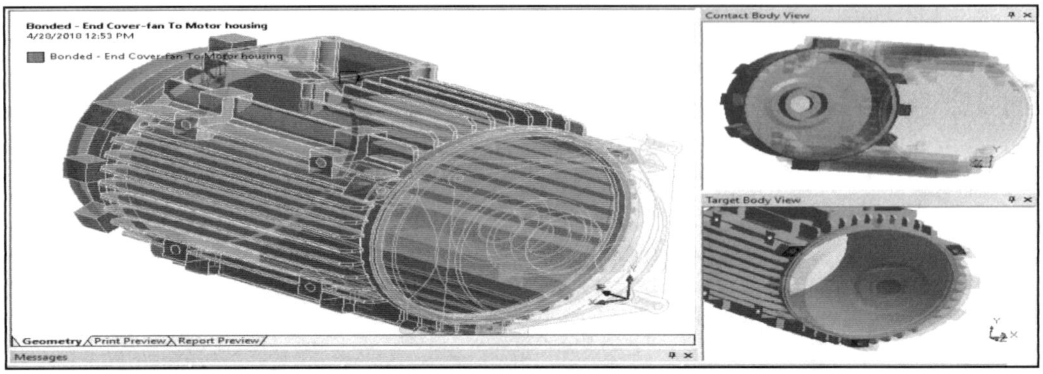

Figure 4. End cover fan to motor housing

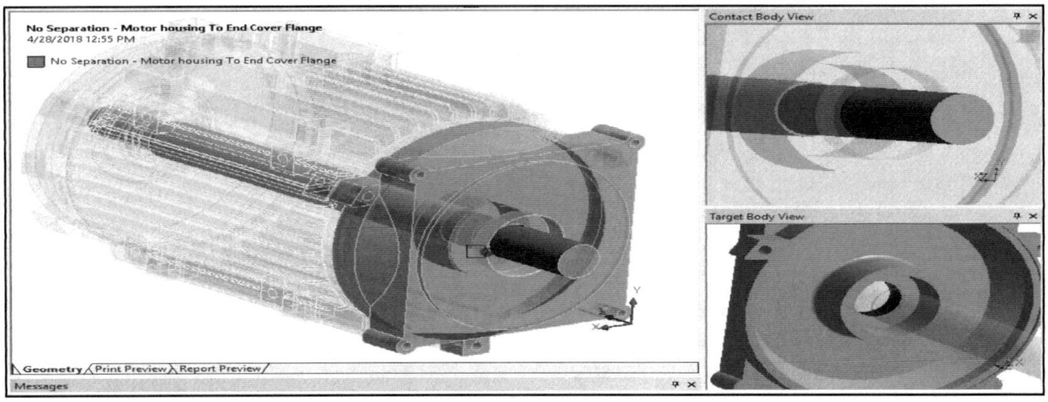

Figure 5. Motor housing to end cover flange

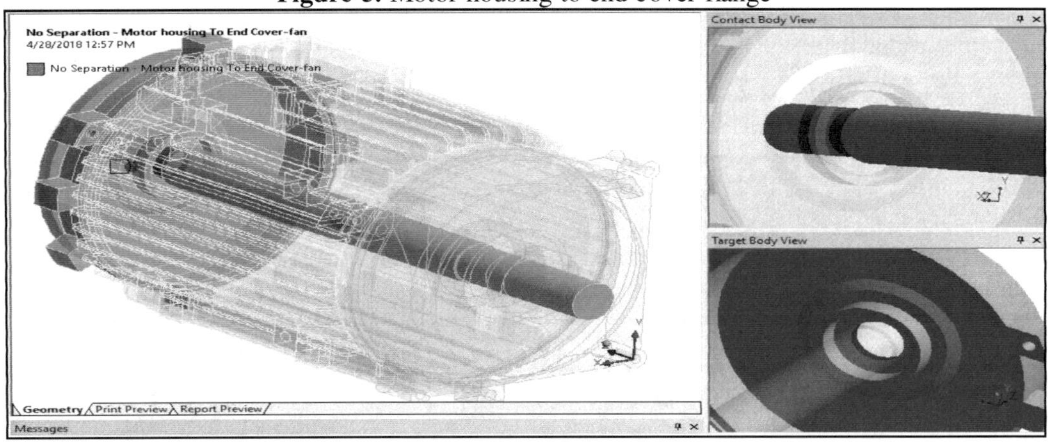

Figure 6. Motor housing to end cover fan

7. Modal Analysis

Modal Analysis is mostly dispensed to seek out natural frequency of the system and completely different mode shapes at different Eigen values [12,14]. In this work, totally different modes and Eigen values square measure calculates and premeditated between mode shapes versus natural frequencies [15,20,

21]. This shows the stiffness and rigidity of the material toward totally different modes. Using Modal analysis it can be captured the first 10 natural frequencies and related mode shape of each frequency for the 3 cases. Figure 7 shows the modal analysis results for case A. Each figure represents the frequency of each mode and displacement related to the frequency. Due to rotational velocity the model is exhibiting rotational vibration.

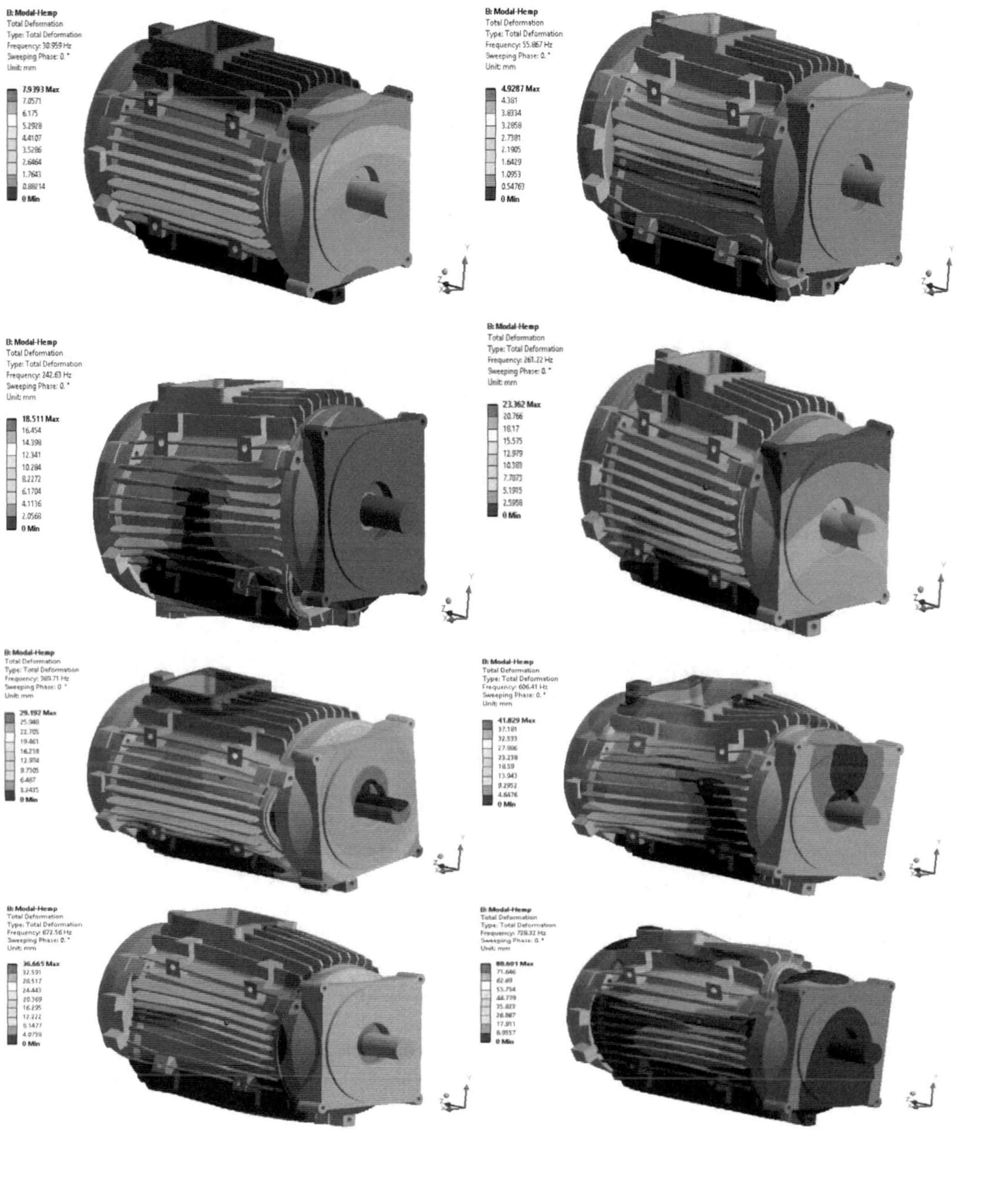

Figure 7. Modal analysis for 10 modes for case A

IOP Conf. Series: Materials Science and Engineering 1112 (2021) 012013 doi:10.1088/1757-899X/1112/1/012013

Table 1 shows the value of modal analysis for case A, B and C. According to this table Case C is having more natural frequency than other 2 cases.

Table 1. Modal analysis for case A, B and C

Modes	Case -A	Case -B	Case -C
1	30.959	45.103	49.496
2	55.867	79.428	86.464
3	242.63	306.88	327.84
4	261.22	330.73	351.68
5	369.71	496.33	515.18
6	606.56	747.08	753.8
7	672.56	751.02	791.26
8	728.32	907.61	911.07
9	754.32	910.27	931.84
10	880.19	986.57	1021

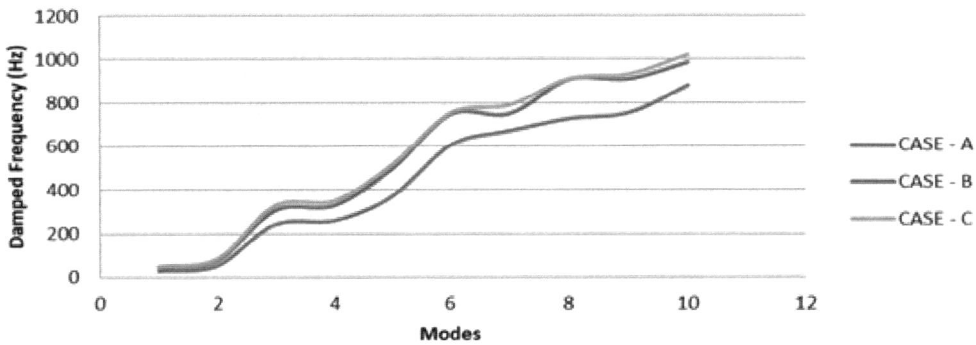

Figure 8. Modes vs Damped frequency

Figure 8 shows the relation between modes and damped frequency. Based on modal analysis results Case C (Hybrid Hemp - Flax fibre) has additional stiffness (due to high natural frequency) while compared to other two cases.

Conclusion

In this analysis, a study for three totally different composite materials particularly Hemp fibre, Flax fibre and Hybrid of Hemp and Flax fibre for Finite Element Analysis of electrical motor casing has been carried out. The motor casing should be rigid and better in stiffness to resist operative vibrations.

3D modelling of motor casing is built using Parametric Creo one among the fine 3D modelling software. These 3 materials are applied to motor casing as a material properties of the model and three types of analysis namely, Modal Analysis, Rotor Dynamics and Structural Transient Analysis carried out to check the stiffness, reliability towards operative frequency and durability of motor casing.

As determined from modal analysis Case C (Hybrid Hemp Flax fibre) has better stiffness compared to two different fibres as shown in graph fig 7 (Modes vs Damped Frequency).

Campbell diagram was premeditated between frequencies vs speed of the motor for predicting critical speed. The operative frequency and operative speed do not match with the critical speed and frequency.

Model analysis was performed for finding the vibrational stability of the composite materials and observation Case C shows low deflection and equivalent stresses compared to different fibres. Force vs deflection curve and stress vs strain curve are premeditated to indicate the stiffness of the materials and energy vs deflection curve premeditate show of the strain energy observed, when deformation happens. As determined from all above studies Case C has higher stiffness and strength compare from different two fibres.

Reference

[1] Bajuri F, Norkhairunnisa Mazlan, Mohamad Ridzwan Ishak and Junichiro Imatomi, 2016, "Flexural and compressive properties of hybrid kenaf/silica nanoparticles in epoxy composite", Procedia Chemistry, Vol.19, pp:955-960.
[2] Korniejenko. K, Frczek. E, Pytlak. E and Adamski. M, 2016, "Mechanical properties of geo polymer composites reinforced with natural fibres", Procedia Engineering, Vol.151, pp:388-393.
[3] Lut Pil, Farida Bensadoun, Julie Pariset and Ignaas Verpoest, 2016, "Why are designers fascinated by flax and hemp fibre composites?", Composites: Part A, Vol.83, pp:193-205.
[4] Shuhimi F, Mohd Fadzli Bin Abdollah, Kalam M.A and Hilmi Amiruddin, 2016, "Tribological characteristics comparison for oil palm fibre/epoxy and kenaf fibre/epoxy composites under dry sliding conditions", Tribology International 101.
[5] Raman Bharath, Vijaya Ramnath B and Manoaran N, 2015, "Kenaf fibre reinforced composites: A review", ARPN Journal of Engineering and Applied Sciences, Vol.10, No.13, pp:5483-5485.
[6] Zamria M, Hazizan Md Akil and Zainal Ariffin Mohd Ishak, , 2016 "Pultruded kenaf fibre reinforced composites: Effect of different kenaf fibre yarn tex", Procedia Chemistry, Vol.19, pp:577-585.
[7] Bensadoun. F, Depuydt. D, Baets. J, Verpoest. I, Van Vuure. A.W, 2017, "Low velocity impact properties of flax composites", Composite Structures, Vol.176, pp:933-944.
[8] Changlei Xia, Jason Yu, Sheldon Q. Shi, Ying Qiu, Liping Cai, Felix Wu. H, Han Ren, Xu Nie and Hualiang Zhang, 2017, "Natural fibre and aluminum sheet hybrid composites for high electromagnetic interference shielding performance", Composites Part B, Vol.114, pp:121-127.
[9] Ming Liu, Andreas Baum, Jurgen Odermatt, Jens Berger, Liyun Yu, Birgitte Zeuner, Anders Thygesen, Jesper Holck and Anne S. Meyer, 2017, "Oxidation of lignin in hemp fibres by laccase: Effects on mechanical properties of hemp fibres and unidirectional fibre/epoxy composites", Composites: Part A, Vol.95, pp:377-387.
[10] Bambach. M.R, 2018, "Geometric optimisation and compression design of natural fibre composite structural channel sections", Composite Structures, Vol.185, pp:549-560.
[11] Ravi Prakash Babu K., Rao P.K.V., Ali M.A., Raghu Kumar B., 2019, 'Crack simulations in shaft-hub-blade system using modal analysis', International Journal of Recent Technology and Engineering, 8(1), PP.2646-2650.
[12] Bala Subramanyam P.N.V., Rao B.N., Prakash R.L., Rajavardhan D.S.K., Sankarudu B.S.K., Ganesh P.,2019, 'Modal analysis on go-kart chassis', International Journal of Engineering and Advanced Technology, 8(4), PP.1701-1705.
[13] Syed K., Mannam H., Koya M.K., Meduri S.K., Laxmi V.S., Kasturi S.P., 2019, 'Dynamic simulation of gas turbine blade using finite element analysis', International Journal of Mechanical and Production Engineering Research and Development, 9(), PP.745-751.

[14] Dama K.K., Surya S., Babu C., Pavan, Kiran R., 2018, 'Modal analysis of an automotive subsystem using cae techniques', International Journal of Mechanical Engineering and Technology, 9(3), PP. 1157-1162.

[15] Deepak Kumar G., Kommuri N., Kota H., Prasanth V., Hameed S, 2018, 'Vibrational analysis of a human body on an automobile', Journal of Advanced Research in Dynamical and Control Systems ,10(6), PP. 39-45.

[16] Devi S.K., 2018, 'The optimization of layer densities for multilayered insulation systems by using ansys', International Journal of Mechanical and Production Engineering Research and Development ,8(4), PP. 1087-1098.

[17] Balasubramanyam P., Nageswara Rao B., Banerjee M., Lakshmana Swamy B., 2019, 'Impact analysis on Go-Kart chassis with variable speeds using ansys 19.0', International Journal of Engineering and Advanced Technology, 8(6), PP.2614-2620.

[18] Kancheti N., Reddy Vemula A., Reddy Gudibandla G., Krishna H., Bala Subramanyam P.N.V. 2019, 'Modeling and analysis of wheel rim using ansys', International Journal of Innovative Technology and Exploring Engineering, 8(8), PP.415-418.

[19] Bharath E.S., NagaSree Harsha O., Bala Subramanyam P.N.V., Reddy G.S.S., 2019, 'Impact analysis on honey comb structured go-kart bumper using ansys R19.1', InternationalJournalof Innovative Technology and Exploring Engineering, 8(6), PP.396-400.

[20] Ravi Prakash Babu K., Gupta S.K., Phaneendra Sai Sri D., Gautam K., Kumar C.M. 2017,'Significance of modal analysis to detect and diagnose misalignment fault in turbine rotor',Journal of Advanced Research in Dynamical and Control Systems,9(Special issue 14),PP.2637-2648.

[21] Babu K.R.P., Prasanth K.F.V., Sai V.V., Kumar M.V.S., Kumar B.N.S., Chaitanya B.S. 2017,'Effect of pre-twist on free vibration characteristics of a cantilever beamwithexperimental validation',International Journal of Mechanical Engineering and Technology,8(5),PP:389-399.

[22] T. Srinivasan, CH.V.L.N. Ram Gopal, P. Sasidhar, P. Vineeth Kumar and Shaik Sohail, 2020, Effect of change of cross section and Frequency Analysis of Anti-Roll Bar Using FEA, IOP Conf. Series: Materials Science and Engineering 954 (2020) 012027 IOP Publishing, doi:10.1088/1757-899X/954/1/012027

IOP Conf. Series: Materials Science and Engineering 1112 (2021) 012014 doi:10.1088/1757-899X/1112/1/012014

Comparison of Effect of Orientation on SS304 and SS316 Material Samples Fabricated with DMLS Technology

Dr. M. Kedar Mallik, D. V. Seshagirirao*, K. J. Padmaja, B. Aditya Mani Sai Pavan, P. Bhagyasri
Department of Mechanical Engineering, Vasireddy Venkatadri Institute of Technology, Guntur, A.P., India
* Corresponding author: dvsrao@vvit.net, +91 99511 59363

Abstract
SS316 and SS304 are most widely used materials in mechanical engineering. Many times the products are subjected to tensile load during work. The knowledge of tensile properties is essential for the mechanical engineer working with materials. SS304 and SS316 samples were fabricated using Direct Metal Laser Sintering (DMLS) as per JIS standards in different orientations. The samples were then tested for their tensile strength. The same has been simulated on the NX platform for prediction of variations in the tensile strength with change in orientation. The results obtained from the simulation are observed to be in good agreement with the experiment results when the process parameters are same in both the cases. Further, the simulation has been used to predict the tensile strength of the samples that were not experimented.

Key words: SS304, SS316, Direct Metal Laser Sintering, Tensile Strength

Introduction

DMLS (Direct Metal Laser Sintering) is an additive manufacturing method. It creates parts additively by sintering fine metal powder particles, to fuse them together locally. Like in the SLS process, the metal part will be fabricated layer by layer, according to the 3D model. A major difference is the sintering temperature. Polyamide needs to be sintered at a temperature of 160°C to 200°C, when metals melt at a temperature around 1510°C which requires high-wattage laser power to reach that temperature. Basically the DMLS machine sinters the metal powder by focussing a laser beam on the thin layer of metal powder on a platform. The high power laser beam moves as per the CAD model/digital data submitted to the machine in the form of layers. A roller applies a layer of metal powder, on platform in a chamber. Then the laser sinters the powder according to the 3D file followed by the downward movement of the platform and the process repeats until the desired part is created. Once it's finished, the metal part needs to cool down before being extracted. If the metal part needed supports, these supports will then be removed manually.

The basic principle of the Direct Metal Laser Sintering (DMLS) Technology is to melt down thin layers (200μm) of Metal Powder with an electronically driven LASER beam (200W). Layer by layer, it is possible to build any kind of shape and geometry, even those which are impossible to obtain with any other kind of technology. The accuracy is ± 0.05mm.

In this technology it is desirable to have metal powders in the form of spherical and nearly spherical geometry and in tens of micron size to achieve good bonding, homogeneity and crack free products [9]. Among the 50 different atomized powders within reach, most widely used metal powders in AM technology are Ti based alloys, Nickel based super alloys, Al based alloys, Co-Cr alloys, Stainless steels (316L and17-4PH), Precious metals (Au, Ag), Refractory metals(W, Ta), Cu based alloys, inter metallic and low alloy steels [10].

The formation of reinforced TiN -SS316 metal matrix composites using fiber laser sintering through DMLS process were studied with varying parameters like laser power, scanning speed, spot diameter, layer thickness and hatching distance . The fabricated specimens were investigated and the experimental results states that the density of the sample increases with increase in the laser power. Also the suitability of the above samples for various engineering applications was discussed. [6]

In DMLS process a porous internal structure can be used to lessen the weight and assuring the strength at the same time. This results in considerable redeeming in powder material and weight as well as energy consumption during processing. [7]

Paul F. Kelley et al. [8] studied the effect of parameters and build orientation on the properties of DMLS Inconel 718. There was an improvement of surface finish by post processing (micro machining). The fatigue strength in build orientation in Z is approximately 14% lesser than X/Y axes. The tensile and yield strengths were found to

be 10% less in Z direction than X direction. The Z and X directions with respect to built platform were as shown in Fig.18. Onome Scott-Emuakpor et al. [9]
The DMLS process is highly beneficial for those who need to produce their metal parts for prototyping or low-volume production by eliminating time-consuming tooling. It also allows to create complicated and highly detailed designs that would not be possible with any other technology, due to the limitations of the traditional manufacturing processes.
The present article discusses about the effect of orientation on tensile strength of SS304 and SS316 materials.

Experimentation
The experimentation is carried out in two stages, one is simulation of tensile strength using NX software and the second one is physical experimentation.
CAD models of the tensile specimens have been created using NX platform. The.stl files are imported in to the software user interface, the material properties are appended and the loads are applied in longitudinal directions and are virtually simulated and analysed using NX-CAE.
Commercially available SS316 and SS304 powders of particle size 50 -140 microns have been used for fabrication of samples using Direct Metal Laser Sintering technology. As per the JIS standards, six sets of the tensile specimens, each five in number, in the dumbbell shape of rectangular cross section are fabricated. Three different orientations are selected and are shown in Table 1. All the other process parameters like scan speed, laser power etc. are kept constant.

Table 1: Details of the samples fabricated using DMLS technology

Sl. No.	Sample Identification No.	Material	Orientation
1	ZL1	SS316	Length of the specimen in Z-direction
2	ZW1	SS316	Width of the specimen in Z-direction
3	ZT1	SS316	Thickness of the specimen in Z-direction
4	ZL2	SS304	Length of the specimen in Z-direction
5	ZW2	SS304	Width of the specimen in Z-direction
6	ZT2	SS304	Thickness of the specimen in Z-direction

The fabricated samples are separated from the substrate, using EDM. Standard procedure was followed to polish the specimens using sand papers of grit size 200 to 1200 followed by polishing by single disc polishing machine under wet condition in the presence of Alumina paste of 0.1mm particle size. The samples that are mirror finished are shown in Fig. 1.

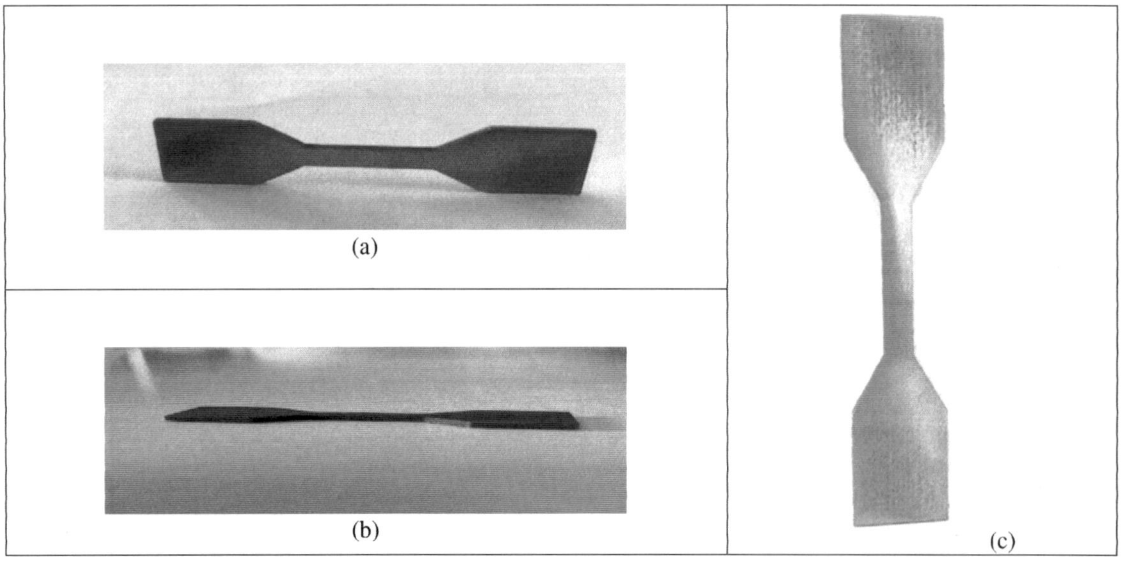

Fig. 1: Samples fabricated using DMLS technology (a) Width along Z-Direction, (b) Thickness along Z-Direction and (c) Length along Z-Direction

The fabricated specimens of SS316 and SS304 are tested under tensile loading by using micro tensometer. The specimens are placed in the work holding device of the instrument and load is gradually applied. Due to gradual increase in the load, the specimens break at a particular point and value is recorded.

Results and Discussion

The results obtained from the simulation and physical experimentation of tensile tests conducted on SS304 and SS316 stainless steels that are fabricated using DMLS technology are shown in the Table 2.

Table 2: The tensile properties of SS304 and SS316 obtained from Simulation and experimentation

Sl. No	Sample No.	Simulation results		Experiment results	
		Yield Strength MPa	Ultimate Tensile Strength MPa	Yield Strength MPa	Ultimate Tensile Strength MPa
1	ZL1	270	560	277	564
2	ZW1	283	586	281	587
3	ZT1	310	580	313	591
4	ZL2	205	490	199	488
5	ZW2	215	505	208	497
6	ZT2	210	488	205	499

The analysis on the obtained results reveal that the tensile properties of both SS304 and SS31 steels are almost at par with the as cast samples of the same material [6] and are in agreement with the results obtained from simulation. The maximum YS and UTS are obtained from the samples that are fabricated with SS316 when the orientation is 'thickness in Z-direction' (sample No. ZT1). Similarly the minimum YS and UTS are obtained from the samples fabricated with SS304, in 'length in Z-direction' (Sample No. ZL2). Evidently, the samples SS316 are stronger in nature than SS304. Further, during fabrication in DMLS process, as explained earlier in this article, the curing of the powder takes place layer by layer manner i.e. the powder is cured as a thin layer on a platform, and further layers are cured over the previous such that the new layer is bonded to the previous. When compared with the other directions, the bonding between the two consecutive layers is weaker in Z-direction. Hence, both the materials have lesser strengths in Z- direction. The results are also compared with the results obtained from the tests conducted on the samples when deposited with Laser Engineered Net Shaping (LENS^{TM}) technology, an energy dispersion technique [10]. Eventually, the samples from DMLS technology have shown lesser strength, since the samples of LENS technology are deposited by spraying the powder on the platform at the point where the laser beam hits the platform or previous layer, with high pressure.

Conclusions

The conclusions drawn from the experiments conducted on SS304 and SS316 steel samples fabricated using DMLS technology for obtaining their tensile strength are that the samples fabricated with SS316 when oriented in 'thickness in Z-direction' have shown highest Yield strength and Ultimate Tensile strength where as the samples that are oriented as 'length in Z- direction', fabricated with SS304 are resulted in minimum YS and UTS. The results are compared with as cast and LENS deposited samples and the reasons are discussed.

References:

[1] Mantrala Kedar Mallik, Chalamalasetty Srinivasa Rao and Vaddi Venkata Sundara Kesava Rao, "Effect of heat treatment on hardness and wear behaviour of weld deposited Co-Cr-Mo alloy", Revista Materia, Vol. 20, No. 02 , pp.544-549, 2015

[2] Mantrala Kedar Mallik, Ch. Srinivasa Rao, V. V. S. Kesava Rao and Abdul Munaf Sk. "Effect of Heat Treatment on Corrosion Behaviour of Weld Deposited Co-Cr-Mo Alloy, ARPN Journal of Engineering and Applied Sciences- Vol. 11, No. 20, October 2016

[3] Ganzi Suresh, K. L. Narayana, M. Kedar Mallik, V. Srinivas, G. Jagan Reddy and I. Gurappa "Electro Chemical Behaviour of LENS^{TM} Deposited Co-Cr-W Alloy for Bio-Medical Applications" International Journal of Mechanical and Production Engineering Research and Development, Special Issue, pp. 41-52, Jun 2018

[4] Stephane Gorsse, Christopher Hutchinson, Mohamed Gouné & Rajarshi Banerjee, "Additive manufacturing of metals, a brief review of the characteristic microstructures and properties of steels,

Ti6Al-4V and high-entropy alloys", Science and Technology of Advanced Materials,vol.18(1), 2017, pp.584-610.

[5] Available Materials for Metal Additive Manufacturing, Characteristics & Applications. Farinia Group. https,//www.farinia.com/additivemanufacturing/3dmaterials/characteristics-and-applicationsofavailable-metals-for-additive-manufacturing

[6] VijayMandal, ManowarHussain, VikasKumar, AlokKumar Das, N.KSingh "Development of reinforced TiN-SS316 metal matrix composite (MMC) using direct Metal laser sintering (DMLS) and its characterization" Materials Today: Proceedings Volume 4, Issue 9, 2017, Pages 9982-9986

[7] Liciane Sabadin Bertol, Wilson Kindlein Júnior, Fabio Pinto da Silva, Claus Aumund-Kopp, "Medical design, Direct metal laser sintering of Ti–6Al–4V", Materials and Design, vol.31, 2010, pp.3982–3988.

[8] Paul F. Kelley, Anil Saigal, "Tensile and fatigue behaviour of direct metal laser sintered (DMLS) INCONEL 718", Proceedings of the ASME 2015 International Mechanical Engineering Congress and Exposition, Houston, Texas, IMECE2015November 13-19, 2015.

[9] O. Scott-emuakpor, J. Schwartz, T. George, C. Holycross, C. Cross and J. Slater, "Bending fatigue life characterization of direct metal laser sintering nickel alloy 718".Wiley Publishing Ltd. Fatigue Fract Engng. Mater Struct., vol.38, 2015,pp.1105–1117.

[10] K M Mantrala, C S Rao, VVSK Rao and S A Munaf "Effect of process parameters on the tensile properties of LENS deposited Co-Cr-Mo alloy", International Journal of Manufacturing Technology and Management Vol.32, No. 6, pp. 509-516, 2018.

Influence of Heat Treatment on Mechanical Characteristics of Al7075/Al_2O_3/TiC Hybrid Metal Matrix Composite

E Sai Baba*[1], M Vinitha[2] and Dr D V V Krishna Prasad [3]

[1,2] Assistant Professor, Department of Mechanical Engineering, Vasireddy Venkatadri Institute of Technology, Guntur, AP, India

[3]Professor, Department of Mechanical Engineering, RVR & JC College of Engineering, Chowdavaram, Guntur, AP, India

* Corresponding author: amaranath.sai@gmail.com

ABSTRACT: The present paper deals with the study and characterization of the physical and mechanical behaviors of Al-7075/ Al_2O_3 / TiC hybrid metal matrix composite (HMMC) materials. The assimilation of dual particles like self-lubricated ceramic particles (Al_2O_3) along with hard particles (TiC) in the Al-7075 matrix is done using liquid stir casting method. The hybrid composite is produced with varying weight percentages of (8, 6, 4, 2, 0) Al2O3 and TiC (0, 2, 4, 6, 8) particulates. The produced Al HMMC materials are subjected to heat treatment (T6). The behavior of properties like density, hardness, impact and tensile strengths are analyzed based on experimentation. The obtained properties of cast Al-7075 HMMC materials are compared with heat treated Al-7075 composite materials. The dispersion of TiC hard particles in ductile Al-7075 matrix, predominates Al_2O_3 in improving mechanical properties because of its superior properties.

***Keywords**: Hybrid Metal Matrix Composite, Al-7075, Al_2O_3, TiC, T6 Solution heat treatment.*

1. INTRODUCTION

The Aluminum metal matrix composite (AMMC) materials are low in density, good in castability, dimensionally stable and they exhibit excellent mechanical properties, tribological behavior, thermodynamic stability and high strength to weight ratio[1,2]. Among various Al-alloys, Al-7075 is strong and it has good fatigue strength. Compared to other Al alloys, the 7075-aluminum alloy is most preferably used for aircraft structural fittings[3].
Demand for further advanced materials, researches focused on the development of hybrid metal matrix composites. It consists of two different materials reinforced in the metal matrix. The comprehensive reports show improved properties of a hybrid than single reinforced metal matrix composites[4, 5].
As a comparison with metal matrix composites, less literature is available with Aluminum HMMCs. Different Al alloys are combined with commonly used hard reinforcement particles like SiC, Al_2O_3, B_4C, Cr_3C_2, TiB_2, TiC in addition to RHA, beryl, fly ash, zirconia, graphite, MgO etc[6,7]. These reinforcements improve mechanical properties in the composites[8]. Among various reinforcements, TiC was preferred more due to better properties at high temperatures [9,10].
The discontinuous composites are fabricated using cost-effective stir casting technology[4]. From the literature studies, it was identified that most of the composites are produced with a simple stir casting method which is more cost effective and used for mass production[11]. This casting technique avoids damage to reinforcement particles compared to other techniques[12]. The composites produced by stir casting illustrate better dispersion of particulates, good wettability between the materials and reduced porosity [13].
The Al-7075/ TiC MMCs are used as advanced materials in aircraft applications. The addition of hard filler material TiC up to 8% in weight with the Al-7075 matrix alloy through stir casting strengthened the mechanical properties and tribological behavior in comparison with untreated Al-7075 alloy.

Researches preferred TiC as reinforcement material at ageing temperatures due to the elevated modulus of elasticity, low chemical reactivity, hardness and wettability [14]. The improvement in properties of Al/TiC metal matrix composites is due to better TiC particle bonding with Aluminum matrix compared with other reinforcements[15].
The Al-7075/Al_2O_3 matrix composite also has exhibited excellent mechanical and tribological properties compared to the base Al alloy[16]. The priority for Al_2O_3 as reinforcement particles is due to its high thermal stability and good electrical conductivity along with hardness. The results show an increase in wear resistance and a decrease in the coefficient of friction by the addition of Al_2O_3 with Al-7075 alloy. The characteristics like impact strength also improved by the reinforcement of Al_2O_3 with Al-7075 matrix alloy using stir casting method [17]. The tensile strength, hardness, and thermal properties show superior results at 10% reinforcement of Al_2O_3 with Al-7075alloy [18].
From the Literature study, it is found that researchers worked on Al-7075 with Al_2O_3 and also Al-7075 with TiC combinations as single and hybrid composite materials and gained advantages by improving mechanical properties. The present work is an attempt to combine the advantages of both Al_2O_3 and TiC with Al-7075 alloy. The novelty of current work is to prepare hybrid metal matrix composite materials by Al-7075 as base metal and Al_2O_3 and TiC as reinforcements and the physical and mechanical properties are investigated. Also, the effect of heat treatment on the composite material is studied.

2. EXPERIMENTATION

The Al matrix alloy is reinforced with dual Al_2O_3 and TiC with variations in weight to percentage ratios. The hybrid metal matrix composite is produced using a liquid stir casting method. This investigation is to study and characterize the influence of heat treatment on physical properties like microstructure and mechanical characteristics like hardness, tensile and impact strength of Al-7075/ Al_2O_3/TiC composite materials.

2.1 Preparation of Al-7075/ Al_2O_3 /TiC hybrid composite specimens

The raw Al-7075 received in the form of 160mm cylindrical rods with 20mm diameter. The average particle size of Al_2O_3 is 40-50microns and TiC is 44microns. The elemental composition of the Al-7075 alloy has mentioned in the table 1[19]. Weight percentage ratio of Al_2O_3 & TiC is mentioned in table 2.

Table 1: Elemental Composition of Al-7075

Element	Al	Si	Cu	Fe	Zn	Mg	Ti	Cr
Weight %	Base	0.20	1.71	0.23	5.29	2.46	0.54	2.21

Table 2: Samples with various percentages

Sl. No	Sample Name	Percentages of samples
1	Sample 1	Pure Al-7075
2	Sample 2	92%Al 7075 + 8% Al_2O_3 + 0% TiC
3	Sample 3	92%Al 7075 + 6% Al_2O_3 + 2% TiC
4	Sample 4	92%Al 7075 + 4% Al_2O_3 + 4% TiC
5	Sample 5	92%Al 7075 + 2% Al_2O_3 + 6% TiC
6	Sample 6	92%Al 7075 + 0% Al_2O_3 + 8% TiC

Stir Casting
In the stir casting process, Al-7075 was melted in a graphite crucible at a temperature of 700^{0}C for 20 minutes. In the intermediate time reinforcement particulates are preheated in another furnace to a temperature of 300^{0}C for 1hr. The preheating of matrix and reinforcement materials before mixing them helps in removing the moisture content and trapped air in between the particles to avoid agglomeration [20]. When the temperature of the melt has crossed the pouring temperature, the stirrer which was already preheated, is introduced into the melt. A brisk stirring of the melt was started and parallelly the preheated Al_2O_3 and TiC particles are poured into the melt at a rate of 10g/ min. The mixture is stirred at 600rpm for 20 min using graphite impeller and the temperature was maintained at 700^{0}C. A degassing tablet is used to remove the impurities. To improve the wettability of molten aluminum to mix with solid reinforcement particles, a small quantity of magnesium as a catalyst is added[21]. Later the molten mixture is poured into the preheated die and allowed for natural cooling at room temperature. The same process was carried out for the preparation of 2 sets of 6 specimens with different weights to percentage ratios of Al_2O_3 and TiC particles. One set of specimens is used to study the properties before heat treatment while the other set for comparison. The experimental set-up of stir casting is shown in figure 1.

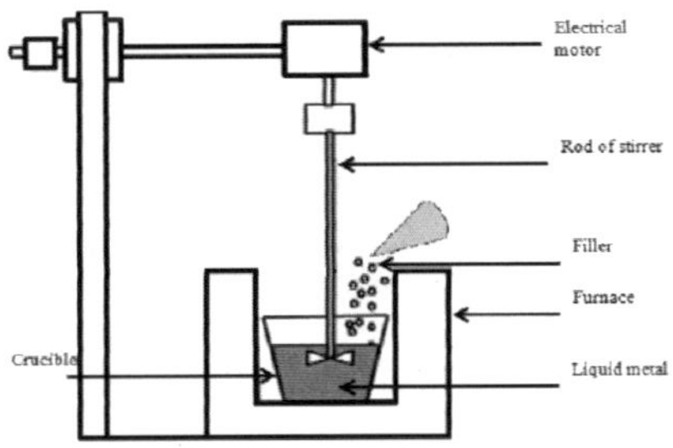

Figure 1: Experimental set-up of stir casting

2.2 Solution Heat Treatment
The Al-7075 HMMC specimens are subjected to T6 tempering (solution heat treated and artificially aged) [22]. For the tempering treatment, the specimen materials are heated in a furnace to a temperature value of 478^{0}C for 1 hr and immediately quenched in the water for cooling purposes [23]. Later aging of the specimens is done by heating in a furnace at a temperature of 127^{0}C for 24 hrs. The specimens are cooled by natural air at room temperature. The effect of solution heat treatment, quenching and age hardening processes on the mechanical properties of Al-7075 HMMC are investigated.

3. Physical and Mechanical Tests

3.1 Density and Porosity
The density is determined to analyze the effect of Al_2O_3-TiC Wt% propositions in Al-7075 hybrid composites. The theoretical density was calculated by using the formula rule of the mixture and experimental density was measured by volume of water displacement.

3.2 Metallographic Test
The test pieces are sectioned to a convenient size from the casted specimen and are mounted in an epoxy material to facilitate handling during grinding and polishing steps. Grinding is
done by grazing the surface of specimens under a series of operations using gradually finer abrasive grits. Grit sizes ranging from 100 to 1200 mesh are used. For fine finish grit sizes of 1/0, 2/0, 3/0 and 4/0 are also used. The persistence of grinding is actually to eradicate the oxide layer or uneven surfaces that might have occurred during the last segmenting operation.

Polishing is the final phase of a surface to get the flat, free from scratches and mirror-like appearance. A velvet cloth is lapped over the disc of polishing machine and Al_2O_3 paste is used as a finishing agent. The disc is rotated with an electric motor and the specimen is held on the disc by hand. Once the mirror finish is obtained, Keller's (composition) reagent is used to etch the samples for microscopic study. The specimens are now examined under an optical microscope to study the microscopic structure of the Al-7075/ Al_2O_3/TiC hybrid composite samples.

3.3 Hardness Test:

The Vickers hardness tester is used to evaluate hardness values for Al-7075/ Al_2O_3/TiC hybrid composite materials. A load of 150kg is applied over the specimen for 15 seconds. Four values at different locations on each specimen are taken and the average is considered for evaluation.

3.4 Impact Test:

As per ASTM A370 standards specimens are prepared. The impact load in terms of Joules is noted from the scale. The energy absorbed area and impact resistance are calculated based on the impact load.

3.5 Tensile Test:

The tensile test is conducted on the Tensometer. The specimens are prepared as per ASTM E8 standards. The tensile strength is assessed at crosshead speed of 0.3 mm/min. The true displacement–load curves are obtained.

4. RESULTS AND DISCUSSIONS

4.1 Density and Porosity

The percentage of porosity obtained from theoretical and experimental densities mentioned are shown in table 3.

Table 3: Calculated Density and estimated percentage Porosity

Sl. No.	Reinforcement Compositions	Theoretical Density(g/cm^3)	Experimental Density(g/cm^3)	Percentage Porosity
1	Sample 1	2.81	2.79	0.07
2	Sample 2	2.9	2.83	2.47
3	Sample 3	2.92	2.86	2.09
4	Sample 4	2.94	2.89	1.73
5	Sample 5	2.96	2.91	1.71
6	Sample 6	2.98	2.94	1.36

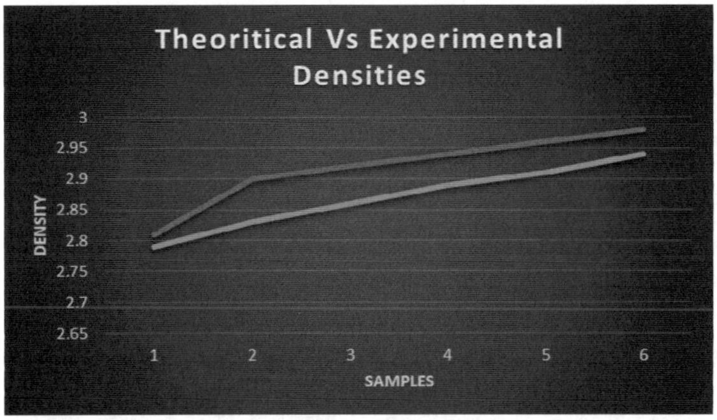

Figure 2: Graphical representation of theoretical & experimental densities with different compositions of Al-7075/Al_2O_3/TiC hybrid composites

Figure 2 indicate that the values of density are varying linearly. The density increased with an increase in TiC along with a decrease of Al_2O_3 particles in the composition. The porosity levels are within the limits (3%). The porosity is less and does not affect the mechanical behaviors of Al-7075/ Al_2O_3/TiC hybrid composite materials [24, 25]. The casted aluminum hybrid matrix composites are acceptable using stir casting technique.

4.2 Microscopic Images

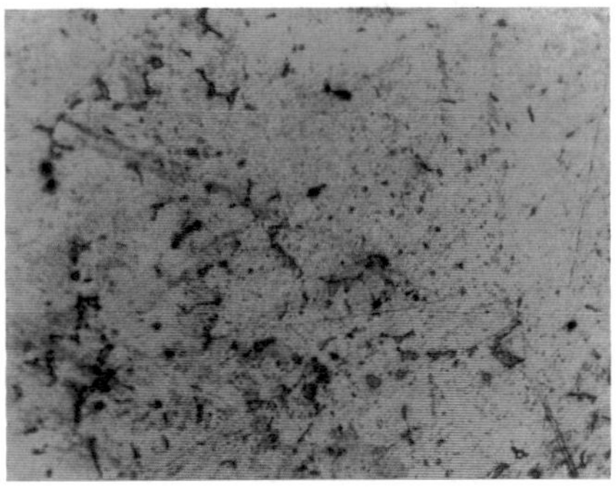

Figure 3: Microscopic image of Al-7075 with 2% Al_2O_3 and 6% TiC composite before heat treatment at 200X

Figure 4: Microscopic image of Al-7075 with 2% Al_2O_3 and 6% TiC composite after heat treatment at 200X

Figure 3 demonstrates the distribution of Al_2O_3 and TiC particles in the Al-7075 matrix. It has evident that there are no voids, cracks or dendrites in Al7075 HMMC material.
After completion of heat treatment, due to the formation of grain boundaries, the movement of grains was restricted. Such restriction is essential for increment in the strength of composite. Comparison of figure 3, figure 4 shows a fine dispersion of reinforcements into the Al matrix with the impact of heat treatment (T6). The fine dispersion of reinforcements improves the mechanical properties of Al HMMC.

4.3 Hardness

Figure 5 shows a comparison in hardness value before and after the heat treatment of Al-7075 HMMC materials. The characteristics and functionality of a material especially depends on the material hardness. The hardness of a material increase resistance to plastic deformation and wear resistance. The Al-7075 hybrid composite specimens are subjected to Vicker's test and an optimized weight percentage of reinforcement is analyzed. Also, the heat treatment influence on the specimens has examined.

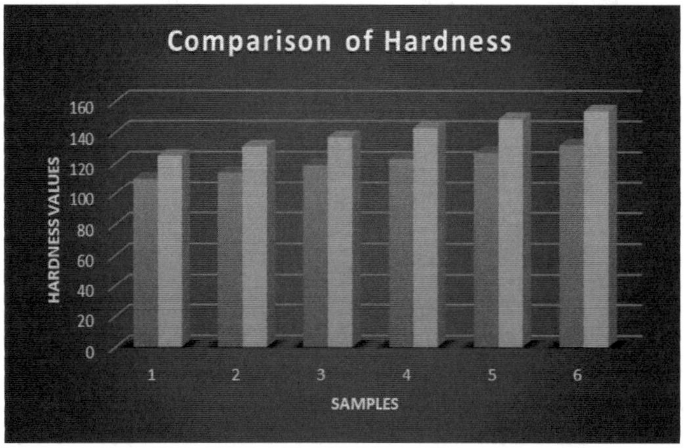

Figure 5: Comparison of hardness values before and after heat treatment on Al-7075/Al_2O_3/TiC hybrid composites

From figure 5, the sample composition with 8% Al_2O_3 and 0% TiC achieved high hardness value than pure Al-7075 material because of presence of hard ceramic particles (Al_2O_3). The TiC hard particles (%) improved hardness in the Al composite than Al_2O_3. Also, the uniformly distributed reinforced particulates in Al matrix improved the hardness as the load is shared by matrix and the hard particles. A rise in hardness is observed for the Al-7075 cast condition from 110 to 132 VHN.
The specimens are subjected to solution and age heat treatment. From figure 5 it is noticed that the hardness has increased after heat treatment process.

The hardness value of heat-treated specimens increased from 125 to 154 VHN due to refinement of grains. The maximum hardness is obtained at 8% of TiC and 0% of the Al_2O_3 composition. Figure 5 shows the overall improvement in hardness for aged Al-7075/ Al_2O_3/TiC composite specimens than the cast condition composites. Hardness is improved by 40% after heat treatment.

4.4 Impact test

Table 4 shows the influence of heat treatment on impact load and figure 6 represents impact resistance before and after heat treatment on Al-7075/ Al_2O_3/TiC hybrid specimens.

Table 4: Comparison of Heat Treatment on Impact Load

S. No	Samples	Before Heat Treatment		After Heat Treatment	
		Impact Load (N)	Energy absorbed (J/cm2)	Impact Load (N)	Energy absorbed (J/cm2)
1	Sample1	5	0.472	6	0.51
2	Sample2	6	0.519	7	0.54
3	Sample3	7	0.543	8	0.58
4	Sample4	9	0.628	11	0.69
5	Sample5	10	0.615	12	0.68
6	Sample6	9	0.602	11	0.69

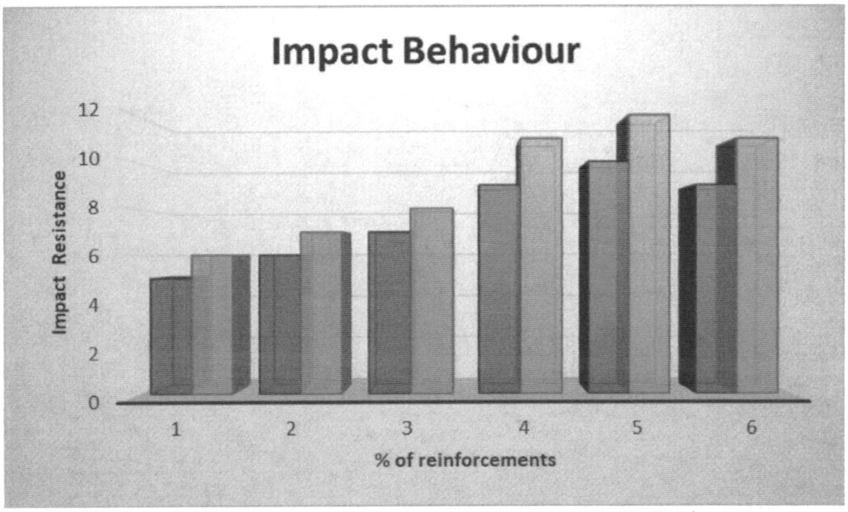

Figure 6: Comparison of Heat Treatment on Impact Resistance

The table 4 indicates a rise in impact energy for the Al 7075 cast condition. The cause might be due to the high toughness of Al_2O_3 and TiC and load distribution capacity of TiC particles that restrict crack or fracture propagation in the Al matrix[26]. Thereby, impact strength is improved. From figure 6 it is observed that at 2% Al_2O_3 and 6% TiC, the maximum impact strength is obtained. But, the increase in TiC percentage increases brittleness and decrease the impact strength at 8% TiC composition.
Figure 6 reveals that the impact resistance has been increased by 65% in aged specimens when compared with cast samples. The heat treatment may cause fine dispersion of reinforcements (Al_2O_3 and TiC) in the Al matrix. This fine dispersion improves bonding strength and the specimens can withstand the impact loads. The Al-7075 reinforced with 2% Al_2O_3 and 6% TiC specimens exhibited more impact strength.

4.5 Tensile strength

The table 5 exhibits the results of the tensile tests conducted on the samples prepared as per the standards and heat treated.

Table 5: Percentage Elongation and Tensile Strength before and after T6 heat treatment

Sl. No.	Samples	% Elongation		Tensile Strength (Mpa)	
		Before T6 Heat Treatment	After T6 Heat Treatment	Before T6 Heat Treatment	After T6 Heat Treatment
1	Sample1	9.32	7.13	314.6	336.4
2	Sample2	8.75	6.78	352.1	376.2
3	Sample3	8.42	6.51	365.8	391.8
4	Sample4	8.13	6.12	391.3	419.6
5	Sample5	8.05	5.92	415.7	437.2
6	Sample6	7.52	5.21	390.4	428.6

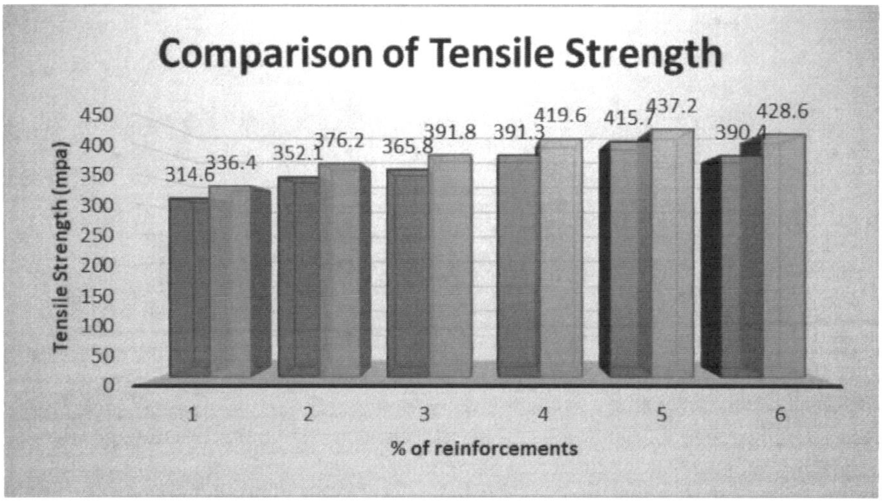

Figure 7: Graphical representation of tensile strength with different compositions of Al-7075/Al_2O_3/TiC hybrid composites

From Table 5, it is evident that the tensile strength at 2% Al_2O_3 and 6% TiC composition is maximum. The ductility is percentage elongation obtained from a tensile test. Thereby brittleness is increased and the tensile strength is decreased when the samples have 8% of TiC since the reduction in percentage elongation of material reduces ductility due to percentage increase in TiC hard particles.

The heat-treated specimens (Figure 7) exhibit greater tensile strength compared to cast Al alloy specimens. The cause may be with increase in strengthening property by the dispersion of filler materials like Al_2O_3 and TiC. The percentage of TiC hard particles reinforced into a soft matrix alloy increases the tensile strength called as direct strengthening[27].

The tensile strength increased by 38% further with T6 solution heat treatment. The heat treatment may cause thermal mismatch between matrix alloy and reinforced particles. This increases dislocation densities around reinforcement particles. The hard particles Al_2O_3/TiC restrict dislocation movement and increase tensile strength called as indirect strengthening, called Orowan mechanism [28].

5.CONCLUSIONS

The mechanical properties of Al-7075 HMMC materials have been considerably improved by adding Al_2O_3 and TiC as reinforcement. The improvement in properties is further enhanced with the combination of solution and ageing heat treatments. The study of microscopic images revealed the fine distribution of particles by the effect of T6 heat treatment than cast Al-7075 composites. The sample 6 with Al-7075 and 8 % TiC, 0% Al_2O_3 results in high hardness and sample 5 with 2 % Al_2O_3 and 6 % TiC resulted in high impact and tensile strength. The hardness increased by 40%, impact strength by 65% and tensile strength increased by 38% for reinforced and T6 heat treated Al-7075 composites.

6. REFERENCES

[1] Himanshu Kala S, Mer KK, Sandeep Kumar 2014 A Review on Mechanical and Tribological Behaviors of Stir Cast Aluminum Matrix Composites *Procedia Materials Science* **6** 1951-1960
[2] Uvaraja V, Natarajan N, Sivakumar K, Jegadheeswaran S and Sudhakar S 2015 Tribological behavior of heat-treated Al 7075 aluminium metal matrix composites *Indian journal of Engg. & Material Science* **22** 51-61
[3] Utkarsh Pandey, Rajesh Purohit, Pankaj Agarwal, Swadesh Kumar Singh 2018 Study of Fabrication, Testing and Characterization of Al/TiC Metal Matrix Composites through different Processing Techniques *Materials Today* **5** 4106–4117
[4] Ramkumar K R, Sivasankaran S, Fahad A. Al-Mufadi, Siddharth S, Raghu R 2019 Investigations on microstructure, mechanical, and tribological behavior of AA 7075–x wt.% TiC composites for aerospace applications *Archives of Civil and Mechanical Engineering* **12** 428-438
[5] Rajendra S K, Ramesha C M 2015 A survey of Al7075 aluminium metal matrix composite *International Journal of Science and Research* **4** 1071-1075
[6] Jaswinder Singh, Amit Chauhan 2015 Characterization of hybrid aluminium matrix composites for advanced applications- A Review *j mater res techno.* **170** 1-11
[7] Sreenivasa Reddy M, Dr. Soma.V. Chetty, Dr. Sudheer Premkumar, Reddappa H N 2014 Influence of reinforcement and heat treatment on mechanical and wear properties of Al 7075 based hybrid composites *Procedia Materials Science* **5** 508 – 516
[8] MK Mallik, CS Rao, VVSK Rao 2014 Effect of heat treatment on hardness and wear behavior of weld deposited Co-Cr-Mo alloy *Procedia Engineering* **20**, 544-549
[9] Mohanavel V, Rajan K, Suresh Kumar S, Chockalingam A, Aquila Roy, Adithiyaa T 2018 Mechanical and characterization of stir-cast Al-SiCp composite *Materials Today: Proceedings* **5** 1740–1746
[10] Anitha1P, Shrinivas Balraj U 2017 Dry Sliding Wear Performance of Al/7075/Al2O3p/Grp Hybrid Metal Matrix Composites", *Materials Today: Proceedings* **4** 3033–304
[11] Sambathkumar M, Navaneetha krishnan P, Ponappa K, Sasikumar K S K and Arunkumar P 2016 Analysis of Al7075 Hybrid Metal Matrix Composite Using Two- Dimensional Microstructure Model Based Finite Element Method *International Journal of Advanced Engineering Technology* **7** 180-183
[12] Rajeswari B, Amrithagadeswaran K S and Ramya R 2014 Microstructural studies of Aluminum 7075- Silicon Carbide-Alumina Metal Matrix Composite *Advanced Materials Research* **984** 194- 199
[13] Mohanavela V, Rajan K, Senthil P V, Arul S, 2017 Mechanical behaviour of hybrid composite (AA6351+Al2O3+Gr) fabricated by stir casting method *Materials Today: Proceedings* **4** 3093– 3101
[14] Hashim J, Looney L, Hashm M S Ji, 1999 Metal matrix composites: production by the stir casting method, Processing Technology *Journal of Materials Processing Technology* **92** 1-7

[15] Udoye N E and Inegbenebor a O, 2019 The study on improvement of aluminium alloy for engineering applications *International Journal of Mechanical Engineering and Technology* **10** 380-385

[16] RamakoteswaraRao V, Ramanaiah N, Srinivasa Rao M, Sarcar MMM, and Kartheek G, 2016 Optimization of process parameters for minimum volumetric wear rate on AA 7075- TiC Metal matrix composite *International Journal of Automotive and Mechanical Engineering* **13** 3669 - 3680

[17] Kok M, 2005 Production and mechanical properties of Al2O3 particle reinforced 2024 Aluminium alloy composites *Journal of Materials Processing Technology* **161** 381-387

[18] G Suresh, KL Narayana, MK Mallik, 2018 Characterization and Wear Properties of Co-Cr-W alloy Deposited with Laser Engineered Net Shaping" *International Journal of Recent Technology and Engineering* **7** 151-155

[19] Indumathi Deshmanya and GK Purohit, 2012 Development of models for predicting impact strength of Al 7075/ Al2O3 composites produced by stir-casting *Journal of Composite Materials* **46** 3247-3253

[20] Jacob S, 2017 Thermal analysis on Al7075/Al2O3 metal matrix composites fabricated by stir casting process", *IOP Conf. Ser.: Mater. Sci. Eng.* **183** 1-9

[21] Devinder Priyadarshi, Rajesh Kumar Sharma, 2016 Porosity in aluminium matrix: cause, effect and defence *MaterialScience an Indian Journal* **14** 119-129

[22] Hanumanthe Gowda, Rajendra Prasad P, 2016 Influence of Heat Treatment on Corrosion Resistance of A356/RHA/Al2O3 Based Hybrid Composites *Materials Today: Proceedings* **4** 10870-10878

[23] Clark Jr T, Coughran Ba, Traina I, Hernandez A, Scheck T, Etuk C, Peters J, Lee b E W, Ogren J, Es- Said O S,2005 On the correlation of mechanical and physical properties of 7075-T6 Al alloy *Engineering Failure Analysis* **12** 520–526

[24] Veeresh Kumar G B, Rao C S P, Selvaraj N, Bhagyashekar M S, 2010 Studies on Al6061-SiC and Al7075-Al2O3 Metal Matrix Composites *Journal of Minerals and Materials Characterization and Engineering* **9** 43-55

[25] Kannan C, Ramanujam R, 2017 Comparative study on the mechanical and microstructural characterisation of AA 7075 nano and hybrid nanocomposites produced by stir and squeeze casting", *Journal of Advanced Research* **8** 309–319

[26] KM Mantrala, CS Rao, VVSK Rao, SA Munaf, 2018 Effect of process parameters on the tensile ____properties of LENS deposited Co-Cr-Mo alloy *International Journal of Manufacturing* Technology *and Management* **32** 509-516

[27] Adeyemi Dayo Isadarea, Bolaji Aremob, Mosobalaje Oyebamiji Adeoyec, Oluyemi John Olawalec, Moshood Dehinde Shittu, 2013 Effect of Heat Treatment on Some Mechanical Properties of 7075 Aluminium Alloy", *Materials Research* **16** 190-194

[28] Baradeswaran A, Elaya Perumal A, 2014 Study on mechanical and wear properties of Al 7075/Al2O3/graphite hybrid composites *Composites: Part B* **56** 464-471

IOP Conf. Series: Materials Science and Engineering 1112 (2021) 012016 doi:10.1088/1757-899X/1112/1/012016

Multi Response optimization of drilling parameters in Glass Fiber Reinforced Thermoplastic composites

Srinivasan.T[1] , Arunkumar. R[2], Meghanathan. S[2] and Ramu.P[2]

[1]Professor, Department of Mechanical Engineering, Koneru Lakshmaiah Education Foundation ,Vaddeswaram,Guntur, Andra Pradesh,India

[2]Asst Professor, Department of Mechanical Engineering, Sri Sai Ram institute of science,Technology, Tambaram, Chennai.

Corresponding author: seenusreeja@gmail.com

Abstract: Optimization of drilling parameters in Glass Fiber Reinforced Thermoplastic composites (GFRTP) has been analyzed using grey relational analysis in this research.Thegoal of this thesis was to investigate the effect of drilling parameters on composites of Glass Fiber Reinforced Thermoplastic composites (GFRTP) using hand layup method. The circularityand surface roughness are diminished. In drilling, tests were carried out on Glass Fiber Reinforced Thermoplastic composites with HSS drill bits of 6 mm, 7 mm and 8 mm diameter used in dry conditions using the Taguchi L9 orthogonal array process. ANOVA analysedtheeffectonoutput parameters such as surface roughness and circularity of multiple drilling input parameters such as cutting speed, feed, and cutting depth. For finding the best machining state, Grey relational analysis equations were used. For the drilling of hybrid Glass Fiber Reinforced Thermoplastic composites, the most critical parameter, namely the cutting speed, was found to have affected the thrust power, circularity and surface roughness.

1. Introduction

Drilling is a method of making circular holes in a concrete material or widening existing holes with the use of drills or drill bits called multi-point cutting equipment. Drilling is a method of continued machining. For drilling, various cutting tools are available, but the twist drill is the most common. To serve numerous purposes, a broad range of drilling processes is available. The main problem is the rapidly increasing efficiency and competitiveness of technology. Productivity is concerned with theSR and the product characteristics apply to circularity during machining process and efficiency. By parameter optimization, efficiency and productivity can also be increased. There are a range of research works relating to optimization of drilling parameters to obtain output responses. The primary reactions are surface roughness and circularity of the drill bit

Raj Mohan et al. [1] performed push drive drilling experiments. In the experiment, the feed was found to be the major parameter controlling the thrust force. The aim is to investigate the impact of the parameter machining for newly reinforced surface MMCs machinability and increased roughness [2-5]. Ramprasad et al [6] analyzed the effect of machining parameters on MRR and SR in wire EDM process parameters of Titanuium alloys. The optimal levels are Peak current-20, Pulse on time-105, Pulse off time-95 and Servo voltag-50. Basha et al [7] deliberate the Host alloy C276 using Taguchi Techniques and found the lower values of discharge current and pulse on time, higher values of pulse off time the optimum condition for surface roughness. Venkataiah et al [8] calculated the design model and helped to select optimized parameters with a minimum number of experiments for new lightweight technical applications in the production of grain-refined ZE41 Mg alloy. In order to obtain the optimal MRR and minimum surface roughness (Ra) of the Inconel718, Dharmandera et al [9] analysed the Taguchi orthogonal series. L9 and L16 orthogonal arrays were selected and signal to noise ratios were computed by Suresh et al [10]. The input parameters of velocity, feed, cut depth, drill depth and outcome

IOP Conf. Series: Materials Science and Engineering 1112 (2021) 012016 doi:10.1088/1757-899X/1112/1/012016

parameters of surface roughness, material removal rate and running time were optimized at the end. Using zinc-coated brass wire electrode, A Kumar et al [11] analyzed the material removal rate, surface roughness and corner deviation, where investigated using Inconel 718. The other others are applied the some other optimization techniques and implemented in different materials and the dissimilar results [12-16].

2. Materials and Methods

In this study, Thermoplastic composite laminates of 4 mm thickness is prepared by film stacking process. Six layers of E-glass fibre woven mat (200 gsm) are used as the reinforcing material and the seven layers of polypropylene thermoplastic sheets having 0.5 mm thick is used as the matrix. The polypropylene film sheet and glass fiber woven mat are prepared with a length of 250 mm and width of 250 mm. alternately arranged glass fiber woven mat and polypropylene film sheet have been stacked in a hot compression molding machine in between the moving and fixed platens of the hydraulic press. Suitable releasing agent (silicone spray) is used during the fabrication.

The specimen is heated electrically in all the sides up to the glass transition temperature and Pressure is applied to form the shape (50 kg/cm^2). The laminate is hot cured in platens at a temperature of 220°C. Finally, the composite plate is allowed to cool at room temperature under the same pressure. Drilling experiments are carried out with the aid of Vertical Machining Centre (VMC) controlled by the computer by using L9 orthogonal array. Based on the experience of the authors and discussed literature, the drilling variables are chosen. The parameters are taken into account and their values are tabulated in Table 1. The experimentation was carried out with three different step twist HSS drill bits (6 mm, 7 mm and 8 mm diameter). The measurement of surface finish of each drilled hole was done withtheassistance of a Video Measuring System (VMS-2010 F), and the measurement of circularity of each drilled hole was done with the help of a computer-controlled surface roughness tester, Kosaka (SE3500), JAPAN. The Machining parameters and experimental values are tabulated in Table 1 and 2.

Table 1. Machining parameters

Parameters	Unit	Level 1	Level 2	Level 3
Speed (A)	rpm	750	1000	1250
Feed (B)	mm/min	15	25	35
Drill diameter (C)	mm	6	7	8

3. Grey Relational Analysis

Grey contextual analysis is the best theory used to examine the unpredictability of multiple inputs and discrete information [17, 18]. The estimation of the cumulative value between the two is an evenly weighted grey relational analysis. Sequences that are based on the separation of details. It is being used for measuring a reasonably precise series association by substituting 0.5 for equal weight. The GRG steps explained in the Figure and formulas given from Eq 1-4.

Table 2. Experimental values

Speed (rpm)	Feed Mm/min	Drill diameter mm	SR (µm)	Circularity (mm)	S/N SR	S/N Circularity
750	15	6	6.52341	0.132	-16.289	17.589
750	25	7	6.52362	0.129	-16.290	17.788
750	35	8	6.52383	0.125	-16.290	18.062
1000	15	7	6.17632	0.137	-15.815	17.266
1000	25	8	6.17653	0.134	-15.815	17.458
1000	35	9	6.40801	0.123	-16.134	18.202
1250	15	9	5.82923	0.152	-15.312	16.363
1250	25	6	6.06071	0.1323	-15.650	17.569
1250	35	7	6.06092	0.1391	-15.651	17.133

1	Identify the response variables and corresponding S/N ratio
2	Identification of controllable parameters, interactions and their levels
3	Identify orthogonal array and assign parameter levels to each colum
4	Conduct the experiment and collect the data for response variables
5	Data preprocessing by normalizing the S/N ratio for all sequences
6	Calculate grey relational coefficients for each sequence
7	Determine the grey relational grade by averaging the grey coefficients
8	Determine the optimum sequence, from the higher grey relational grade
9	Determination of optimum parameters
10	Predication of GRG for optimal parameters

Figure 1. Steps for Grey relational Analysis

Optimization Steps of GRG and formulas

Step 1

Transform the experimental value into S/N ratio using the appropriate formulae depending on the type of quality characteristics. S/N ratio of s surface roughness and circularity consider as a smaller the better.

$$S/N\ ratio= -10*\log_{10}* 1/n\ \Sigma Y_i^2 \quad (1)$$

Step 2

Normalized the value for i^{th} experimental trial for j^{th} dependent variable/response

$$Z_{ij} = \text{Max}\,(Y_{ij},\, i=1,2,3,\ldots n)-\, Y_{ij}\,/\, \max(Y_{ij}, i=1,2,3\ldots n)-\text{Min}\,(Y_{ij},\, i=1,2,3\ldots n) \quad (2)$$

(to be used for S/N ratio with smaller the better)

Step 3

Compute the grey relational coefficients GC for the normalized S/N ratio values.

$$GC_{ij}= (\Delta_{min}+\lambda\Delta_{max}) \,/\, (\Delta_{ij}+\lambda\Delta_{max}) \quad (3)$$

GC_{ij}- grey relational coefficient.

Δ- Absolute difference between Y_{oj} and Y_{ij}.

Y_{oj}-optimal performance value.

Y_{ij}- Normalised values.

Δ_{Min}- Minimum value of Δ.

Δ_{Max}- Maximum value of Δ.

λ- Distinguishing coefficients.

Steps 4 Compute the GRG

$$GRG= 1/m*\Sigma GC_{ij} \quad (4)$$

Step 5 Use response graph method or ANOVA and select optimal level for the factor based on maximum average GRG value.

4 Results and discussion

Optimization of Drilling Conditions Using Taguchi Analysis

As mentioned above for better surface roughness and circularity, the optimization of drilling parameters in this work is done using Taguchi grey relational analysis. Therefore the experimental results are then converted into an S/N ratio after a series of experiments provided by orthogonal array used in data analysis, and the lower the better is used to evaluate the quality characteristics, as given in equations 1-**4.** The OA with control factors, experimental results, and the corresponding normalised S/N ratio values. GRC and Grades are shown in Table 3.

SR and circularity enhancement is one of the key features of any machining process. Due to machining procedures, SR and circularity reveals defects in the surface of each workpiece. A high SR and circularity machined component results in unnecessary wear, fatigue, lower ability of material to resist corrosion, and lower corrosion resistance. Production of a substance. The drilling parameters, tool geometry, and relative vibration caused by the tool could affect SR and circularity. The analysis of grade values are shown the Figure 2, as the best combination of parameters can be obtained at speed at level 1 (750 rpm), feed (15mm/min) level 1, and drill diameter is level 3 (8mm).

Table 3. Normalized, grey relational coefficient and grade values

Normalised S/N		GRC		Grade
SR	Circularity	SR	Circularity	
0.999	0.334	0.999	0.429	0.714
1.000	0.225	0.999	0.392	0.696
1.000	0.076	1.000	0.351	0.676
0.514	0.509	0.507	0.505	0.506
0.514	0.405	0.507	0.456	0.482
0.841	0.000	0.759	0.333	0.546
0.000	1.000	0.333	1.000	0.667
0.346	0.344	0.433	0.433	0.433
0.346	0.581	0.433	0.544	0.489

Table 4. Mean values of Grade

Factors	Level 1	Level 2	Level 3	max-min
A	**0.695**	0.511	0.529	0.18386
B	**0.629**	0.537	0.570	0.0919
C	0.564	0.563	**0.608**	0.04457

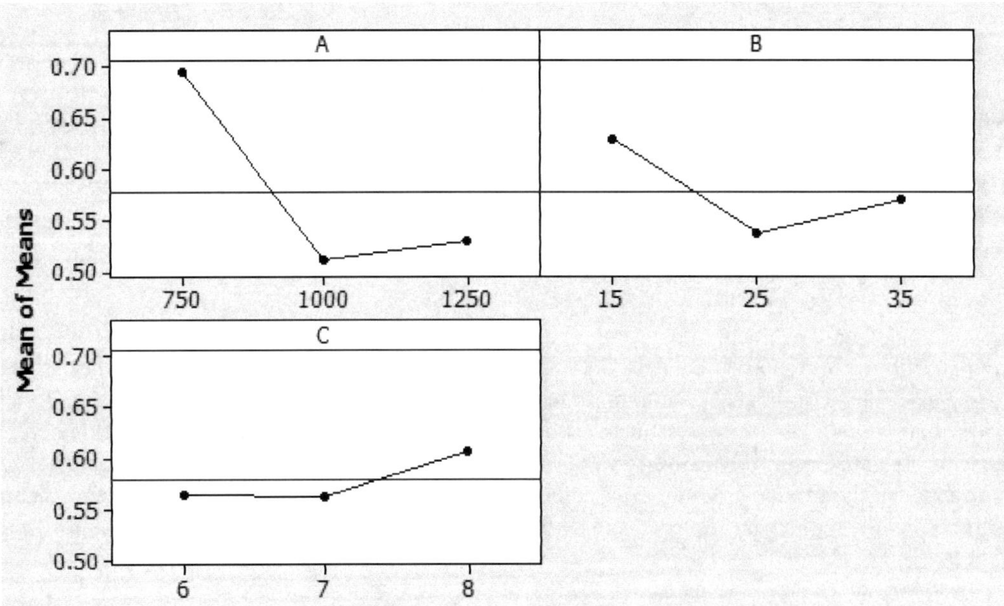

Figure 2. Optimal level of Grade

Analysis of variance

In this analysis, ANOVA was used to calculate the significance of each of the method parameters for SR and circularity. The ANOVA results for the SR and HS are shown in Tables 5, respectively. The percent contribution in ANOVA is used to explain how much the output responses are influenced by each process parameter. The P-value will validate the effect on responses of method parameters and indicates that values smaller than 0.05 do not have an effect.

Table 5. ANOVA parameters contribution

Parameters	SS	Dof	MS	% Contribution
Speed	0.061638	2	0.0308	65.36
feed	0.013017	2	0.0065	13.80
Drill bit	0.003932	2	0.002	4.17
Error	0.004	2	0.002	16.67
Total	0.09431	8		83.33

SS- sum of the square, Dof –Degree of freedom, MS- Mean square.

Table 7 thus indicates that speed contribution is 65.36 percent and feed contribution 13.80 percent respectively, were the statistic effects of speed, feed and drill bit size. This illustrates that the has greater effect than the sped on hole roughness. However, in the case of speed, the influence of speed was more than the doc and drill size value given in Table 5. Feed percentage share is 13.80 percent, which is higher than drill bit size contribution with a 4.17 percent statistical meaning. The model adequacy checking was conducted after performing an ANOVA analysis to verify the normality assumption of the residual. Figs.3 and 4 shows normal probability plots of the residuals and these figures reveal that almost all the residuals follow a straight line pattern and this agrees well with the results reported. This work will be useful for industries while the selection of process parameters in the drilling of GFRP composite materials, to improve the quality of the drilled holes by reducing the delamination.

The regression equation is surface roughness is given Eq 5 and circularity is given Eq 5.
SR = 7.34 - 0.00106 *A + 0.00882*B - 0.0328 *C

The regression equation is circularity is
Circularity = 0.113 + 0.000024*A - 0.000618*B + 0.00159*C (5)

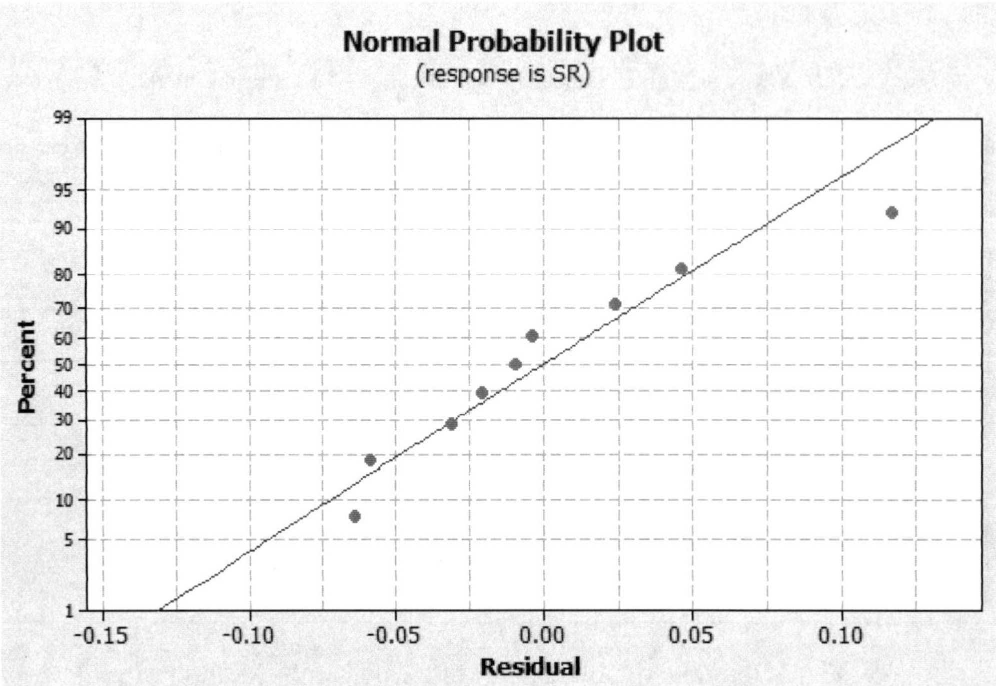

Figure.3 Normal probability plot of residuals for SR

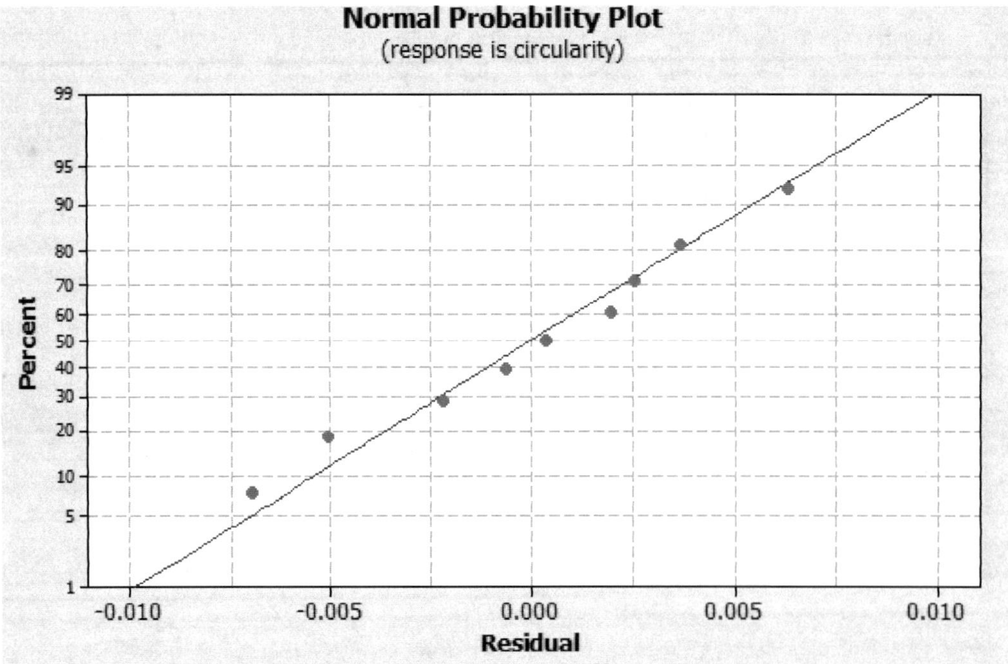

Figure.4 Normal probability plot of residuals for circularity

Conclusions

A research on the optimal selection of glass fiber reinforced thermoplastic composites, especially in this article, the optimization of surface roughness and circularity deviation of drilled holes is carried out for the automobile and aerospace industries. In this regard, the GRA approach is recommended for decision-making on the collection of suitable material, which yields optimum surface roughness values and drilled holes circularity deviation. The output from the Taguchi process was fed to the GRG as an input. Finally, the result generated in GRA suggests the suitable alternative of glass fiber reinforced thermoplastic GRA methods.

References

[1] T. Rajmohan and K. Palanikumar, "A mathematical model to predict thrust force in drilling hybrid metal matrix composites," *Frontiers in Automobile and Mechanical Engineering -2010*, Chennai, India, 2010, pp. 1-3, doi: 10.1109/FAME.2010.5714788.

[2] Hayajneh MT, Hassan AM, Mayyas AT (2009) Artificial neural network modeling of the drilling process of self-lubricated aluminum/alumina/graphite hybrid composites synthesized by powder metallurgy technique. *J Alloys Compd* , 478, pp::559–565.

[3] Basavarajappa S, Chandramohan G, Mahadevan A, Thangavelu M, Subramanian R, Gopalakrishnan P (2007) Influence of sliding speed on the dry sliding wear behavior and the sub surface deformation on hybrid metal matrix composite. *Wear,* 262, pp:1007–1012.

[4] Gultekin D, Uysal M, Aslan S, Alaf M, Guler MO, Akbulut H (2010), The effects of applied load on the coefficient of friction in Cu-MMC brake pad/Al-SiCp MMC brake disc system. *Wear,* 270, pp:73–82.

[5] Basavarajappa S, Chandramohan G, Davim J (2008) Some studies on drilling of hybrid metal matrix composites based on Taguchi techniques. *J Mater Process Technol* 196(1–3), pp:332– 338.

[6] Ram Prasad A.V.S., Ramji K., Raghu Kumar B.(2018), Study of wire-electrical discharge machining parameters of titanium alloy by using taguchi method, *International Journal of Engineering and Technology*, 7(2), pp:10-12.

[7] Basha S.K., Kolli M., Jagannadha Raju M.V (2018), Parametric optimization of Edm on Hastelloy C-276 Using Taguchi L18 Technique, *International Journal of Engineering and Technology*,7, pp:714-716.

[8] Venkataiah M., Kumar T.A., Rao K.V., Kumar S.A., Sunil B.R. (2019), 'Role of friction stir processing parameters on the microstructure and hardness of Ze41 Mg alloy: A Taguchi approach', *Materials Performance and Characterization*, 8(1), pp:582-593.

[9] B.V D., Kodali S.P., Boggarapu N.R. (2019), Multi-objective optimization for optimum abrasive water jet machining process parameters of Inconel718 adopting the Taguchi approach, *Multidiscipline Modeling in Materials and Structures*, 16(2), pp:306-321.

[10] Suresh A., Diwakar G. (2019), 'Optimization of machining process parameters in turning and drilling by using design of experiments with aluminum 6061-O alloy and austenitic stainless steel', *International Journal of Innovative Technology and Exploring Engineering*, 8(11), pp:625-633.

[11] Kumar A, Vivekananda K, Abhishek K. (2019), Experimental Investigation and Optimization of Process Parameter for Inconel 718 Using Wire Electrical Discharge Machining, *Journal of Advanced Manufacturing Systems*, 18(3), pp:339-362.

[12] Mugutkar H., Tamiloli N, Kohir V.V. (2019), The processing techniques and behaviour of aluminum metal matrix with different reinforcement materials, *International Journal of Mechanicaland Production Engineering Research and Development*, 9(4), pp:793-804.

[13] Venkatarao K, Anup Kumar T. (2019), An experimental parametric analysis on performance characteristics in wire electric discharge machining of Inconel 718', *Proceedings of the Institution of Mechanical Engineers, Part C: Journal of Mechanical Engineering Science*, 233(14), pp:4836-4849.

[14] Suresh A, Diwakar G. (2019), Optimization of process parameters in turning operation of austenitic stainless steel rod using taguchi method and anova, *International Journal of Mechanical and Production Engineering Research and Development*, 9(3), pp:379-386.
[15] Ram Prasad A.V.S., Ramji K., Kolli M., Vamsi Krishna G. (2019), Multi-Response Optimization of Machining Process Parameters for Wire Electrical Discharge Machining of Lead-Induced Ti-6Al-4V Alloy Using AHP-TOPS64V IS Method', *Journal of Advanced Manufacturing Systems*, 18(2), pp:213-236.
[16] Ramya K., Koneru S., Santhosh Kumar R .(2018), Multi variable optimization of surface grinding process using genetic algorithm, *International Journal of Mechanical and Production Engineering Research and Development*,8(2), pp:77-486.
[17] Uthayakumar M (2013) Dry sliding friction and wear studies of fly ash reinforced AA-6351 metal matrix composites. *Adv Tribol.* https ://doi.org/10.1155/2013/365602.
[18] Pannerselvan R (2012) Design and analysis of experiments. *PHI Learning Private Limited*, NewDelhi

Optimization of EDM Process Parameters on Machining Characteristics of SiC and Graphene Reinforced Al 6061-T6nano-Composites

Saikiran Ammisetty[1, *], DhanunjaykumarAmmisetti[2], K.Satyanarayana[3], Suresh Chitturi[4], N.Sunil Naik[5]

[1]JNTUA College of Engineering, Anantapur, INDIA

[2,5]Lakireddy Bali Reddy College of Engineering, INDIA

[3]Vasireddy Venkatadri Institute of Technology, INDIA

[4]Vignan's Institute of Engineering for Women

[*] Corresponding Author

`dhanaammisetti@gmail.com

Abstract. Aluminum composites play an important role in various applications due to their low weight to strength ratio, high corrosive resistance, etc. As reinforcement materials, Graphene and SiCnano particles were used. Since, they havea high melting point, young modules, strong electrical and thermal properties etc. Electrical Discharge Machining (EDM) is one of the non-conventional machining techniques widely used for the machining of hard materials such as composites. The aim of current work is to evaluate the influence of EDM process parameters such as peak current, pulse on-time, voltage and flushing pressure on machining of Aluminium 6061 reinforced with 0.5 wt% SiC and 0.5 wt% Graphene nanoparticles. Al-based Metal Matrix Composite was casted by using liquid processing technique i.e. Stir Casting. Taguchi's Design of Experiments was used to evaluate the effect of process parameters on machining characteristics i.e. Material Removal Rate (MRR) and Tool Wear Rate (TWR) of developed composite. From the results it was observed that peak current is the most influencing factor on MRR and TWR followed by other parameters.
Keywords: Electrical Discharge Machining, Metal Matrix Composite, Design of Experiments, Material Removal Rate, Tool Wear Rate

1. Introduction

Metal matrix composites are used in many fields due to their high specific strength, strong corrosion resistance and improved mechanical properties. Currently, industries are looking for light weight material with improved properties such as defence, automotive, sports, electronic, aerospace etc. For that AMMCs plays a major role due to its improved physical and mechanical properties. This improvement in properties was done because of reinforcement addition to the base material. Various types of reinforcements are available such as SiC, TiC, Al_2O_3, MoS_2, B_4C, TiB_2, ZrO_2, Graphite, Graphene, Carbon Nano Tubes, etc.Machining of this newly developed composite material viz. metal matrix composites, nano composites, and ceramic composites, found difficult by existing manufacturing industries / conventional machining processes [1]. Because it is hard and difficult to machine, reduces surface finish and accuracy and increased machining cost. To machine these outlandish composites non -conventional methods are preferable such as Electric Discharge Machining (EDM), Electro Chemical Machining (ECM) etc.[2,3]. In this, no direct contact between cathode and anode, no external pressure on work piece and dielectric fluid is used to dissipate the generated heat at sparking zone and to flush out the debris from the machining surfaces.Many authors reported their work on EDM of composites. N. Radhika et.al [5] developed the Aluminium hybrid metal matrix composite. They reported the impact of process parameters on MRR, TWR and SR by using Taguchi's Design of Experiments. It states that SR was influenced by peak current, MRR was influenced by flushing pressure and TWR was influenced by pulse on-time mostly.Sarivastavaet. al [6] studied the machinability behavior of developed Al6063/SiC MMC using EDM. Developed composite shows improved mechanical properties. Machinability was enhanced by 37.5% at optimum condition of process parameters. B.C. Kandpalet.al [7] reported the optimization of process parameter of EDM of AMMCs. Optimization was done using Taguchi method in combination with S/N ration and ANOVA test. They found the MRR increase with increasing peak current and pulse on time but decreases with increasing duty factor. Optimum parameters for getting higher MRR are 14A, 200μs and 50%. K. G. Maniyaret. al [8] described the effect of process parameters, weight percentage of reinforcement on MRR, TWR and SR with the help of L27 orthogonal array and its significance on output responses was studied by ANOVA. The process parameters were designed to get maximum MRR and minimum TWR and SR. Contribution current for MRR was 83.5%, for SR was 70.22% when compared to other parameters. C. Karet. al [9] fabricated the aluminium - red mud reinforced MMCs by stir casting method and machined using EDM.

From the literature survey, it can be concluded that so much of research work was done on fabrication and machining of MMCs. However, there is a gap in the development of AMMCs with SiC/GNPs as reinforcement. In this work, influence of EDM process parameters such as flushing pressure, pulse on-time and peak current on MRR, TWR and Surface Roughness of SiC/GNPs reinforced AMMCs by DOE.

2. Materials and Method

Any composite material contains two different phases. One is called as matrix phase which acts as supporting material and the other is reinforcement phase which gives strength to the composite. In the present work Silicon Carbide (SiC) and Graphenenano particles were mixed with aluminium 6061 alloy.

2.1 Methodology

Initially AA6061 ingots were loaded in the electrical furnace assisted graphite crucible and melted at 800^0C. After melting, to extract the trapped gases from the melt, 3 grams of C_2Cl_6 was added to the melt. Reinforcement in the respective proportions was mixed with 2 grams of Potassium hexaFluoroTitanate (K_2TiF_6) and then preheated in preheater die for 2 hours so that it is free from foreign particles, moisture content and also increases the wettability. K_2TiF_6 is called as flux powder or molten salts used for improving wettability between matrix and reinforcement materials. Laterpreheatedreinforcementhas mixed with the AA6061 melt in 4 steps while rotating the stirrer at 400 rpm. Addition of reinforcement in steps improves the distribution in the matrix and it leads to better properties. The stirrer has rotated for 10 min in the melt to get uniform distribution of reinforcement in the melt. After stirring the molten mixture was held in the furnace for 5 minutes to settle down the particles then it has poured into the preheated die.The developed casting was shown fig 1. From the developed casting 40*10*10(mm^3) billets were prepared for EDM machining and shown in fig 2.

IOP Conf. Series: Materials Science and Engineering 1112 (2021) 012017 doi:10.1088/1757-899X/1112/1/012017

Fig. 1: Casted Aluminium Composite

Fig.2: Work- pieces before EDM Machining

2.2 Design of Experiments

In any industry, high quality and high productivity is a challenging task. For that, Design of experiments (DOE) mainly used for generating effective combination of process parameters by reducing number ofexperiments.In this study, Taguchi based mathematical model with 27 experiments was developed to study the process parameter influence on machining characteristics. This is one of the best methods to create the mathematical model. From the literature, it was observed that the most effective parameters on machining characteristics are pulse on–time (T_{on}), flushing pressure (P), peak current (I_p) and voltage (V). The other parameters such as pulse off-time, duty factor, polarity and spark gap also influence EDM performance with less effect. T_{on}, P, I_p and V are taken as input variables. MRR and TWR are taken as output variables. The input variables and their levels are shown in table 1. Based on machine capacity the level values are taken into consideration. In this work, L_{29} orthogonal array is considered with27 rows and 9 columns. In any optimization process number of experiments should be more to get accurate results.

IOP Conf. Series: Materials Science and Engineering 1112 (2021) 012017 doi:10.1088/1757-899X/1112/1/012017

Table 1: Machining parameters and Levels

S.No	Parameters	Symbol	Units	Level A	Level B	Level C
1	Pulse on-time	T_{on}	µs	100	200	300
2	Peak Current	I_p	A	10	20	30
3	Voltage	V	V	40	50	60
4	Flushing Pressure	P	kg/cm^2	0.3	0.5	0.7

For each parameter three levels of values are selected based on the minimum and maximum values of the parameters. MINITAB18 has used for generating the relation between parameters. This is specifically used for DOE applications. The interactions of input variables on output variables also investigated. After finding the MRR and TWR for each experiment, the entire data is converted to signal-to-noise (S/N) ratio. It is the ratio of the mean of the signal to the standard deviation of noise. Rank of input variables to be determined with this ratio. It contains 3 different quality characteristics, known as smaller the better, larger the better and nominal the best [10]. For TWR, smaller the better characteristic is suitable because tool life should be high in any machining process. But it is quite opposite in case of MRR where larger the better is considered.

Larger the better characteristics-

$$\frac{s}{N} = -10 \times Log10\left(\frac{su\left(\frac{1}{y2}\right)}{n}\right) \text{......................Eq (1)}$$

Smaller the better characteristics-

$$\frac{s}{N} = -10 \times Log10\left(\frac{su(y^2)}{n}\right) \text{......................Eq (2)}$$

Analysis of Variance (ANOVA) is used to find the relative effect ofoneparameter on the other. To analyse the experimental data this statistical tool was used.The mean square of the result with deviation from sample mean was compared and given the optimized results i.e. minimum TWR and maximum MRR.

3. Experimental Setup:

ELECTRONICA-SMART ZNC kick the bucket sink EDM machine was utilized for experimentation on created Aluminium nano composite. Aluminium composite examples as appeared in fig. 2 were utilized as work-pieces with length, width and thickness being 10mm, 40mm, and 10 mm individually. Copper anode with 8 mm distance across is utilized as instrument material to make pockets on the aluminium billets. A few materials can be utilized as device materials however the vast majority of the scientists selected copper as apparatus material since it has high electrical conductivity and financially accessible. Dielectric liquid utilized in this examination is EDM Oil. The reason this oil is to chill off the created heat among apparatus and work-piece and to flush out the eliminated chips during machining. Morphology of machining zone and test arrangement was appeared in fig.3. The parametric arrangement with 4 components (input factors) and 3 levels (Level1, 2 &3) and their reactions are appeared in Table 2. Most importantly, step in EDM in the wake of turning on the force flexibly is setting the boundaries in the control board. The instrument will perform as indicated by the qualities entered in the control board. In the current work, 8 mm pockets with 3 mm profundity were made on the work piece surface.

The instrument moved opposite to the work surface for example in Z-heading. The device is empowered with servo instrument so it can keep up appropriate hole between the apparatus and work surface and return to the first situation after finish of machining. The machined work pieces are

indicated fig 4. To figure MRR and TWR both apparatus work-piece loads are estimated when machining then determined utilizing condition (3) and (4). This was rehashed for all examinations.
For experiments on the produced Aluminium matrix nano composite, the diesink EDM machine was u sed. Composite aluminium specimens, as shown in Fig. 2 were used as workpieces of 10 mm, 40 mm, and 10 mm in length, width and thickness. The copper electrode was used to manufacture pockets on t he aluminium billets as a tool material. Many materials may be used as tool materials, but most researc hers have chosen copper as a tool material because it has high electrical and economic conductivity.E DM Oil is the dielectric fluid used in this experiment. The purpose of this oil is to cool down the heat produced between the instrument and the work-piece and flush out the chips removed during machining. The parametric configuration with 4 factors (input variables) and 3 levels (Level A, B & C) and their responses are shown in Table 2. The parameters in the control panel are first and foremost set in EDM after switching on the power supply. The tool will work in compliance with the values entered on the control panel. 10 mm pockets with a depth of 5mm were formed on the work-piecesurface in the present work.
The instrument shifted in the Z- direction perpendicular to the work surface. The tool is fitted with a servo mechanism so that it can maintain a correct distance between the tool and the work surface and return to the original location after machining is done. The machined work-pieces are shown in Figure **4.** Both work-piece weights are measured before and after machining and then determined using equations (3) and (4) to calculate MRR and TWR. For all tests, this was replicated.

$$MRR = \frac{\text{I.W.W} - \text{F.W.W}}{MT} \left(\frac{gm}{hr}\right) \text{................ Eq. (3)}$$

$$TRW = \frac{\text{I.W.T} - \text{F.W.T}}{MT} \left(\frac{gm}{hr}\right) \text{.................. Eq. (4)}$$

I.W.W- Initial Weight of Work-piece
F.W.W-Final Weight of Work-piece
I.W.T-Initial Weight of Tool
F.W.T-Final Weight of Tool

Fig.3: Experimental setup of EDM machining

IOP Conf. Series: Materials Science and Engineering 1112 (2021) 012017 doi:10.1088/1757-899X/1112/1/012017

Fig .4: Work piece after machining

4. Results and Discussions:

The influence of input variables on response was found out from the data generated by S/N ratio and ANOVA tools. All the responses for respective experimental conditions were shown in the table 2 : data generated by S/N ratio andANOVA tools.

S. No	I	T_{on}	V	P	MRR	S/N for MRR	TWR	S/N for TWR
1	10	100	40	0.3	5.9550	13.9008	0.1011	19.9050
2	10	100	50	0.5	9.8301	18.9193	0.0949	21.4218
3	10	100	60	0.7	9.7326	18.8228	0.2338	13.3998
4	10	200	50	0.3	12.0117	20.8370	0.2652	11.2070
5	10	200	60	0.5	10.8526	19.8710	0.1794	14.4524
6	10	200	40	0.7	12.475	21.1950	0.1225	18.9769
7	10	300	60	0.3	9.3835	18.4685	0.0986	20.1225
8	10	300	40	0.5	9.3601	18.4442	0.1806	15.3604
9	10	300	50	0.7	10.16	18.2338	**0.09**	21.9382
10	20	100	40	0.7	12.6821	21.3504	0.0953	20.4181
11	20	100	50	0.3	22.1204	26.8959	0.1301	17.7145
12	20	100	60	0.5	19.4157	25.7631	0.2831	10.9612
13	20	200	50	0.7	27.0869	28.3284	0.2086	13.6137
14	20	200	60	0.3	23.95	27.2157	0.135	17.3933
15	20	200	40	0.5	21.2	26.5267	0.24	12.3958
16	20	300	60	0.7	23.6712	27.4844	0.2958	10.5800
17	20	300	40	0.3	26.4705	28.4552	0.4294	5.5243
18	20	300	50	0.5	25.44	28.1103	0.524	4.0963
19	30	100	40	0.5	10.54	20	0.424	7.1309
20	30	100	50	0.7	32.124	21.7556	0.212	10.1169
21	30	100	60	0.3	29.5522	29.4118	1.3032	-2.5628
22	30	200	50	0.5	**46.2857**	33.3089	1.0142	-0.9393
23	30	200	60	0.7	40.1276	32.4913	0.4829	8.3383

24	30	200	40	0.3	27.1818	28.3600	1.0185	-0.2441
25	30	300	60	0.5	33.5172	30.5054	0.4844	7.2422
26	30	300	40	0.7	35.3571	30.9695	0.2928	14.2979
27	30	300	50	0.3	37.6545	31.2825	0.6854	2.0982

4.1 Effect of Machine Parameters on MRR

From the S/N ratios response table of MRR shown in Table 3, it is observed that peak current (I) is influencing more on MRR, followed by T_{on}, V and P.
Table 3: Response Table for S/Noise Ratios- MRR (Larger is better)

Level	I (A)	P (kg/cm^2)	T_{on} (μs)	V (V)
1	18.74	24.98	21.87	23.24
2	26.68	24.61	26.46	25.30
3	28.68	24.51	25.77	25.56
Delta	9.93	0.47	4.59	2.31
Rank	1	4	2	3

From the MRR main effects plot shown in fig.5, it was observed that, MRR increases with improved current and voltage values. So, enhancement in I value causes the improvement in intensity of electric discharge. Whenever this high electric discharge pulses impinges on the work piece surface, the K.E. of pulses converts to heat energy. Owing to this, temperature increases at both tool and work piece surfaces and erosion of metal will happen by melting and evaporation andincreasesMRR.Butincreasing of flushing pressure reduces MRR. As at high pressure velocity is low. So, it will not flush out the debris from the tool and work surfaces. Clustering of debris takes place in the gap and reduces the MRR. MRR is directly proportional to T_{on} in that sense MRR should increase with increased T_{on}. But in fig.5(T_{on}) MRR increases and then decreases. From this it can be observed that MRR not only depends on T_{on} but also requires certain I value to remove material on work surface.So, from the main effects plot for S/N ratios we can found the optimal machining conditions and those are I= 30 A, *P* = 0.3 kg/cm2, *Ton* = 200 μs and V = 60 V.

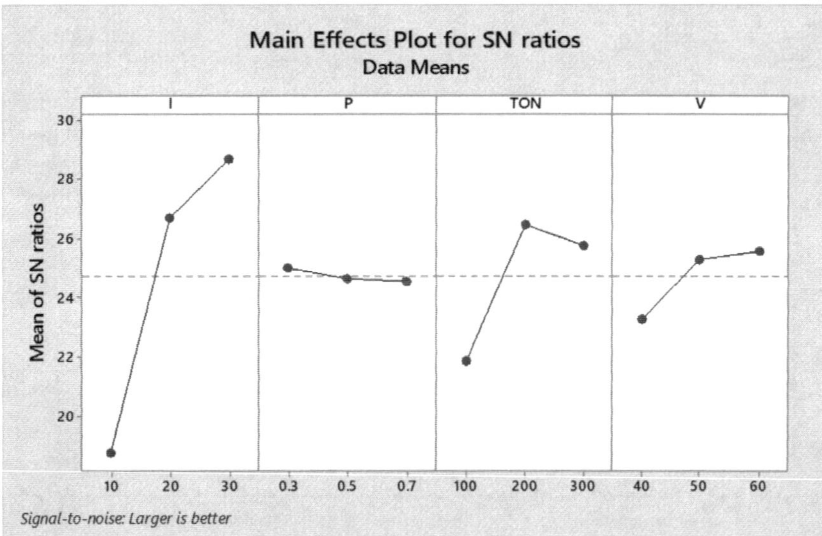

Fig. 5: Main effects plot for MRR

4.2 Effect of Machine Parameters on TWR

Table 4 shows the rankings of process parameters on TWR. It is concluded that I is the most effecting factor followed by P, T_{on} and V.

Table 4: Response Table for S/Noise Ratios-TWR (Smaller is better)

Level	**I (A)**	**P (kg/cm^2)**	**T_{on} (μs)**	**V (V)**
1	17.419	10.127	13.166	12.639
2	12.520	10.235	10.251	11.251
3	5.052	14.629	11.250	11.101
Delta	12.367	4.502	2.590	1.538
Rank	1	2	3	4

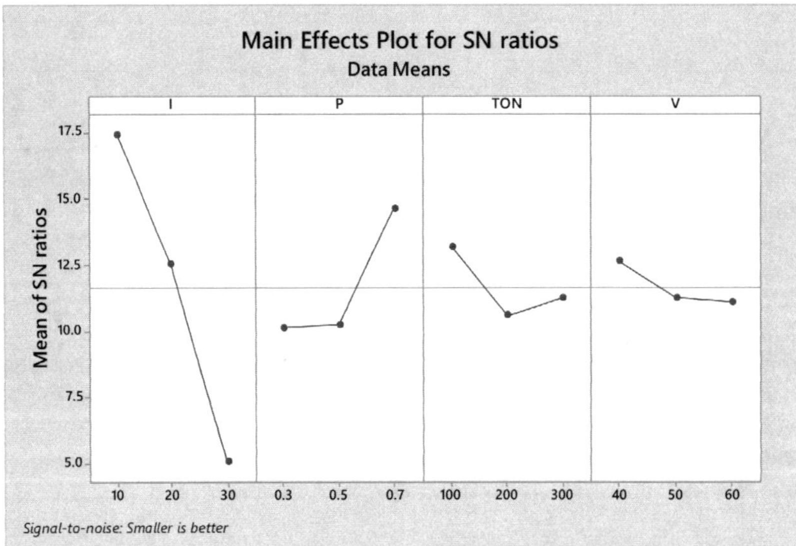

Fig. 6: Main Effect Plot for TWR

Main effects plot for TWR was shown in Fig.6, it can be noticed that TWR decreases linearly with the enhancement in I. Similarly, TWR decreases linearly by increasing voltage factor. Whereas by increasing flushing pressure TWR also increases. But in case of Ton, TWR decreases up to level 2 then it increases TWR. Optimum machining conditions were found from the main effects plot and those are I=30 A, T_{on}=200 μs, P=0.3 kg/cm^2 and V=60 V.

4.3 Analysis of Variance

The most significant factor which affects the output response was found by using ANOVA tool.

ANOVA for TWR

Source	**DF**	**Seq SS**	**Adj SS**	**Adj MS**	**F**	**P**	**Contribution**
I	2	9.2543	9.2543	4.62715	58.92	0.000	52.87%

Ton	2	0.4308	0.4308	0.21542	2.74	0.143	2.46%
P	2	1.5748	1.5748	0.78738	10.03	0.012	9.00%
V	2	0.1716	0.1716	0.08579	1.09	0.394	0.98%
I*Ton	4	2.6422	2.6422	0.66056	8.41	0.012	15.10%
I*V	4	0.2289	0.2289	0.05723	0.73	0.604	1.31%
Ton*V	4	2.7285	2.7285	0.68212	8.69	0.011	15.59%
Error	6	0.4712	0.4712	0.07854			2.69%
Total	26	17.5023	17.5023				100.00%

ANOVA for MRR

Source	**DF**	**Seq SS**	**Adj SS**	**Adj MS**	**F**	**P**	**Contribution**
I	2	6.58647	6.58647	3.29323	71.75	0.000	66.62%
Ton	2	1.46268	1.46268	0.73134	15.93	0.004	14.79%
P	2	0.01458	0.01458	0.00729	0.16	0.857	0.15%
V	2	0.38325	0.38325	0.19163	4.18	0.073	3.88%
I*Ton	4	0.49561	0.49561	0.12390	2.70	0.133	5.01%
I*V	4	0.15429	0.15429	0.03857	0.84	0.547	1.56%
Ton*V	4	0.51424	0.51424	0.12856	2.80	0.125	5.20%
Error	6	0.27537	0.27537	0.04590			2.79%
Total	26	9.88649					100.00%

5.Conclusion

Aluminium MMCs are fabricated successfully by means of stir casting process and machined on Electric Discharge Machine at predefined machining conditions. The main objective of this work is to evaluate the impact of process parameters on MRR and TWR. From this experimental study it is observed that Current (I) and T_{on} are the most influencing parameterson MRR of fabricated Al composites while in the case of TWR, current (I) and interaction of I and T_{on} are the most influencing factors. Optimal parameters for maximizing the MRR and minimizing the TWR are found.

References

[1] Manish Vishwakarma, Vishal Parashar, V.K.Khare, "Advancement in electric discharge machining on metal matrix composite materials in recent: AReview", International Journal of Scientific and Research Publications, Volume 2, Issue 3, 2012.
[2] A. Nag, A.K. Srivastava, A.R. Dixit, S. Chattopadhyaya, A. Mandal, P. Hlavacek, D. Klichova, M. Zelenak, S. HlochLect, Notes in Mech. Eng., Springer, Cham, 2018, https://doi.org/10.1007/978-981-10-5329-0_36.
[3] A.K. Srivastava, A. Nag, A.R. Dixit, S. Tiwari, J. Scucka, M. Zelenak, S. Hloch, P. Hlavacek, J. Manuf. Process. 28 (2017) 11–20.
[4] Tomadi, S.H.; Hassan, M.A.; Hamedon, Z.; Daud, R.; and Khalid, A.G. (2004). Analysis of the influence of EDM parameters on surface quality, material removal rate and electrode wear of tungsten carbide. *Proceeding ofthe International Multi Conference of Engineers and Computer Scientists*, Hong Kong, 1803-1808.
[5] N. Radhika, A. R. Sudhamshu, G. Kishore Chandran, Optimization of Electric Discharge Machining Parameters Of Aluminium Hybrid Composites using Taguchi Method, Journal of Engineering Science and Technology, Vol. 9, No. 4 (2014) 502 – 512
[6] Ashish Kumar Srivastava, Assessment of mechanical properties and EDM machinability on Al6063/SiC MMC produced by stir casting, Materials Today: Proceedings (2019)
[7] Bhaskar Chandra Kandpala, Jatinderkumar, Hari Singh, Machining of aluminium metal matrix composites with Electrical discharge machining - A Review, Materials Today: Proceedings 2 (2015) 1665 – 1671
[8] Kamalkishor G. Maniyar, Dr.Dilip S. Ingole, Investigation of EDM Process parameters for Hybrid Metal Matrix Composites, IOP Publishing, **377** (2018) 012204 doi:10.1088/1757-899X/377/1/012204
[9] ChinmayeeKar, B. Surekha, Hemalatha Jena, s V Choudhury. (2018). Study of influence of process parameters in Electric discharge Machining of Aluminium- red mud metal matrix composites.Procedia Manufacturing 20 (2018) 392-399.
[10] Puri, Y.M.; and Deshpande, N.V. (2004). Simultaneous optimization of multiple quality characteristics of wedm based on fuzzy logic and Taguchi technique.*Proceedings of the Fifth Asia Pacific Industrial Engineering andManagement Systems Conference*, Gold Coast, Australia, 14.18.1- 14.18.12.

IOP Conf. Series: Materials Science and Engineering 1112 (2021) 012018 doi:10.1088/1757-899X/1112/1/012018

MECHANICAL PROPERTIES OF PARTIAL REPLACEMENT OF CEMENT WITH SUGARCANE BAGGASE ASH

Ch Ravi Teja[1], G Nipun[2], A Monica[3]

[1]Assistant Professor, Department of Civil Engineering, V R Siddhartha Engineering College, Kanuru, Vijayawada, India
[2]Structural Engineer, SriSahitha Engineers and Developers, Vijayawada, India
[3]Assistant Professor, Department of Civil Engineering, V R Siddhartha Engineering College, Kanuru, Vijayawada, India
*Corresponding author: chrt@vrsiddhartha.ac.in

Abstract. Current study focuses on mechanical properties of Sugarcane Baggase Ash (ScBA) as a cementetious material and as pore filling material in concrete. With depleting natural resources these industrial waste materials not only helps as replacement material but also provides in numerous alternate benefits such as protecting environment, land fill compounds, saving energy etc. Difficulty of disposing of ScBA is already an environmental hazard. ScBA imparts high strength in early replacements and also reduces concrete permeability, also presence of silica in chemical composition imparts high chloride resistance and corrosion resistance. Mechanical properties were studied for 5%, 10%, 15% and 20% replacement with cement, compressive strength, split tensile and workability were studied and optimum was found to be at 10% replacement. Also durability studies were performed by conducting Rapid Chloride Penetration Test (RCPT) and durability is found to be higher than conventional concrete.

Keywords. Soft soil, coconut coir fibre, Epoxy resin, Stabilization.

1. Introduction

With growing economy and rapid increase in construction, there is high requirement of alternate materials in construction inorder to save the natural resources and thereby reducing impact on environment. In engineering materials applicable in construction concrete is considered as the most dynamic material. In this context there is high need for replacement in concrete mix with natural wastes such as Fly Ash, ScBA, Rice-husk etc[1]. ScBA is considered as waste material in current study as it is widely used in construction industry. it can be used in variety of building materials such as ceramic tiles, cement replacement, fine aggregate replacement etc. due to good pozzolanic properties of ScBA it can be chosen as alternate material in replacement in concrete mixes [2]. India is rich in its agricultural produce and natural fibers, has a large production of sugarcane, these natural fibers provide excellent increase in physical and mechanical properties upon replacement [3]. Baggase waste generated from indian sugar mills is produced in huge proportions and several ways of converting waste to usable material is being studied [9]. Researchers all around world are focusing on various ways of utilizing either industrial or agricultural wastes as raw materials, these waste reusing, ScBA is among such wastes in sugar refining [4]. Baggase Ash was already used to improve the strength parameters of flexible pavements [8] although results may differ when applied the same for structural concrete. Chloride penetration of concrete specimens can be determined by RCPT test, based on Faraday's laws of electrolysis depositions are directly proportional to quantity of electricity considered in coulombs [7]. RCPT tests were conducted to assess the durability aspect of present study as per [6, 7].

Current study aims to evaluate mechanical properties with main focus on strength parameters such as compressive strength, split tensile strength. The workability aspect is also presented in the work. It is focused on 5%, 10%, 15%, 20% replacement of ScBA. Finally durability aspect was studied by performing RCPT on cylindrical specimens.

2. Material Tests

Preliminary tests on all materials Water, Fine Aggregate, Course Aggregate, ScBA were conducted as per codalprovisons.

2.1 Cement &ScB

Ordinary Portland Cement 43 grade is used in current study for replacement studies. Tests on cement were conducted as per guidelines followed by IS 269-2015. Similar tests were conducted on ScBA for obtaining parameters. Table 1 represents the material properties of ScBA where Specific gravity, mean particle size, density, min specific surface area, particle shape and color determined in the preliminary laboratory tests.

Table 1. Material properties of Sugarcane Baggase Ash

S.No	Property	Value
1.	Density	575 kg/m^3
2.	Specific gravity	2.2
3.	Mean particle size	0.1-0.2
4.	Min sp. surface area	2500m^2/kg
5.	Particle shape	Spherical
6.	Colour	Black

Chemical composition of ScBA is taken from [2],Table 2 represents the Chemical composition of ScBA. Main composition is Silicon dioxide comprising of 78.34% of total volume, and other compounds as in Table 2. OPC 43 grade cement when mixed with ScBA with various percentages (5%, 10%, 15%, 20%) with particle size distribution represented in form of graph shown in Figure 1. with all proportions.Tests on cement for OPC 43 grade were performed and the results were presented in table 3.

Table 2. Chemical Composition of ScBA

S.No	Description of properties	Percentage(%)
1.	Silicon dioxide	78.34
2.	Aluminium	8.55
3.	Iron Oxide	3.61
4.	Calcium oxide	2.15
5.	Sodium oxide	0.12

IOP Conf. Series: Materials Science and Engineering 1112 (2021) 012018 doi:10.1088/1757-899X/1112/1/012018

6.	Potassium oxide	3.46
7.	Manganese oxide	0.13

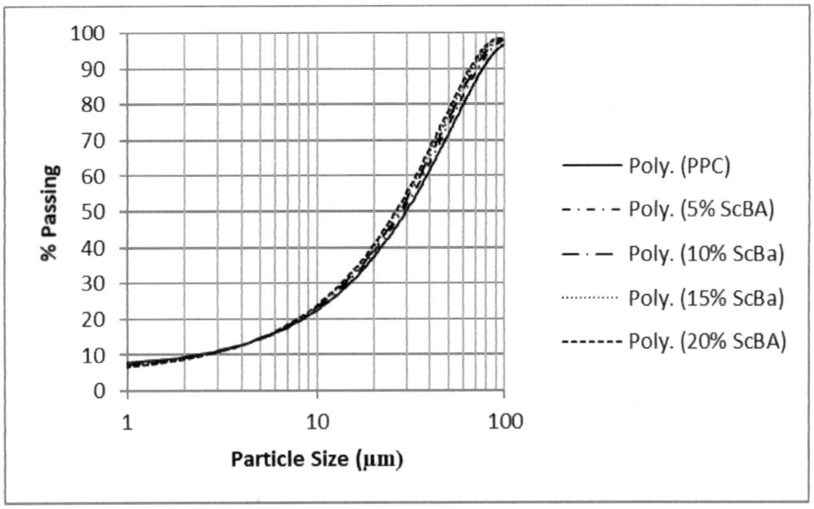

Figure 1. Gradation curve of cement with different % of ScBA

Table 3. Material properties of cement

Test Parameter		**Result**	**Limits as per (IS 269: 2015)**
1.Type of cement chosen		PARASAKTHI CEMENT – OPC 43	
2.Fineness:-		2.2%	< 10%
3.Standard Consistency:-		28.00%	---
4.Setting Time	Initial	120 min	≥30 min
	Final	245 min	≤ 600 min
5.Compressive Strength	3 Days	24.67 MPa	≥ 27.00 Mpa
	7 Days	35.00 MPa	≥ 37.00 Mpa
	28 Days	44.33 MPa	≥ 53.00 Mpa
6.Soundness:-		0.1 mm	< 10mm

2.2 Coarse Aggregates

Coarse aggregates passing through 12.5mm sieve and 20mm sieve were considered accordingly. These respective aggregates are cleaned with water and sun dried before testing. The test procedure was conducted by following guidelines of IS 383-2016 & IS SP 23:1982 the values are represented as shown in Table 4.

Table 4. Sieve analysis and material properties of Coarse aggregates 12.5mm

IS Sieve	**Wt. retained (gms)**	**% of Wt. Retained**	**% of Cumulative Wt. Retained**	**% of wt. passing**	**Grading Limits (IS 383 : 2016)**
63	0	00.00	00.00	100.00	100
40	595	11.90	11.90	88.10	85 - 100
20	4340	86.80	98.70	1.30	0 - 20
10	65	1.30	100.00	0.00	0 - 5
Pan	0	0.00	0.00	0.00	0
2. Aggregate Impact Value :-			21.2%	<30%	IS SP 23 : 1982
3. Aggregate Crushing Value :-			22.8%	<30%	
4. Flakiness :-			11.8%	<30%	
5. Elongation Index :-			13.6%	<30%	
6. Specific Gravity :-			2.71	> 2.6	
7. Water Absorption :-			0.38 %		
8. Bulk density		Loose :-	1430 Kg/Cum		
		Compacted:-	1526 Kg/Cum		

2.3 Fine Aggregates

Fine aggregates are tested as per IS 383 – 2016 and the results were represented in table 5. Fine aggregate was confined to Zone III based on sieve analysis tests. Also the bulk density (loosely compacted and closely compacted), Specific gravity, Fineness modulus were given and values are found to be within limits of codal provisions. It can be seen that maximum percentage of weight passing lies in Zone III.

Table 5. Sieve analysis of Fine Aggregates considered

1. Gradation (Sieve Analysis)									
S.No.	**Sieve Size (mm)**	**Wt. Retained (gms)**	**Wt. Retained (%)**	**Cumulative Wt. Retained (%)**	**% of weight passin g**	**Zone - I**	**Zone-II**	**Zone - III**	**Zone - IV**
1	10	0	0.00	0.0	100.0	100	100	100	100
2	4.75	27	2.7	2.7	97.3	90-100	90-100	90-100	95-100
3	2.36	72	7.2	9.9	90.1	60-95	75-100	85-100	95-100
4	1.18	88	8.8	18.7	81.3	30-70	55-90	75-100	90-100
5	0.6	132	13.2	31.9	**68.1**	15-34	35-59	60-79	80-100
6	0.3	568	56.8	88.7	11.3	5--20	8--30	12--40	15-50
7	0.15	76	7.6	96.3	3.7	0- 10	0- 10	0- 10	0-15
8	0.075	37	3.7	100.0	0.0	0-5	0-5	0-5	0-5
Grading Zone					Zone-III				
2.Fineness Modulus:-					2.70				
3.Specific Gravity:-					2.64		> 2.6		
4.Bulk Density:-			Loose		1620 Kg/Cum				
			Compaction		1756 Kg/Cum				
5.Silt Content:-					1.20%				
6.Average Moisture Content:-					0.62				

Standard consistency of cement with various percentages of ScBA was tested and the values were given in Table 6.

Table 6. Standard consistency of cement with % replacement of ScBA

S.No	**Replacement of bagasse ash in cement (%)**	**Consistency(%)**
1.	0	33
2.	5	34
3.	10	38
4.	15	42
5.	20	46

Tests on water used in mix were tested as per IS10500-2012 and same was utilized in the mix.

2.4 Concrete Mix Preparations

As per 10262-2019 mix for M30 concrete was prepared, cubes of 150x150x150mm dimensions, cylinders of 150mmdia and 300mm height for split tensile strength and durability studies were conducted in accordance with ASTM C1202 (2012). Mix was designed with 5%, 10%, 15%, 20% ScBA to evaluate the strength parameters. Concrete proportions for all five mixes were shown in Table 7. Water cement ratio and other weights were obtained from weight batching from mix specifications.

Table 7. Mix quantities by weights

Type of Mix	w/c ratio	%ScBA	Water (Lt)	Cement (Kg)	FA (Kg)	CA12mm (Kg)	CA20mm (Kg)	ScBA (Kg)
ScBA0	0.30	0	2.43	8.15	10.36	7.77	7.77	0
ScBA5	0.30	5	2.39	7.74	10.30	7.50	7.50	0.407
ScBA10	0.30	10	2.40	7.33	10.32	7.62	7.62	0.815
ScBA15	0.30	15	2.41	6.93	10.35	7.72	7.72	1.225
ScBA20	0.30	20	2.42	6.52	10.34	7.74	7.74	1.630

Note: FA – Fine Aggregate, CA – Coarse Aggregate, ScBA0, 5, 10, 15, 20 – SugracaneBaggase Ash Percentage

3. Results and Discussions

3.1 Workability

Slump for conventional concrete is found to be higher results in higher workability and with different percentages of ScBA slump decreased with increase in ScBA content which results in lower workability for higher percentages of ScBA. Table 8 represents the values of Slump obtained with corresponding replacement of ScBA.

Table 8: Slump values (Workability)

% ScBA	0	5	10	15	20
Slump (mm)	98	92	88	76	60

3.2 Compression Test:

Concrete cubes of 150mm dimensions were cured for 7, 28, 56 days and tested for different proportions of ScBA (upto 20%). Tests were conducted in Compressive Testing Machine (CTM) by following the guidelines provided in IS 516: 1959. From the maximum load bearable by concrete cube specimens compressive strength obtained.

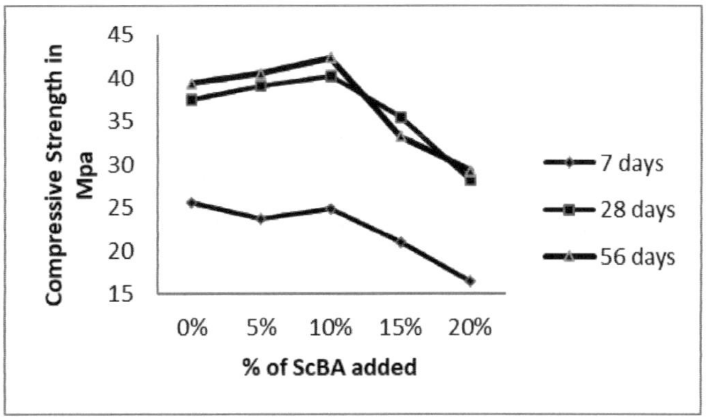

Figure 2. Compressive Strength of concrete with varying percentages of ScBA

From Figure 2 of Compressive strength, it is observed that maximum strength with respect to 7 days curing was found with no replacement of ScBA whereas with age of concrete increasing it is found that strength at 10% replacement of ScBA showed significant improvement and immediate declination in strength with greater percentages of ScBA.

3.3 Split Tensile Strength

Split Tensile Strength was determined by casting cylindrical specimen of 150mm diameter and 300mm height. Results obtained were represented in Figure 3.

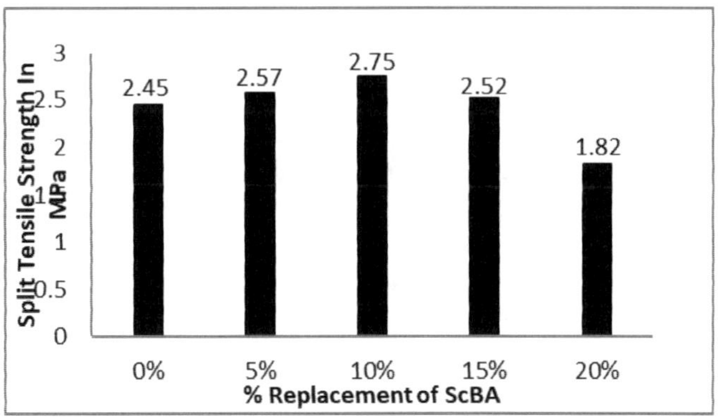

Figure 3: Split Tensile Strength with replacement

From the figure 3 it can observed that maximum tensile strength is at 10% replacement of ScBA which is greater than conventional concrete.

3.4 Rapid Chloride Penetration Test

Rapid Chloride Penetration Test (RCPT) was conducted on 28 and 56 days old concrete in accordance with ASTM C1202 (2012) codal specifications.Test results were represented in below, which the overall chloride ion permeability decreases with increase in ScBA content, from the table it can observed that penetration of chloride ions decreases with increase in ScBA content. It might be due to high amount of Silicon dioxide present in the composition. Also it can be seen that durability is slightly better than conventional concrete. Table 9 represents the chloride ion charge passed.

Table 9. RCPT values of ScBA replaced

ScBA %	28 Days		56 Days	
	Charge-Passed (Coulombs)	**Chloride Ion Permeability**	**Charge-passed (Coulombs)**	**Chloride Ion Permeability**
0	1598	L	1266	L
5	1420	L	1156	L
10	1335	L	1122	L
15	1235	L	1058	L
20	1153	L	1011	L

Note: L – Low, 1000-2000 – Low (According to ASTM C1202 (2012)

4. Conclusions

Mechanical properties and durability based on chloride ion permeability were studied for 0, 5%, 10%, 15%, 20% replacement of ScBA. High compressive strength, Split tensile strength were observed at 10% replacement of ScBA and strength immedietly declined at higher percentages of ScBA. Workability decreases with higher replacement of ScBA although strength decreased. Durability also increases with higher ScBA content compared with durability of Conventional concrete, durability is said to be slightly higher than conventional concrete.

Although it is established that ScBA has decent durability and higher strength at 10% replacement further need to study with closer percentages around 10% which gives the optimum replacement percentage.

IOP Conf. Series: Materials Science and Engineering 1112 (2021) 012018 doi:10.1088/1757-899X/1112/1/012018

References

[1] Akshaydhawan, Nakulgupta, Rajesh goyal, K Ksaxena,2020, Evaluation of mechanical properties of concrete manufactured with fly ash, bagasse ash and banana fibre, *Materials Today: Proceedings*

[2] PoojaJha, A K Sachan, R P Singh, 2020, Agro-waste sugarcane bagasse ash (ScBA) as partial replacement of binder material in concrete, *Materials Today: Proceedings*

[3] G Hemath Kumar, HBabuvishwanath, Rajesh Purohit, PramodSahu, R SRana, 2017 Investigations on Mechanical Properties of Glass and Sugarcane Fiber Polymer Matrix Composites, *Materials Today: Proceedings,* **Volume 4,** Issue 4, Part D, 2017, Pages 5408-5420,

[4] Prashant O Modani, M RVyawahare, 2013, Utilization of Bagasse Ash as a Partial Replacement of Fine Aggregate in Concrete, *Procedia Engineering,* **Volume 51**, Pages 25-29

[5] Elias MolaeiRaisi, JavadVaseghiAmiri, Mohammad Reza Davoodi,2018, Mechanical performance of self-compacting concrete incorporating rice husk ash, *Construction and Building Materials,* **Volume 177**, Pages 148-157

[6] Chikkala RaviTeja, GorantlaNipun, SajjaSatish, 2019, Durability of Robust Self Compacting Hybrid Fiber Reinforced Concrete, *International Journal of Recent Technology and Engineering,* **Volume 8,** Pages 1749-1755

[7] Kang-Shiun Huang, Chung-Chia Yang, 2018, Using RCPT determine the migration coefficient to assess the durability of concrete, *Construction and Building Materials,* **Volume 167**, Pages 822-830.

[8] BManjula Devi, Hemant S Chore, 2020, Feasibility study on bagasse ash as light weight material for road construction, *Materials Today: Proceedings,* **Volume 27**, Part 2, Pages 1668-1673.

[9] DeepankarVarshney, Prasad Mandade, YogendraShastri, 2019, Multi-objective optimization of sugarcane bagasse utilization in an Indian sugar mill, *Sustainable Production and Consumption*, **Volume 18**, Pages 96-114.

COMPARATIVE STUDY ON ASPHALT MIXTURE WITH NANO MATERIALS AND POLYMER

Akhila Bobbili [1], Sai Krishna Kollipara [2], Dr. V. Mallikarjuna[3] and Malathi Narra[4]

[1]Underdraduate student, Department of Civil Engineering, V.R. Siddhartha Engineering College, Vijayawada – 520 007, Andhra Pradesh, India.
[21]Underdraduate student, Department of Civil Engineering, V.R. Siddhartha Engineering College, Vijayawada – 520 007, Andhra Pradesh, India
[3]Associate Professor, Department of Civil Engineering, V.R. Siddhartha Engineering College, Vijayawada – 520 007, Andhra Pradesh, India.
[4]Assistant Professor, Department of Civil Engineering, V.R. Siddhartha Engineering College, Vijayawada – 520 007, Andhra Pradesh, India

ABSTRACT: Pavement materials are crucial factors affecting pavement durability. Now-a-days, there is dire necessity for roads which are more stable and stronger. Due to weathering conditions and heavy traffic, the pavement surfaces are getting deteriorated by rutting, pot holes etc. Among them asphalt is the most sustainable pavement material for construction pavements and can be used for many applications including highways, airport runways, parking lots and drive ways. In order to provide effective durability than that of asphalt, in its original form it has been modified using Nano-materials known as modified asphalt. In this paper we are gone deal with the advances in Nano-materials in hot mix asphalt and also addition of recronfibre and the comparison was made. With the addition of this Recronfibre, there is an improvement in the properties of bitumen like increase in stability value and decrease in the flow value, % of air voids etc. Recronfibre is an artificial material obtained from the polyester and which is also used as a secondary reinforcement for attaining tensile strength. It helps to resist the cracks obtained by the improper laying of pavement surface and heavy loaded vehicles. But whereas, the clay Nano-particles are the primary materials applying in asphalt construction adding Nano- particles like Nano clay, Nano silica and nanotubes in asphalts normally increase the viscosity of asphalt binders and improves the rutting and fatigue resistance of asphalt mixtures. From this the performance of asphalt when treated with Nano-particles and its sustainability compared to that of other pavement materials is examined and studied.

Key words: Modified Bitumen, Flexible Pavement, Polyester [Recron 3s] Fibber, Nano-materials, Rutting, and fatigue resistance.

1. INTRODUCTION:

Generally, roads are basic requirement for transportation facilities. The pavement should be stronger and more stable. But flexible pavements are generally affected to heavy traffic, weathering and geological conditions of the pavement which causes a reduction of quality and performance. In order to overcome from the effects like rutting, pot holes, shrinkage cracks etc., properties of bitumen are improved with addition of fiber and Nano silica.

The fibers used is Polyester (Recron-3S) Fibber. This is an artificial material obtained from polyester. This Fibre helps to resist the cracks obtained by heavy loaded vehicles and any changes occurred due to varying temperatures. This also helps to increase in flexural strength and tensile strength to the pavement. Bitumen is viscous fluid material which consists of binding and adhesive property (which binds all the components in it without any changes in their properties and it is insoluble and acts as a sealant). During the construction of flexible pavements, the bitumen binder is added to increase the life span of the pavement surface. While layingof road, the bitumen and coarse aggregate are mixed together providing good bonding and frictionbetween vehicle wheels and road surface. But the major problem in the bitumen pavements is due to rising high temperatures, the volatile compounds present in the bitumen are evaporated and the bitumen will become hard.

Due to this, when the traffic loads are heavy, the pavement will get deteriorated. During winter rainy seasons, due to presence of moisture content, the pavement may be damaged. To prevent this, Polyester (Recron-3S) fiber is added at certain proportions (3%, 6%, 9%, and 12%) to the bitumen mix. Generally, asphalt pavements can absorb water and losses its strength. But, with the addition of this fiber to these pavements, the strength is not decreased even with the absorption of water and also it with stands to abrasion. Polyester (Recron-3S) fiber in bitumen protects from the fatigue, flow value, cracking and deformation on flexiblepavements.

In recent years, nanotechnology has gradually been incorporated into the field of modified asphalt with various kinds of nanomaterials being used to modify asphalt. There are various nanomaterials which have potential to be used in asphalt modification; such as Nano clay, Nano silica, Nano-hydrated lime, Nano-sized Nano-silica powders, or polymerized powders, fibers, and nanotubes. The recent practices of nanotechnologies are those using Nano-sized particles in so-called nanomaterials and nanometer-size features on integrated circuits. These materials frequently show properties that are completely different from those demonstrated by the same products with larger dimensions. A reason for this can be found in the increased relative surface area of minute particles.

GENERAL:

In 2013, M. Faramarzi et al. Attempted to promote technical characteristics of asphalt mixture using carbon nanotubes as an additive material for bitumen. In this study, marshal test parameters of hot mix asphalt, modified with 0.1, 0.5, and 1% carbon nanotubes, are investigated and compared to conventional asphalt mix. Wet and dry process methods are most practical ways of mixing CNF in AC. It was decided that the best method to adopt for this investigation was dry process. According to the results, the more carbon nanotubes increase, the better asphalt concrete specifications will be. Thus sample containing 0.01 carbon nanotubes by weight of bitumen, has the best results. This sample regarding Marshal Stability 32.53 percent and Marshal Ratio 44.71 percent is higher than the control sample. Also Marshal Flow 8.4 percent and specific gravity 0.68 percent is lower than control sample. It should be noted that despite the decrease in flow, it is still within the permitted regulation. The initial cost of both samples 0.005 and 0.01 is higher than the control sample but for total cost, the amount and type of work should be investigated. When using modified mix, due to its high stability, the lower layer thickness will be less than the control mix and then the amount of total costs willdecrease.

Yadykina, V.V et al. made a guess based on theoretical researches that inserting CNT into bitumen will make a positive impact on its stress-strain properties. In researchers' opinion it may be possible because of aromatic compounds presence in bitumen connected with π-conjugated links system. It will cause the CNT and carbon amorphous particles dispergating improvement and forming theirs Table suspensions in organic solvents. These suspensions will consist of separated carbon amorphous particles and carbon nanotubes, this will cause forming a grid from nanotubes and carbon nanoparticles at asphalt-concrete and asphalt binder with improved stress-strain properties.

Ziari Hasan et al. discussed the mixing conditions of CNT with asphalt and the effect of various mixers on the CNF mixing with asphalt cement (AC). In the first stage, multi wall carbon nanotube, which its weight proportion to asphalt is three percent, and asphalt are mixed together. This mixture is divided into three smaller mixtures (samples). Each sample is mixed by using a different mixer and in different conditions. These mixers are mechanical, high shear and ultrasonic mixers similar results were observed in almost all samples mixed by the mechanical mixer. While there was a concentration of Nano materials in some parts of samples there was almost no Nano materials concentration in some other samples. This indicates that the use of the mechanical mixers, to mix Nano materials, leads to forming a heterogeneous mixture based on the materials dispersion in asphalt. In addition, it cannot prevent Nano materials from becoming agglomerated. In contrast to the provided samples by the mechanical mixer, the dispersion of Nano materials in the whole sample mixed by the high shear mixer was homogeneous. But unfortunately, this mixer, like the previous one, could not disperse Nano materials in Nano scale and these materials were dispersed in micro scale. Therefore, agglomeration of Nano materials in the sample was observed. Consequently, they normally become agglomerated and approach micro scale. Therefore, this mixer cannot separate these Nano materials from each other. For samples mixed by an ultrasonic mixer, Nano materials are observed homogeneously and separately dispersed in asphalt (without agglomeration). Also, CNT particles are separated from each other and easilyrecognizable.

*MojtabaGhasemi*et al. reported the potential benefits of nano-SiO2 powder and SBS for the asphalt mixtures used on pavements. Five asphalt binder formulations were prepared using various percentages of SBS and Nano- SiO2 powder. Then, Marshall Samples were prepared by the modified and unmodified asphalt binders. The results of this investigation indicated that the asphalt mixture modified by 5% SBS plus 2% nanoSiO2 powder could give the best results in the tests carried o6ut in the current study so that this modification can increases physical and mechanical properties of asphalt binder and mixtures. The modified bitumen's were prepared by a high shear mixer. The physical properties of the modified bitumen (such as softening point, penetration and ductility) were measured. The obtained optimum bitumen content for the control mixtures was 6.3% which was used for preparing all other modified mixes in order to maintain consistency throughout the study.

VangariManikanta, Gopi Sai Reddy (2018), states that, the permeable Asphalt Pavements are suited for environment in order to percolate the water into the ground for raising the ground water table which is suited for India. In this process, the pavement loses its strength. Further, it is mentioned that, in such cases, addition of fibers to asphalt mix may give the efficient results like additional strength, resistance to shrinkage splits and stability to the pavement.

P.Rajendhra Kumar & P. Archana (2017), got good results on the performance of expansive soils while adding the Recron-3s fiber. By adding this fiber, there is a reduction in maximum dry density & optimum moisture content and increase in the tensile strength.

G. Pradeep Reddy, G. Tarun Krishna (2017), presented a paper giving clear information about the nature of bitumen with the addition of Recron fibers. And they show the comparison of nominal bitumen and bitumen with Recron fiber based on strength, flow value, voids filled with mineral aggregates, bulk density and unit weight.

G. Jenitha, M. Shenbagavalli(2016), in their work gave the clear information about properties of Recron fiber and advantages. They analyzed that with the addition of Recron, there is a change in strength and workability of concrete.

Muhammad Nawazish Husain, Praveen Agarwal (2015), explained about the effect of adding recron fiber to the silty soils which is situated in Kurukshetra. By adding 0.15% of Recron fiber, the CBR value had increased by 3.5% to 20.2% and with further addition of Recron fiber of 0.3%, 0.4%, and 0.6%, there is a little improvement in CBR value. With 0.15% fiber the UC value also has beenincreased.

FarhadZafari, Mohammed Rahi(2014), exposed their results about by adding Nano silica to the asphalt pavement. This improves the resistance of pavement by the effect of rutting and rheological properties of asphalt binder.

KishanKhunt(2013), concluded about the strength of the soil which may increase to certain extent by using additive materials. Especially, Recron-3s mixed with soil &fly ash mixture gave a wonderful result. Fiber absorbs everything and keeps the road surface intact and many problems are solved like pot holes, cracking and failure of pavement. Fiber plays an important role in increasing the efficiency of pavement and helps to bind the soil under theroad.

2. OBJECTIVES

- To study the properties of bitumen by adding different proportions of Polyester(Recorn-3S).
- To modify the strength of the bitumen mix by using Polyester (Recorn-3S)Fiber.
- To obtain the optimum value of Recron fiber to be mixed with the nominalbitumen.

3. MATERIALSUSED

Aggregate

It is a term for the mineral materials, for example, sand, rock and pulverized stone that are utilized with a coupling medium for such as water, bitumen, Portland Concrete, lime and so forth to shape compound materials. The properties of the aggregates are shown in the Table: 1

Table 1. Aggregate Test Results & Accepted Values

S.NO	Test	Results	Acceptable value
1	Aggregate crushing value	17.95 %	30% (maximum)
2	Impact value	12.725	10-20(strong) 20-30(good)
3	Specific gravity	2.91	2.5-3.0

Polyester (Recron-3S)fiber

Recron fiber is a modified polyester fiber and helps to resist the micro shrinkage cracks caused due to hydration in the bitumen pavement. It also helps to increase flexural strength of the pavement.

Properties:

- Tensile Strength: 4000 – 6000 kg/cm^2
- Cut Length: 6mm or12mm
- Melting Point:>250˚C
- Color:White
- Source: RelianceIndustries
- Cross section:Triangle
- Diameter: 35- 40 micron
- Ignition temperature: >450^0 C

Advantages:

- Improves homogeneity of the concrete by reducing segregation ofaggregates.
- Reduces shrinkage cracks/microcracks
- Abrasion resistance increases by more than25%.
- Impact and shatter resistance increase by100%.
- Increases ductility, compressive, flexural and tensilestrength.
- Reduces water permeability which helps to prevent correction ofprimarysteel.
- Increases energy absorption capability ofconcrete.
- Replaces or reduces non-structural steel in floors, roads andpavement.

Bitumen

The bitumen used in this study is 80/100. This grade is mainly used for good flexible pavements design mixes with chips. This bitumen type has good properties. The main property of this grade is that this has more viscosity as compared to other grades.

Table 2. Bitumen Test Results & Accepted Values

S. No.	Tests	Results	Acceptable Values
1	Ductility	74	65(min)
2	Softening point	47	46(min)
3	Penetration	80/100	80/100
4	Fire	325	235(min)
5	Flash	315	205(min)

Nano-Material

The extremely small things which are studied by Nano technology are known as Nano-materials. Materials with the size of 10^{-9}m order are termed as Nano-materials. They can be of various states such as liquid form and solid. Nano- material used is Nano-silica it is of liquid form and it is brought from the Silica Ceramic Pvt Ltd, a tiles manufacturing company in DubacherlanearChebrolu, Eluru, Krishna district and there Nano-silica is used for fine polishing given to the tiles during manufacturing to make it glossy and give texture to the tiles and helps in increasing the service life ofthem.

Properties:

- A Nano-fluid is intended for use in high voltage engineering is a heat transfer fluid, containing small fraction of Nano-sized fillermaterials.
- These nanoparticles exhibit unique properties, compared to those of the same material at the bulkscale.
- The term Nano-fluid is defined as to a colloid fluid, composed of a liquid phase and dispersed Nano particles insuspension.
- The thermal conductivity was measured in the temperature range 10^0C to 80^0C with up to 0.1% silica Nano particles.

4. METHODOLOGY:

4.1 Using Polyester (Recron-3S)fiber

The Polyester (Recron-3S) fiber is collected and made into pieces of 2 cm length. Then aggregates properties are obtained by conducting various tests such as crushing test, impact test, specific gravity, water absorption and abrasion test.

4.2 Aggregate Impact Test

The aggregates that pass through 12.5 mm and 10 mm with pan below the sieves are taken and the aggregate which retained on 10 mm sieve is collected. 1/3rd of the cylinder is filled with aggregate and tampered with tamper rod for 25 times. Again the cylinder is filled up to 2/3rd level with aggregate and tampered. The cylinder is placed at the bottom of the impact machine and hammer is lifted up to 300 mm and it was dropped freely until 15 blows. After this, the aggregate is passed through 2.36 mm sieve and readings are noted.

4.3 Crushing Test

Aggregates which retain on 10 mm sieve are collected and the cylinder is filled by 1/3 proportion giving tampering of 25 times. Then it is placed in the crushing machine and the load of 50tonnes, 5tonnes per minute is applied. Then, the aggregates are sieved on 2.36mm sieve and readings are noted.

SpecificGravity

Aggregates are collected and placed in the wire basket and the aggregates are immersed into water and kept for 24 hours. After that, the aggregates are cleaned by using cloth and readings are noted.

4.4 Penetration Test

The bitumen sample is heated and then cooled for 24 hours. The bitumen mould is placed on the bottom base of penetration machine and the time is set. The reading is noted from the instrument.

4.5 Softening Point

The bitumen is placed in the rings and fixed well and the ball is placed and the whole apparatus is put in to the water bath and temperature is switched on. At certain temperature, the ball penetrates through the bitumen and touches the bottom the lid.

4.6 Marshall MixDesign

Marshall Mix design is used to find the optimum binder content of the bitumen, Stability, flow value and bulk density for the bitumen content.

- Different types of aggregates are selected for grading by using MORTHtable.
- First, we need to assemble the mould with a base plate and we need to apply sumlubricate.
- Before that we need to sieve the aggregate as per the MORTHtable.
- Then, sample of aggregate is taken as per the MORTH table and heated up to certain temperature by using pan.
- The bitumen content is added to the sample and mixed thoroughly and taken into the mouldimmediately.
- 75 blows are given on both sides of the sample mechanically ormanually.
- The sample is ideally put for 24 hours and then removed from themould.
- The water bath is taken at 60 degrees for the sample and the surface of the sample iscleaned.

Then marshal stability test is done and readings are noteddown.*UsingNano-silica*

- This project aims to study the improvement in properties of modified bitumen, by adding nanomaterial as a percentage ofweight.
- The effect of Nano-materials (Nano-silica) is examined. To Examining the Performance of Hot Mix Asphalt Using Nano-Materials achieve the optimum content, five different percentages of Nano-silica (1, 3, 5, 7, and 9%) will be mixed withbitumen.
- Nano-silica will be mixed with bitumen using two different methods: (i) using the mechanical mixer and/or (ii) using the hand mix as shown in Figure (1); the better method later determined through results.
- A variety of tests are carried out on all samples to study potential improvement. Tests that will be conducted are the penetration test, viscosity test and softening point, ductility test and flash and fire point test. The Marshall test will be carried out on specimen prepared by finding the optimum bitumen content of modified bitumen of dry aggregateweight.
- Unconfined compression tests are also carried out on the optimum percent of the Marshall specimen to obtain the stress-strain curve,accordingly.
- The test procedure will be carried out after mixing bitumen with Nano material (Nano-silica) and filling it in the mould and the properties will bechecked.
- Then the results will be compared for that of normal asphalt to modified bitumen and properties that are enhanced and useful to increase the service life of bituminous pavements will beknown.

5. RESULTS ANDDISCUSSION:

Using Polyester (Recron-3S)fiber

5.1.1 NominalMix

With the utilization of various proportions of aggregates of various sizes, the voids in the mix got reduced giving more stability to the mixture. The various sieve sizes taken for getting the required quantity of aggregate and weights are as follows: e.g.: for 5% of bitumen.

Table 3. Sieve Sizes & Aggregate

Sieve sizes	**Aggregate weight taken**		
	5% of	5.5% of	6% of
10	60	120	60
8	120	120	120
6.3	240	180	120
4.75	240	180	240
2.36	300	300	360
Filler	240	300	300
Total weight	1200	1200	1200

5.1.2 Weights ofMaterial

The bitumen content is taken as per the total weight of the aggregate 1200 grams.

5% of Bitumen = (1202.327*5)/100= **60**grams.
5.5% of Bitumen = (1200*5.5)/100=**66**
grams. 6% of Bitumen = (1200*6)/100 =
72 grams.

The fiber % is taken as per the total bitumen for each sample that is:

Table 4. Fibre % and weights taken

S.NO	BITUMEN (%)	FIBER (%)	WEIGHT OF FIBER TAKEN (gms)
1.	5	3	1.80
		6	3.60
		9	5.40
		12	7.20
2.	5.5	3	1.98
		6	3.96
		9	5.94
		12	7.92
3.	6	3	2.16
		6	4.32
		9	6.48
		12	8.64

5.1.3 Marshal Tests on VariousMixes

Various Mix types, their propositions and the obtained values of unit weight, stability, flow, percentage of air voids and percentage of voids filled by bitumen are as follows

Table 5. Mix Types and the Values Obtained

Mix	**Bitumen (%)**	**Flow (mm)**	**Marshal stability (KN)**	**Unit wt. (g/cm$^{2)}$**	**% of Air Voids**	**% of voids filled**
C	5	4	10.35	2.56	5.72	68.39

	5.5	3.9	12.07	2.57	5.5	76.46
	6	4.1	11.65	2.56	5.64	71.73
M1	5	3.9	13.36	2.54	4.9	71.59
	5.5	3.6	14.58	2.56	4.2	80.07
	6	3.7	12.76	2.54	4	78.21
M2	5	3.5	11.73	2.53	4	75.68
	5.5	3.3	13.57	2.54	3.8	80.32
	6	3.5	12.67	2.53	3.9	76
M3	5	3.2	15.66	2.5	3.01	80.7
	5.5	3	16.79	2.52	2.66	86.79
	6	3.1	16.16	2.47	2.9	83.83
M4	5	4	14.76	2.52	3.4	78.68
	5.5	3.7	15.33	2.54	3.2	83
	6	4.2	13.2	2.53	3.3	81.13

UsingNano-silica

5.2.1. Nominalmix:

The various sieve sizes required for the mix are taken as follows..,

Materials	Aggregate weight taken (gms)		
	5%	6%	7%
12mm	336	336	336
6mm	336	336	336
Stone dust	480	480	480
Cement	48	48	48
Bitumen	60	72	84

5.2.2. Nano-silica addition:

In this project we had followed wet process i.e. Nano-silica is binded or mixed with the bitumen content at different proportions.

The Nano-silica content is increased by decreasing the bitumen content. For 1200 grams, the 5.55% OBC = 66.6 grams

For 1% Nano-silica content

Required Nano-silica content = 0.66 g and then bitumen content =65.94 g

Requirements	**Sam 1**	**Sam 2**	**Sam 3**	**Avg**
Height (cm)	6	6.1	6	6

IOP Conf. Series: Materials Science and Engineering 1112 (2021) 012019 doi:10.1088/1757-899X/1112/1/012019

Diameter (cm)	10.0	10.2	10.1	10.1
Volume	471 cc	498.19 cc	480.46 cc	483.21
Air weight (gm)	1246	1258	1252	1252
Weight in water (gm)	702	698	710	703
Ut. Weight (g/cc)	2.645	2.525	2.605	2.59
Gt	2.39	2.39	2.39	2.39
Gm	2.29	2.24	2.30	2.27
Vv (%)	4.18	6.27	3.76	4.73
Vb (%)	12.10	11.84	12.15	12.03
VMA	16.28	18.11	15.91	16.76
VFB	74.32	65.38	76.36	72.02
Marshall value	350 div	365 div	370 div	361.6 div
Flow value	370 div	390 div	380 div	380 div

For Marshall Value 116 div = 5 KN
for 361.6 div = 1588.58 kg
Flow value for 380 div = 3.8 mm

For 3% Nano-silica content:

Required Nano-silica content = 1.95g and then bitumen content =65.27 g

Requirements	**Sample 1**	**Sample 2**	**Sample 3**	**Average**
Height	5.9 cm	6.1 cm	6.0 cm	6.0
Dia	10.1 cm	10.1 cm	10 cm	10.05
Volume	472.45 cc	488.47 cc	471 cc	477.3
Air weight	1240gm	1253 gm	1249gm	1247
Weight in water	725gm	706gm	712gm	714
Unit weight	2.62 g/cc	2.56 g/cc	2.65 g/cc	2.61 g/cc
Gt	2.40	2.40	2.40	2.4
Gm	2.35	2.29	2.32	2.33
Vv	2.08%	4.58%	3.33%Z	3.33%
Vb	12.17%	11.86%	12%	12.01%
VMA	14.25%	16.44%	15.33%	15.34%
VFB	85.4%	72.14%	78.28%	78.60%
Marshall value	395 div	360 div	390div	381.66div
Flow value	460 div	390 div	430 div	426.67 div

For Marshall Value 381.66 div =1676.84
kg Flow value for 426.67 div =4.26 mm
For 5% Nano-silica content:

Required Nano-silica content = 3.23g and bitumen content = 64.6g

Requirements	**Sample 1**	**Sample 2**	**Sample 3**	**Average**
Height	5.9 cm	6.2 cm	6.1 cm	6.1
Diameter	10.2 cm	10 cm	10 cm	10.06
Volume	481.86 cc	486.7 cc	478.85 cc	482.47
Air weight	1256gm	1240gm	1250gm	1248.66
Weight in water	725gm	730 gm	709gm	721.33
Unit weight	2.60 g/cc	2.54 g/cc	2.61 g/cc	2.58 g/cc
Gt	2.39	2.39	2.39	2.39
Gm	2.36	2.38	2.31	2.353
Vv	1.25%	0.418%	3.34%	1.67%
Vb	12.09%	12.19%	11.83%	12.03%
VMA	13.34%	12.60%	15.17%	13.70%
VFB	90.62%	96.74%	78%	88.45%
Marshall value	505 div	445 div	520 div	490 div
Flow value	250 div	340 div	242 div	278 div

For Marshall Value 490 div = 2152.80
kg Flow value for 278 div = 2.78 mm
For 7% Nano-Silica Content

Required Nano-silica content = 4.47g and bitumen content = 63.93g

Requirements	**Sample 1**	**Sample 2**	**Sample 3**	**Average**
Height	6.1 cm	6.1 cm	6.0 cm	6.05
Diameter	10.1 cm	10 cm	10.1 cm	10.05
Volume (cc)	488.47	478.85	480.46	482.5
Air weight (gm)	1242	1256	1240	1246
Weight in water (gm)	698	715	705	706
Unit weight (g/cc)	2.54	2.62	2.58	2.58
Gt	2.35	2.35	2.35	2.35
Gm	2.28	2.32	2.31	2.30
Vv %	2.98	1.30	1.82	2.03

Vb %	11.55	11.75	11.07	11.67
VMA%	14.53	13.05	13.52	13.70
VFB%	79.49	90.03	86.53	85.35
Marshall value	590 div	680 div	630 div	633.33 div
Flow value	250 div	260 div	270 div	245 div

For Marshall Value 633.33 div = 2782.52
kg Flow value for 255 div = 2.55 mm
For 9% Nano-Silica Content

Required Nano-silica content = 5.69g and bitumen content = 63.27

Requirements	**Sample 1**	**Sample 2**	**Sample 3**	**Average**
Height	6.1 cm	6.1 cm	6.2 cm	6.133
Dia	10.2 cm	10.2 cm	10.1 cm	10.1
Volume	498.19 cc	498.19 cc	496.48 cc	497.62
Air weight	1291gm	1285gm	1280gm	1285.33
Weight in water	720gm	727gm	732gm	726.33
Unit weight	2.59 g/cc	2.57 g/cc	2.57 g/cc	2.57 g/cc
Gt	2.40	2.40	2.40	2.4
Gm	2.26	2.30	2.33	2.296
Vv	5.83%	4.16%	2.91%	4.30%
Vb (%)	10.78	10.97	11.11	10.95
VMA(%)	16.61	15.03	13.88	15.17
VFB(%)	64.90	72.98	80.04	72.64
Marshall value (div)	480	465	500	481.66
Flow value (div)	390	334	382	368.66

For Marshall Value 481.66 div = 2116.16
kg Flow value for 386.66 div = 3.68 mm
5.2.3: Marshall Stability test

The average values that are obtained are..,

Nano-silica (%)	**1**	**3**	**5**	**7**	**9**
Marshall stability (kg)	1588.58	1676.84	2152.80	2782.52	2116.16
Flow value (mm)	3.8	4.26	2.77	2.55	3.68
Air voids (%)	4.73	3.33	1.67	2.03	4.30
Unit weight (g/cc)	2.59	2.61	2.58	2.62	2.57

VFB (%)	72.02	78.60	88.45	85.35	72.64
VMA (%)	16.76	15.34	13.70	13.70	15.17

6. CONCLUSION:

1. The unit weight is more for nominal mix as compared to the modified bitumen whereas in modified bitumen, the unit weight is least for the mix-3proportion.

2. The void percentage is more for the nominal bitumen mix when compared to the modified bitumen mix. The least percentage value of voids is2.66%.

3. Void in the nominal mix is mostly filled by bitumen where as in modified bitumen; the voids are filled by bitumen and fiber. Hence, the bitumen percentage for filling voids is less when compared to nominalmix.

4. The flow value is more for the nominal mix whereas the modified bitumen is having much lesser flow value. In nominal mix, the maximum flow value is 4.1mm where as in modified bitumen mix flow value is3.1mm.

5. The stability value is higher for the Mix-3 proportion than others. The maximum stability value for modified Bitumen is 16.79 KN, whereas for nominal mix the maximum value is 12.07kN.

6) The optimal percentages for Nano-silica is7%.

7) A 7% Nano-silica content decreases the penetration degree by 7.13% in case of mechanical mixing and by 8.1% in case of high shearmixing.

8) A 7% Nano-silica increases softening by 9.52% in case of mechanical mixing and by 11.9% in case of high shearmixing.A 7% Nano-silica increases ductility by 8.33% in case of mechanical mixing and by 10% in case of high shear mixing.Mechanical mixing is better for Nanosilica.

9) Mixing Nano-silica using a hand mixer yields a heterogeneous mixture; Nano-silica agglomerates on mixing through mechanical means giving a homogenous mixture with agglomerationNano-silica.

10) Mixing Nano-silica using a mechanical mixer or hand mixer gives the same results a homogenous mixture with no agglomeration ofNano.

Among the 2 materials that are added, Nano-silica gave the more stability when compared to the fiber. The optimum Bitumen content obtained is 5.55%. For this OBC, different proportions of Nano-silica is added and compared with the bitumen specimens. Nano-silica will increase the melting point of the bitumen. The use of theinnovativetechnologynotonlystrengthenedtheroadconstructionbutalsoincreasedtheroadlifeaswellas will help to improve the environment and creating a source of income. Use of Nano-silica in pavement construction would be a boon for India's hot and extremely humid climate, where temperatures frequently cross 50°C and torrential rains create havoc, leaving most of the roads with big potholes.

REFERENCES:

[1] J.-Y. Yu, P.-L. Cong, and S.-P. Wu, "*Laboratory Investigation on the Properties of Asphalt Modified with Epoxy Resin," Journal of Applied Polymer Science, vol. 113, pp. 3557–3563,2009.*

[2] M. Sienkiewicz, J. K. Lipka, H. Janik, and A. Balas, *Waste Management, vol. 32,2012.*

[3] L.H. Lewandowski, 67(3): 447, July-August, 1994, *Polymer Modification of Paving Asphalt Binders.* Rubber Chemistry andTechnology.

[4] J.A. Shen, 2011, *Pavement Performance of Asphalt and Asphalt Concrete,* China Communication Press, Beijing, (in Chinese).

[5] J.S. Zhang, Z. Li, M.T. Li, J. Xu, W.J. Yin, L. Liu, Highway 8 (2005) 142-146 (inChinese).

[6] Zhanping You, *Nanomaterials in Asphalt Pavements*, Technological University, Department of Civil and Environmental Engineering, Michigan Houghton, Michigan, 49931-1295, USA, DOI:10.6135/Ijprt.Org.Tw/2013.6 (3). Iv, ISSN 1997-1400 Int. J. Pavement Res. Technol.6(3)

[7] Erik Bakker, april 2008, *Nanotechnology and Human Health in the Construction Industry."* The Short Internship in Industry Performed at IVAM BV Amsterdam as Part of the Larger Curriculum for the Degreeof Msc.Nanoscience,

[8] Changing Fang, Ruien Yu, Shaolong Liu, and Yan Li, 2003, *"Nanomaterials Applied in Asphalt Modification,* A Review" Xi'an University of Technology, Xi'an 710048, China, SciverseSciencedirect' J. Mater. Sci. Technol., 29(7),589-594.

[9] M. Faramarzi, M. Arabani, A. K. Haghi, and V. Motaghitalab, 7-8 March, 2013. *"A Study on The Effectsof CNT's on Hot Mix Asphalt Marshal- Parameters"* University of Guilan, 7thSASTtech 2013, Iran, Bandar- AbbasOrganized by Khavaran Institute of Higher Education.

[10] Yadykina, V.V., Akimov, A.E., Spitsyna, N.G., and Lobach, A.S. *"Prospects for the Application of Nano- Carbon Tubes to Improve the Quality of Bitumen and Asphalt":* Innovative Materials and Technologies: Fri. Reports. Intern. Scientific and Practical. / Belgor. State. Tehnol. Univ. ¬ Belgorod State Technological University Publishing House, 2011. - Part 4. – P.P. 306-309.

[11] Ziari Hasan, Rahim-of Kamran, Fazilati Mohammad, Goli Ahmad, and FarahaniHosein, *" Evaluation of Different Conditions on The Mixing Bitumen And Carbon Nano-Tubes",* Iran University of Science And Technology, International Journal of Civil & Environmental Engineering IJCEE-IJENS Vol:12No:06.

[12] MojtabaGhasemia, SeyedMortezaMarandia,B, Majid Tahmooresib, Reza Jalal Kamalia, and Reza Taherzadec,2012, *"Modification of Stone Matrix Asphalt with Nano-Sio2"* Acivil Engineering Department,ShahidBahonar University, Kerman, Iran Binternational Centre For Science, ISSN 2090-4304 Journal of Basic And Applied Scientific Research, J. Basic. Appl. Sci. Res., 2(2)1338-1344.

[13] VangariManikanta, Gopi Sai Reddy, January 2018,"*Construction of Permeable Asphalt Pavement using Recron fiber*", International Journal of Engineering Technology Science & Research (IJETSR), Vol. 05, pp. 1124-1127.

[14] G. Pradeep Reddy, G. Tarun Krishna, K. Hemantha Raja, S. Shahabas, April 2017, "*Design of Polyester (Recron-3s) fiber Reinforced Flexible Pavement*", International Journal of Civil Engineering and Technology (IJCIET), Vol. 08, , pp.1231-1237.

[15] P. Rajendhra Kumar, P. Archana, November 2017, "*Effect of CBR of Black Cotton Soil Reinforcedwith Recron fiber*", Imperial Journal of Interdisciplinary Research (IJIR), Vol. 03, pp.1093-1097.

[16] G. Jenitha, M. Shenbhagvalli,December 2016, "*Experimental Study on Behaviour of Recron fiber Reinforced Concrete*", InternationalJournal of Science & Engineering Research, Vol. 07, , pp.1137-1140.

[17] Remadevi, M. Anjali, G. Pillai, Elizabeth Baby George, Priya Narayan, Sophiya Sunny, March 2014, "*Study of fiber Reinforced Bituminous Concrete*", International Journalof Engineering Research and Development (IJERD), Vol. 10, pp.49-56.

[18] Muhammad Nawazish Husain, Praveen Agarwal, November, 2015, "*Application of Recron-3s fiber in Improving Silty Subgrade Behaviour*", IOSR Journal of Mechanical and Civil Engineering, Vol. 12, pp. 51-

EFFECT OF EPOXY RESIN ON CBR IMPROVEMENT OF SOFT CLAYEY SUBGRADE MIXED WITH COCONUT COIR FIBRES

Raja Veerendra Yenigalla[1], G. V.Slaney[2]

[1]Assistant Professor in Civil Engineering, V R Siddhartha Engineering College, Vijayawada, A.P, India. E-mail: raja.yenigalla@vrsiddhartha.ac.in.

[2]Masters Student, Geotechnical Engineering, V R Siddhartha Engineering College, Vijayawada, A.P, India E-mail: vijayaslaneyias@gmail.com.

*Corresponding Author: raja.yenigalla@vrsiddhartha.ac.in.

Abstract. Road infrastructure is a key factor for the development of an economy. For construction of road pavements, good sub grade soil is a major feature which influences the design, construction and durability of the pavement. Pavements built on soft subgrade soils experience faster degradation due to the low strength and high compressibility of soft soil. Replacement of soft soil with good soil is not sustainable, since it is expensive and also impacts the environment. Stabilization of soft soil is a sustainable alternative which can be used to improve the properties of soft soil in-situ. The use of various additives can improve the properties of soft soils. In this study, the effect of using coconut coir in combination with polymeric resin on the improvement of properties of soft soil is studied. Coconut coir is a waste material which can be used as bio-fibers, but since they are decomposable, are used in combination with polymeric resin. The effect of random percentages of coconut coir along with polymeric resin on the compaction and penetration behaviour of soil is presented.

Keywords. Soft soil, coconut coir fibre, Epoxy resin, Stabilization.

1. Introduction

Pavement service life depends upon the subgrade conditions among many other factors. Subgrade is the bottom most soil layer on which the pavements are constructed. Subgrade should have sufficient strength to bear the traffic loads coming on to it. Soft soils do not have sufficient strength to withstand heavy loads and hence pavements constructed over such soils are subjected to early degradation. Earlier practice of replacing soft soil with good quality soils is not feasible due to the high costs involved in the procurement and transportation of the soil, in addition to the pollution caused during the transportation. The method of treating soft soils at the field itself using additives is called stabilization of soils. In this treatment, the properties of soil are improved by mechanical or chemical means. Over the years various materials have been used for stabilization of weak soils such as clays. A recent trend is to utilize waste materials for modifying soft soils. The advantage of using waste materials in construction includes its low cost involved and a new solution for the waste disposal.

India is one of the leading countries in the production of coconut. Coconut palm is one of the most useful tropical palms and one of the important cottage industries related to coconut palm, is the coir industry. It is estimated that the coir-processing industries in India produce roughly 0.5 million tonnes of coir waste [Leema Peter et al., (2014)] annually. These coir wastes are accumulated along the sides of the production units thus creating environmental solution problems. Open disposal of coir pith into heaps acts as a medium for bacteria growing resulting in poor hygiene of the surrounding areas. A major portion of the coir is utilized as a fuel in domestic stoves which creates air pollution. Also, dumping of coir waste into water bodies like rivers, lakes etc., may affect the life within water bodies. Thus a proper disposal method of coir waste is to be proposed.

Many studies have been conducted relating to the behaviour of soil reinforced with randomly distributed fibers. Leema Peter et al., (2014)examined the improvement of subgrade characteristics of expansive soil stabilized with coir waste by performing laboratory studies. They concluded that the addition of coir pith and coir fibres combinedlycauses a decrease in the Maximum Dry Density(MDD) value and the Optimum Moisture Content(OMC) increases. Mali and Singh (2013) observed that soft soils can be improved with randomly distributed fibers of natural and synthetic types. They proved experimentally that, when loaded, the fibers mobilize tensile resistance, which gives greater strength to the soil. S.Muthu Lakshmi et al., (2018) studied the utilization of coconut coir fiber at different percentages for improving the subgrade strength of clayey sand.From the test results, they observed that on increasing the percentage of coir fiber, OMC is increasing and MDD is decreasing. The Soaked CBR values have also increased. They concluded that use of coir mixed with soil drastically reduces the thickness of subgrade required for pavement construction.

V. Rama Susheel Kumar and J Vikranth (2014) analyzed the characteristics of black cotton soil mixed with fly ash and coconut coir. They concluded that the addition of coconut coir to black cotton soil mixed with Fly ash will result in an increase in compressive strength of more than 200%. M. Mirzababaeia, et al., (2017) studied the effect of fibres and a non-traditional polymer, Poly Vinyl Alcohol (PVA) added in various percentages to soil. They concluded that the addition of fibres and PVA has increased the unconfined compressive strength of soil significantly. R.R Singh & Shelly Mittal (2014) studied the effect ofcoir fibers mixed with soil on the unconfined compressive strength of soil and the CBR values. They concluded that, addition of 1% fibres to soil gives the optimum UCS and CBR values. Nithin and Sayida (2012) observed the properties of silty sand mixed with fly ash and coir fibres and established that discrete and randomly distributed coir fibres are improving the bearing capacity of soil. Parag M. Chaple et al. (2013), focused on the effect of coir on the bearing capacity and settlement of footing using laboratory model tests on square footings supported on highly compressible clayey soil reinforced with randomly distributed coir fibres. By varying the parameters such as thickness of reinforced layer (B, B/2, B/4) and the coir percentages (0.25%, 0.5%, 0.75% and 1.0%), they concluded that the bearing capacity increases significantly when using coir. The use of natural fibers for ground improvement is often advantageous as they are cheap, locally available and eco-friendly

2. Materials

2.1. Soil

Soil was obtained from TTD temple construction site in Venkatapalem near Amaravati, Krishna District of Andhra Pradesh, India. Soil was collected about 3m deep below the ground level to avoid top soil and humus content. The colour of the soil is Blackish brown.The geotechnical properties of the soil are listed in Table 1.

2.2 Coconut coir fibre

Coconut coir Fiber is obtained from the husk of coconut and belongs to the group of hard structural fibers. It is the fibrous portion of the coconut extracted mainly from the green nut. Coir extracted consists of rotting the husk in water and removing the organic material binding the fiber. The long bristle fibers are separated from shorter mattress fibers underneath the skin of nut, a process known as wet milling. The coir fiber is elastic enough to twist without breaking and it holds a curl as though permanently waved.The coir is purchased from a local market in Chirala town of Prakasam District, Andhra Pradesh, India. The Diameter of the coir fibres is 0.5mm. The coir is cut into pieces with length varying from 3cm to 5cm.

2.3 Epoxy Resin

Araldite standard epoxy resin is used as additive. The resin was used in addition to mixing of coconut coir fibres with soil. The Resin is readily available in the market.

Table1. Geotechnical properties of soil

Soil Property	Test Result
Visual Identification	Blackish Brown Clay
Liquid Limit (%)	69.1
Plastic Limit (%)	28.06
Plasticity Index (%)	41.04
Specific Gravity	2.36
Soil type	CH
Optimum Moisture Content (%)	16.1
Maximum Dry Density (g/cc)	1.63
California Bearing Ratio(%)	2.1 (Soaked)

Table2. Properties of coconut coir fiber

Property	Test Result
Length (cm)	3-5
Density (g/cc)	1.40
Breaking Elongation (%)	30%
Diameter (mm)	0.3-0.5

3. Experimental investigation

The soil was dried in oven at temperatures between 80°C and 105°C, pulverised and then sieved through the 425μ sieve for the consistency limits and 4.75mm sieve for the compaction and CBR tests. The coconut coir fibre were cut into 3-5cmlength pieces and separated as single strands. Coconut coir was first added to the soil with different proportions of the total dry weight of soil for the determination of compaction behaviour and CBR test. The coir fibres were added in four proportions,
0.5%, 0.75%, 1.0% and 1.5%. The fibres and soil were hand mixed thoroughly. For each proportion of fibres added, the compaction characteristics were first determined. From the compaction characteristics, the samples were prepared for the CBR test. After determination of the CBR value for soil mixed with fibres, Epoxy resin was added to further improve the soil properties. The epoxy resin was added in two different methods: First by mixing the resin along with the fibres directly in the soil and second by surface application after mixing the fibres with soil. The CBR test was performed on the soil added with fibres and also Epoxy resin.

4. Results and Discussions

The results of experimental investigation on the effect of coconut coir fibres added to soil along with resin are discussed.

IOP Conf. Series: Materials Science and Engineering 1112 (2021) 012020 doi:10.1088/1757-899X/1112/1/012020

4.1. Soil mixed with Coconut coir fibres only

4.1.1. Compaction Characteristics

Compaction tests were carried out on soil mixed with different proportions of coir fibres. Figure1 shows the variation ofmaximum dry density (MDD) for different percentages of coir fibre.

The optimum moisture content (OMC) was found to be increasing with an increase in the percentage of fibres added to the soil. The OMC obtained from the compaction test was used for preparation of the soil samples for CBR test.

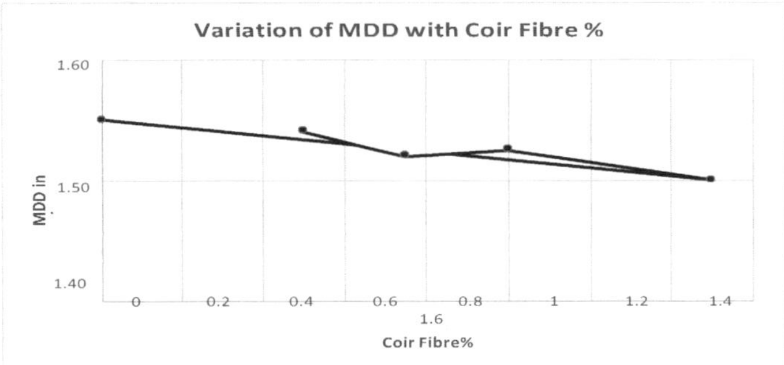

Figure 1. Variation of Soil MDD with different percentages of Coconut Coir Fibre

4.1.2. California Bearing Ratio (CBR)

CBR of the soil is a measure of the strength of the subgrade material to be used as a highway pavement. Soil mixed with coconut coir fibre shows improvement in CBR value of the soil for various percentages of the fibres. The soaked CBR values for soil mixed with different percentages of coconut coir fibres are shown in Figure 2.

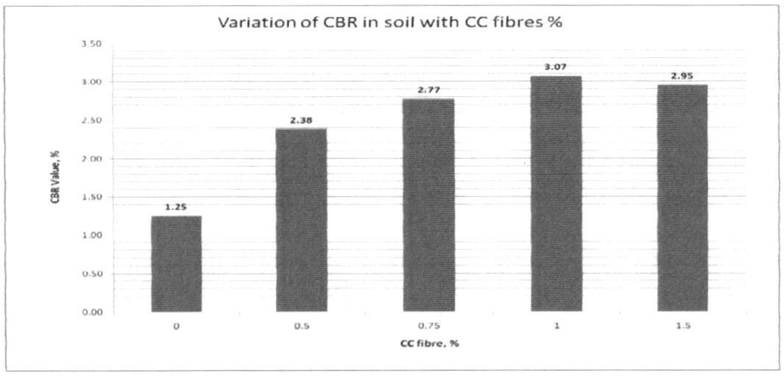

Figure 2. Variation of Soil CBR with different percentages of Coconut Coir Fibre

From the above results it can be seen that, the MDD of the soil has decreased with increase in the percentage of Coconut coir fibres. The CBR value has increased with addition of Coconut coir fibres. The highest increase, in percentage, is seen when 1% coconut coir fibre added to soil

IOP Conf. Series: Materials Science and Engineering 1112 (2021) 012020 doi:10.1088/1757-899X/1112/1/012020

Figure 3. Soaked CBR test performed in the laboratory

4.2. Soil with Coconut coir fibres and Direct Mixing of Epoxy Resin

In the second phase of the experimental investigation, the soil is tested for compaction and CBR values for addition of coconut coir fibres and also epoxy resin. In this phase, Coconut coir fibres percentage is selected as 1%, since the maximum CBR value is obtained at that percentage. The soil is mixed with 1% coconut coir fibres and epoxy resin. The epoxy resin is mixed in three percentages of
0.5%, 1% and 1.5%.

4.2.1. Compaction Characteristics

The results of the compaction tests carried out on soil mixed with 1% coconut coir fibres and different percentages of epoxy resin is are shown in figure 3.

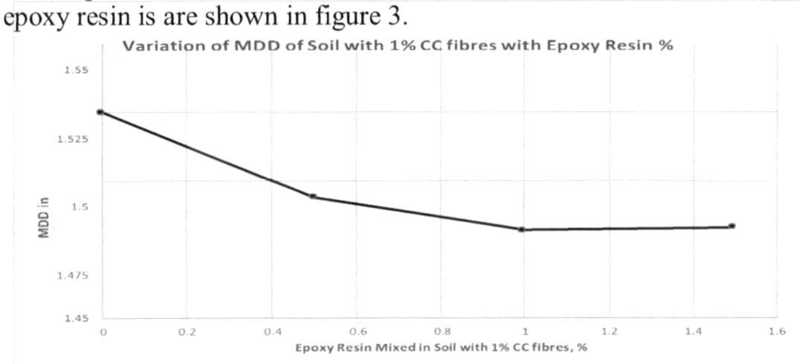

Figure 4: Variation of MDD of soil with 1% CC fibre with Epoxy resin %

4.2.2. California Bearing Ratio (CBR)

The soaked CBR values for soil mixed with 1% coconut coir fires and different percentages of epoxy resin are shown in Figure 4.

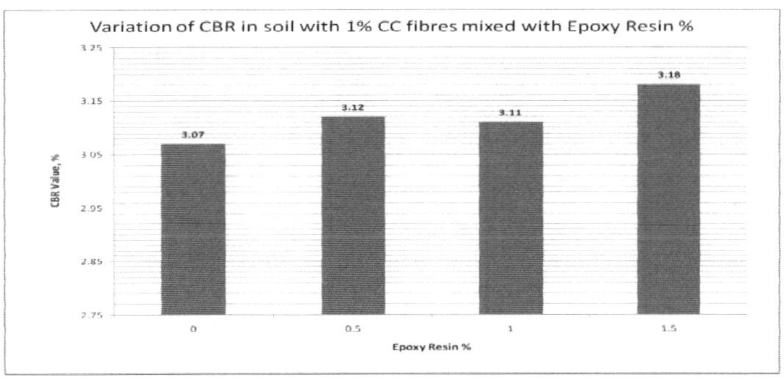

Figure 5. Variation of CBR of soil with 1% CC fibre mixed with Epoxy resin %

From the above results it can be seen that, the MDD of the soil with 1% coconut coir fibres has decreased with increase in the percentage of Epoxy Resin added to the soil by mixing. The CBR value has increased lightly but almost remained the same for increase in percentage of epoxy resin mixed with soil.

4.3. Soil with Coconut coir fibres and Surface Application of Epoxy Resin

In the third phase of the experimental investigation, the soil is tested for compaction and CBR values for addition of coconut coir fibres and surface application of epoxy resin. In this phase, Coconut coir fibres percentage is fixed at 1%, since the maximum CBR value is obtained at that percentage. The soil is mixed with 1% coconut coir fibres and the epoxy resin is applied on the surface. The epoxy resin is applied in the same three percentages of 0.5%, 1% and 1.5%.

4.3.1. Compaction Characteristics

The compaction test was not performed since it is not possible to perform compaction on the sample since the epoxy resin is applied on the top surface. Hence, the compaction test was not performed.

4.3.2. California Bearing Ratio (CBR)

The soaked CBR values for soil mixed with 1% coconut coir fires and different percentages of epoxy resinapplied on the surface are shown in Figure 5.

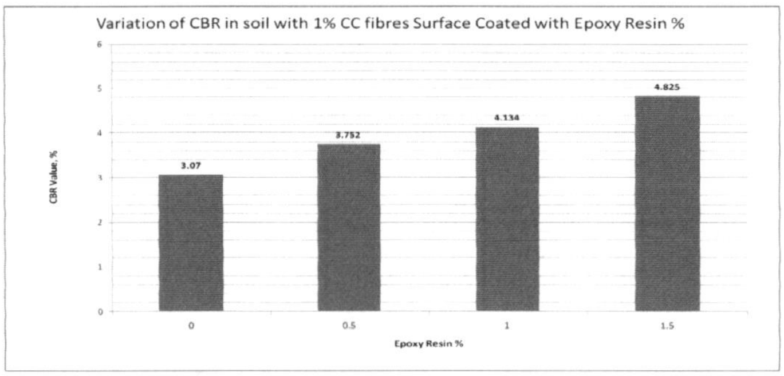

Figure 6. Variation of Soil CBR with 1% CC fibre and Surface Application of Epoxy resin % From

the above results it can be seen that, the CBR value has increased with the increase in the percentage of Epoxy resin applied on the surface of the soil with coconut coir fibres

4.4. Comparison of CBR values of Soil with Coconut fibres and Direct Mixing Vs Surface Application of Epoxy Resin

The CBR values obtained for Soil with 1% Coconut coir fibres and addition of Epoxy Resin via Direct Mixing and Surface Application are compared. The comparison is shown below in Figure 7.

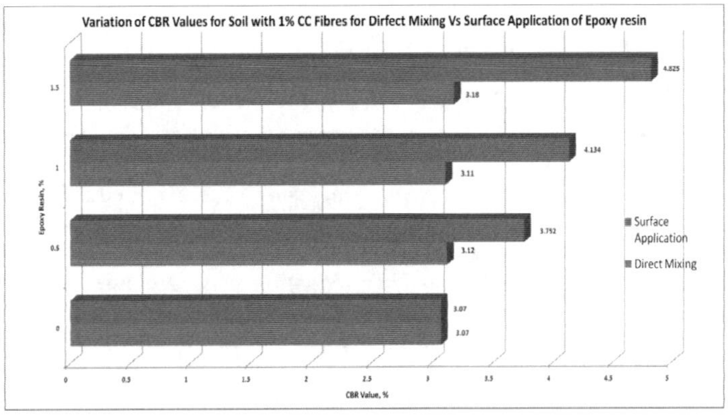

Figure 7. Comparison of CBR values of Soil with Coconut fibres and Direct Mixing Vs Surface Application of Epoxy Resin

5. Conclusions

Laboratory experimentation was done on Soft clayey Subgrade soil to improve the properties. coconut coir fibres and epoxy resin were added to soil in various proportions and the compaction and CBR values of the treated and untreated soil were determined. The Epoxy resin was added in two different methods: Direct Mixing with soil and Surface Application. Based on the test results obtained from the experimental program, the following conclusions are made. Based on the above laboratory investigations conducted on the soil, coconut coir and epoxy resin mixes the following conclusion can be drawn

1. Addition of coconut coir fibres increases the Optimum Moisture Content of the soil but decreases the Maximum Dry Density of the soil.
2. Addition of coconut coir fibres increases the CBR value of the soil. The CBR value has increased by 90%, 122%, 146% and 136% respectively for an addition of 0.5%, 0.75%, 1.0% and 1.5% coconut coir fibres to soil. The highest increase in CBR value was observed when 1% coconut coir fibres are added to soil
3. The Maximum Dry Density values are observed to be decreasing when Epoxy resin is added to soil with coconut fibres by way of Direct Mixing. This can be attributed to the formation of soil balls formed, because of the adhesion properties of the epoxy resin.
4. The CBR value of soil containing 1% coconut coir fibres and different percentages of Epoxy resin added by way of Direct Mixing has almost remained constant. This can also be due to the same reason that the Epoxy resin mixed in soil, forms soil balls and may not be creating bonding between the soil particles. Hence, the CBR value remained same even for an increase in the percentage of Epoxy resin mixed.
5. The CBR value of soil containing 1% coconut coir fibres and different percentages of Epoxy resin added by way of Surface Application has increased. The CBR value has increased by 22%, 35% and 57% respectively for an application of 0.5%, 1% and 1.5% of Epoxy resin on the surface. The highest CBR value was observed when the Epoxy resin applied on the surface is 1.5%
6. From the test results, it can be concluded that Epoxy Resin should be Surface Applied to improve the CBR value rather than Direct Mixing in the soil
7. The combined improvement of the CBR value for 1% coconut coir fibres and 1.5% Epoxy resin applied on the surface is 286%.

References

[1] Leema Peter, P K Jayasree, K Balan, Alaka Raj S, 2016. Laboratory investigation in the improvement of subgrade characteristics of expansive soil stabilised with coir waste, *11th Transportation Planning and Implementation Methodologies for Developing Countries TPMDC,* 10-12 December, Mumbai, India.

[2] Vivi Anggraini *et, al.,* 2016 Reinforcement benefits of Nano modified coir fibre in lime treated marine clay. *Journal of Materials in Civil Engineering.* 28(6), 06016005.

[3] Humphrey Danso, Dorothy Manu, 2020. Influence of coconut fibres and lime on the properties of soil-cement mortar. *Case Studies in Construction Materials.* Volume 12, e00316.

[4] V Rama Susheel Kumar and J Vikranth, 2015 Application of coconut coir and flyash in subgrade strengthening. *International Journal of Engineering and Science.*3:2319-1805, 48-54.

[5] D.Verma & P.C.Gope, 2015. 10 - The use of coir/coconut fibres as reinforcements in composites. *Biofibre Reinforcements in Composite Materials,* 285-319.

[6] S.Muthu Lakshmi *et. al.,* 2018 Utilization of coconut coir fibre for improving subgrade strength characteristics of clayey sand. *International Research Journal of Engineering and Technology.* 5 (4).

[7] Nithin S. and Sayida M. K, 2012 Stabilization of silty sand using flyash and coir fibre. *Advances in civil engineering.*

[8] M.Mirzababaei *et. al.,* 2017 Polymers for stabilization of soft clay soils. *Transportation Geotechnics and Geoecology TGG,* 189 25-32.

[9] B.M Lekha, S.Goutham, A.U.R Shankar, 2015. Evaluation of laterite soil stabilized with areca nut coir for low volume pavements. *Transportation Geotechnics,* 2:20–29.

[10] KR Lekha and BG Sreedevi, 2006. Coir fibre for the stabilisation of weak subgrade soils. *Highway Engineering Lab, NATPAC,* Thiruvananthapuram, 10.

[11] Suchit Kumar Patel & Baleshwar Singh, 2014. Unconfined compressive strength behaviour of fibre-reinforced lateritic soil. *Journal of Civil Engineering and Environmental Technology,* 1 (4), 93–98.

[12] G. L. SivakumarBabu and A. K.Vasudevan, 2008. Strength and stiffness response of coir fibre reinforced tropical soil. *Journal of Materials in Civil Engineering,* 20 (9), 571–577.

[13] Sujatha ER, Geetha A.R, Jananee, R, Karunya, SR, 2018. Strength and mechanical behaviour of coir reinforced lime stabilized soil. *Geomechanics and Engineering,* 16 (6): 627–634.

[14] M.R. Hall, K.B.Najim, and P.KeikhaeiDehdezi, 2012. Soil stabilisation and earth construction: materials, properties and techniques. *Modern Earth Buildings,*222-255.

[15] Dr.Sowmya and Sachin D., 2016. Effect of coconut coir fibre on Black cotton soil Blended with fly ash, *International Journal of Engineering Research and Technology (IJERT),***5**: 2278-0181.

[16] I.Z. Bujang, Awang M. K. and Ismail A. E., 2007. Study on the Dynamic Characteristicof Coconut Fibre Reinforced Composites. *Regional Conference on Engineering Mathematics, Mechanics, Manufacturing & Architecture, Malaysia.*

[17] ChapleParag M and Dhatrak A.I., 2010. Performance of Coir fibre Reinforced Clayey Soil. *International Journal of Engineering and Science (IJES)* 2 (4):54-64.

[18] HejaziSayyed Mahdi, *et. al.,* 2012A simple review of soil reinforcement by using natural and synthetic fibres, *Construction and Building Materials* 30:100-116.

[19] KharadeAmit S, et al, 2014 Waste product 'bagasse ash' from sugar industry can be used as stabilizing material for expansive soils IJRET*: International Journal Of Research In Engineering And Technology (IJRET)* 3: 2319-1163.

[20] R.R Singh and Mittal Shelly, 2014 Improvement of local subgrade soil for road construction by the use of coconut coir fibre. *International Journal of Research in Engineering and Technology (IJRET),* 3 (3).

[21] Muligan L. and Elango R., 2012. Studies on the microbial biodegradation of coir pith. *International Journal of Microbiology and Bioinformatics.*

[22] Singh. H. P. 2013. Strength and stiffness response of Itanagar fly ash reinforced with coir fiber. *International Journal of Innovative Research in Science, Engineering and Technology,* **2**(9):4500-4509.

[23] O. S. Enokela and P. O. Alada, 2012 Strength Analysis of Coconut Fiber Stabilized Earth for Farm Structures, *International Journal of Advancements in Research & Technology,* 1 (2): 1-7.

[24] Kundan Meshram, et al., 2013 "Application of coir geotextile in rural roads construction on black cotton soil subgrade" *International journal of engineering and innovative technology,* 3(4), 264-268.

[25] Dharmendra Kumar, et al., 2015 "California Bearing Ratio Variations in Soil Reinforced with Natural Fibres (A Case Study Bhopal Bypass Road)," *International Journal of Emerging Technologies*, 6 (2): 95-104.

[26] Ajayi,M.A, et al., 1991 "Epoxy –resin based chemical stabilization of fine, poorly graded soil system". *Transportation research record 1295. National research council, Washington DC*

STRENGTHENING OF FLEXIBLE PAVEMENT BY ADDING RECRON FIBRE

Smt. MalathiNarra ,Dr. V. Mallikarjuna[2], Ch. Revathi[3], U. Srikanth[3]

[1]Assistant Professor, Department of Civil Engineering, V.R. Siddhartha Engineering College, Vijayawada – 520 007, Andhra Pradesh, India.
[2]Associate Professor, Department of Civil Engineering, V.R. Siddhartha Engineering College, Vijayawada – 520 007, Andhra Pradesh, India.
[3]Graduate Student, Department of Civil Engineering, V.R. Siddhartha Engineering College, Vijayawada – 520 007, Andhra Pradesh, India.
[4]Graduate Student, Department of Civil Engineering, V.R. Siddhartha Engineering College, Vijayawada – 520 007, Andhra Pradesh, India.
**Corresponding Author: malli.bharghava@gmail.com*

Abstract. Now-a-days, there is dire necessity for roads which aremore stable and stronger. Due to weathering conditions and heavy traffic, the pavement surfaces are getting deteriorated by rutting, pot holes etc. With the addition of this Recronfiber, there is an improvement in the properties of bitumen like increase in stability value and decrease in the flow value, % of air voids etc. Recron fiber is an artificial material obtained from the polyesterand which is also used as a secondary reinforcement for attainingtensile strength. It helps to resistthe cracks obtained by the improper laying of pavement surface and heavy loaded vehicles.
In the present study, various proportions of Recron fiber such as 3%, 6%, 9% and 12%are added to Bitumen to prevent the deterioration of pavement surface andvarious tests areperformed on aggregate and bitumen. From the results, it is found that, the flow value and % of air voids decrease and stability value increases. Hence, it is proved that, it protects the pavement surface from the effect of fatigue, cracking and deformation.
The obtained results show that 9% is the optimum percentage of Recron fiber to be added to the bitumen. For nominal bitumen mix, the maximum stability value is 12.07 kN and for modified bitumen (Recron added) mix the maximum stability value is 16.79 kN.

Key words. Modified Bitumen, Flexible Pavement, Polyester [Recron 3s] Fiber.

1.Introduction

Generally, roads are basic requirement for transportation facilities. The pavement should be stronger and more stable. But flexible pavements are generally affected to heavy traffic, weathering and geological conditions of the pavement. In order to overcomefrom these effectslike rutting, pot holes, shrinkage cracks etc., properties of bitumen are improved with addition of fiber. One of suchfibers isPolyester (Recron-3S) Fiber. This is an artificial material obtained from polyester. This Fiber helps to resist the cracks obtained by heavy loaded vehicles and any changes occurred due to varying temperatures. This also helps to increase in flexural strength and tensile strength to the pavement. Bitumen is viscous fluid material which consists ofbinding andadhesive property (which binds all the components in it without any changes in their properties and it is insoluble and acts as a sealant).
During the construction of flexible pavements, the bitumen binder is added to increase the life span of the pavement surface. While laying of road, the bitumen and coarse aggregate are mixed together providing good bonding and friction between vehicle wheels and road surface. But the major problem in the bitumen pavementsis due to rising high temperatures, the volatile compounds present in the bitumen are evaporated and the bitumen will become hard.Due to this, when the traffic loads are heavy, the pavement will get deteriorated. During winter rainy seasons, due to presence of moisture content,the pavement may be damaged. To prevent this,Polyester (Recron-3S) Fiber is addedat certainproportions (3%, 6%, 9%, 12%) tothe bitumen mix. Generally, asphalt pavements can absorb water and losses its strength. But,with the addition of this Fiber to these pavements, the strength is not

decreased even with the absorption of water and also it withstands to abrasion. Polyester (Recron-3S) Fiber in bitumen protects from the fatigue, flow value, cracking and deformation on flexible pavements.

1.1 General

Vangari Manikanta, Gopi Sai Reddy (2018), states that, the permeable Asphalt Pavements are suited for environment in order to percolate the water into the ground for raising the ground water table which is suited for India. In this process, the pavement loses its strength. Further, it is mentioned that, In such cases, addition of fibers to asphalt mix may give the efficient results like additional strength, resistance to shrinkage splits and stability to the pavement.

P. Rajendhra Kumar & P. Archana (2017), got good results on the performance of expansive soils while adding the Recron-3s fiber. By adding this fiber, there is a reduction in maximum dry density &optimum moisture content and increase in the tensile strength.

G. Pradeep Reddy, G. Tarun Krishna (2017), presented a paper giving clear information about the nature of bitumen with the addition of Recron fibers.And they show the comparison of nominal bitumen and bitumen with Recron fiber based on strength, flow value, voids filled with mineral aggregates, bulk density and unit weight.

G. Jenitha, M. Shenbagavalli (2016), in their work gave the clear information about properties of Recron fiber and advantages. They analyzed that with the addition of Recron, there is a change in strength and workability of concrete.

Muhammad Nawazish Husain, Praveen Agarwal (2015), explained about the effect of adding recron fiber to the silty soils which is situated in Kurukshetra. By adding 0.15% of Recronfiber,the CBR value had increased by 3.5% to 20.2% and with further addition of Recron fiber of 0.3%, 0.4%, and 0.6%,there is a little improvement in CBR value. With 0.15% fiber the UC value alsohas been increased.

FarhadZafari, Mohammed Rahi (2014),exposed their results about by adding Nano silica to the asphalt pavement. This improves the resistance of pavement by the effect of rutting and rheological properties of asphalt binder.

KishanKhunt (2013), concluded about the strength of the soil which may increase to certain extent by using additive materials. Especially, Recron-3s mixed with soil &fly ash mixture gave a wonderful result. Fiber absorbs everything and keeps the road surface intact and many problems are solved like pot holes, cracking and failure of pavement. Fiber plays an important role in increasing the efficiency of pavement and helps to bind the soil under the road.

2. OBJECTIVES

- To study the properties of bitumen by adding different proportions of Polyester (Recorn-3S).
- To modify the strength of the bitumen mix by using Polyester (Recorn-3S) Fiber.
- To obtain the optimum value of Recron fiber to be mixed with the nominal bitumen.

3. Materials Used

3.1. Aggregate

It is a term for the mineral materials, for example, sand, rock and pulverized stone that are utilized with a coupling medium for such as water, bitumen, Portland Concrete, lime and so forth to shape compound materials. The properties of the aggregates are shown in the Table: 1

Table 1. Aggregate Test Results& Accepted Values

S.No.	Test	Results	Acceptable Value
1	Aggregate crushing value	17.95%	30% (maximum)

2	Impact value	12.725	10-20(strong) 20-30(good)
3	Specific gravity	2.91	2.5-3.0

3.2. Polyester (Recron-3S) Fiber

Recron Fiber is a modified polyester Fiber and helpsto resist the micro shrinkage cracks caused due to hydration in the bitumen pavement. It also helps to increase flexural strength of the pavement.

Properties

- Tensile Strength: 4000 – 6000kg/cm^2
- Cut Length: 6mm or 12mm
- Melting Point: >250°C
- Color: White
- Source: Reliance Industries
- Cross section: Triangle
- Diameter: 35-40 micron
- Ignition temperature: >450^0 C

Advantages

- Improves homogeneity of the concrete by reducing segregation of aggregates.
- Reduces shrinkage cracks/micro cracks
- Abrasion resistance increases by more than 25%.
- Impact and shatter resistance increase by 100%.
- Increases ductility, compressive, flexural and tensile strength.
- Reduces water permeability which helps to prevent correction of primary steel.
- Increases energy absorption capability of concrete.
- Replaces or reduces non-structural steel in floors, roads and pavement.

3.3. Bitumen

The bitumen used in this study is 80/100. This grade is mainly used for good flexible pavements design mixes with chips. This bitumen type has good properties. The main property of this grade is that this has more viscosity as compared to other grades.

Table 2. Bitumen Test Results & Accepted Values

S.No.	Tests	Results	Acceptable Values
1	Ductility	74	65(min)
2	Softening point	47	46(min)
3	Penetration	80/100	80/100
4	Fire	325	235(min)
5	Flash	315	205(min)

4. Methodology

The Polyester (Recron-3S) Fiber is collected andmade into pieces of 2cm length. Then aggregates properties are obtained by conducting various tests such as crushing test, impact test, specific gravity, water absorption and abrasion test.

4.1. Aggregate Impact Test

The aggregates that pass through 12.5 mm and 10 mm with pan below the sieves are taken andthe aggregate which retained on 10 mm sieveis collected.1/3rd of the cylinder is filledwith aggregate and tampered with tamper rod for 25 times.Again the cylinder is filled upto 2/3rd level with aggregate and tampered.The cylinder is placed at the bottom of the impact machine and hammer is lifted up to 300 mm and it was dropped freely until 15 blows. After this, the aggregate is passed through 2.36mm sieve and readings are noted.

4.2. Crushing Test

Aggregates which retain on 10 mm sieve are collected and the cylinder is filled by 1/3 proportion giving tampering of 25 times. Then it is placed in the crushing machine and the load of 50tonnes, 5tonnes per minute is applied. Then, the aggregates are sieved on 2.36mm sieve and readings are noted.

4.3. Specific Gravity

Aggregates are collected and placed in the wire basket and the aggregates are immersed into water and kept for 24 hours. After that,the aggregates are cleaned by using cloth and readings are noted.

4.4. Penetration Test

The bitumen sample is heated and then cooled for 24 hours.The bitumen mould is placed on the bottom base of penetration machine and the time is set. The reading is noted from the instrument.

4.5. Softening Point

The bitumen is placed in the rings and fixed well and the ball is placed and the whole apparatus is put in to the water bath and temperature is switched on. At certain temperature, the ball penetrates through the bitumen and touches the bottom the lid.

4.6. Marshall Mix Design

Marshall Mix design is used to find the optimum binder content of the bitumen, Stability, flow value and bulk density for the bitumen content.

- Different types of aggregates are selected for grading by using MORTH table.
- First, we need to assemble the mould with a base plate and we need to apply sum lubricate.
- Before that we need to sieve the aggregate as per the MORTH table.
- Then,sample of aggregate is taken as per the MORTH table and heated up to certain temperature by using pan.
- The bitumen content is added to the sample and mixed thoroughly and taken into the mould immediately.
- 75 blows are given on both sides of the sample mechanically or manually.
- The sample is ideally put for 24 hours and then removed from the mould.
- The water bath is taken at 60 degrees for the sample and the surface of the sample is cleaned.
- Then marshal stability test is done and readings are noted down

IOP Conf. Series: Materials Science and Engineering 1112 (2021) 012021 doi:10.1088/1757-899X/1112/1/012021

5. Results & Discussions

5.1. Nominal Mix

With the utilization of various proportions of aggregates of various sizes, the voids in the mix got reducedgivingmore stability to the mixture. The various sieve sizes taken for getting the required quantity of aggregate and weights are as follows: e.g.: for 5% of bitumen.

Table 3. Sieve Sizes &Aggregates.

Sieve Sizes (mm)	Aggregates Weight Taken		
	5% of Bitumen	5.5% of Bitumen	6% of Bitumen
10	60	120	60
8	120	120	120
6.3	240	180	120
4.75	240	180	240
2.36	300	300	360
Filler	240	300	300
Total weight	1200	1200	1200

5.2. Weights of Material

The bitumen content is taken as per the total weight of the aggregate 1200 grams.
5% of Bitumen=(1202.327*5)/100= 60grams.
5.5% of Bitumen=(1200*5.5)/100=66 grams.
6% of Bitumen = (1200*6)/100 = 72 grams.
The fiber % is taken as per the total bitumen for each sample that is:

Table 4. Fiber % and weights taken

S. No.	Bitumen (%)	Fiber (%)	Weight of fiber taken(gms.)
1.	5	3	1.80
		6	3.60
		9	5.40
		12	7.20
2.	5.5	3	1.98
		6	3.96
		9	5.94
		12	7.92
3.	6	3	2.16
		6	4.32
		9	6.48
		12	8.64

5.3. Marshal Tests on Various Mixes

Various Mix types, their propositions and the obtained values of unit weight, stability, flow, percentage of air voids and percentage of voids filled by bitumen are as follows

Table 5.Mix Types and the Values Obtained

Mix	Bitumen (%)	Flow (mm)	Marshal stability (KN)	Unit wt. (g/cm$^{2)}$	% of Air Voids	% of voids filled
C	5	4	10.35	2.56	5.72	68.39
	5.5	3.9	12.07	2.57	5.5	76.46
	6	4.1	11.65	2.56	5.64	71.73
M1	5	3.9	13.36	2.54	4.9	71.59
	5.5	3.6	14.58	2.56	4.2	80.07
	6	3.7	12.76	2.54	4	78.21
M2	5	3.5	11.73	2.53	4	75.68
	5.5	3.3	13.57	2.54	3.8	80.32
	6	3.5	12.67	2.53	3.9	76
M3	5	3.2	15.66	2.5	3.01	80.7
	5.5	3	16.79	2.52	2.66	86.79
	6	3.1	16.16	2.47	2.9	83.83
M4	5	4	14.76	2.52	3.4	78.68
	5.5	3.7	15.33	2.54	3.2	83
	6	4.2	13.2	2.53	3.3	81.13

5.4. Graphical Representation of Marshall Teston Various Mixes

Comparison of nominal bitumen and fiber added bitumen with various % (i.e., 3%, 6%, 9%, and 12%) are presented in figs 5.4 (a) to 5.4 (e).

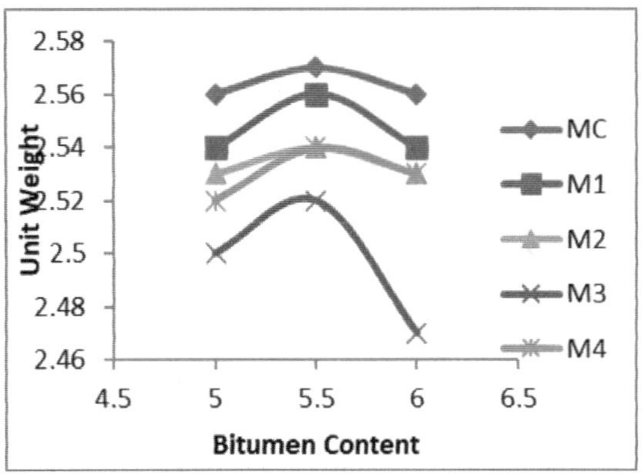

Figure 5.4(a). Unit Weight Vs Bitumen Content

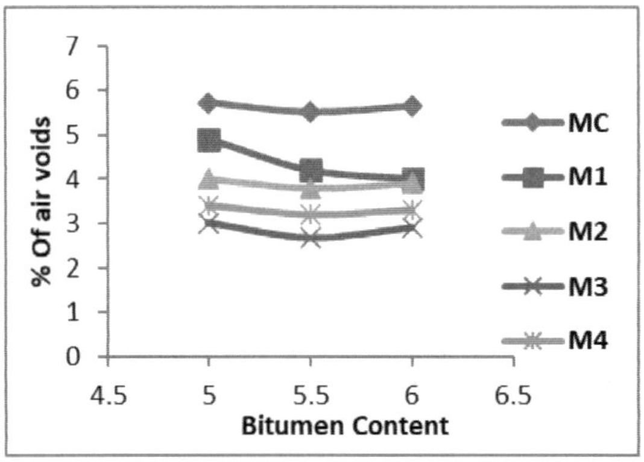

Figure 5.4(b). % of Air Voids Vs Bitumen Content

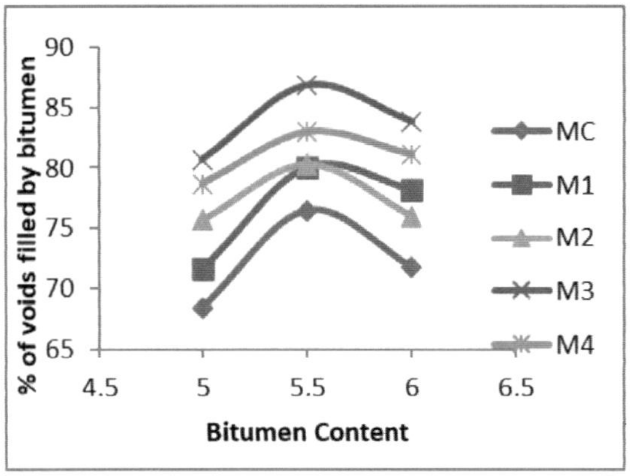

Figure 5.4(c). % of Voids Filled by Bitumen Vs Bitumen Content

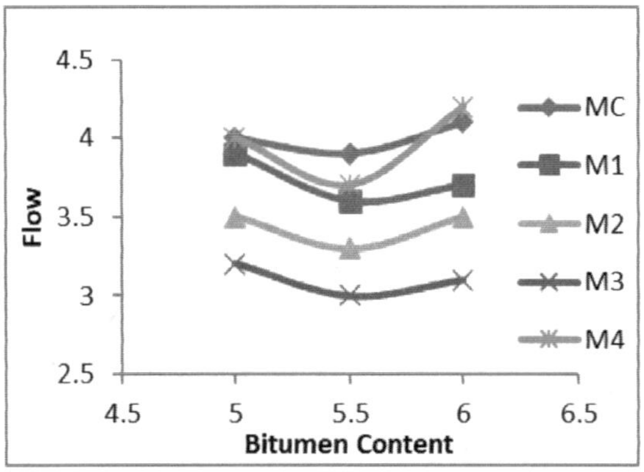

Figure 5.4(d). Flow Vs Bitumen Content

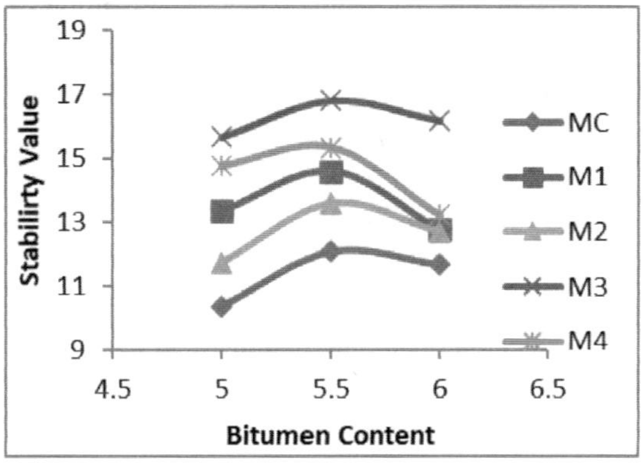

Figure 5.4(e). Stability Vs Bitumen Content

6. CONCLUSIONS

1. The unit weight is more for nominal mix as compared to the modified bitumen whereas inmodified bitumen, the unit weight is least for the mix-3 proportion.
2. The void percentage is more for the nominal bitumen mix when compared to the modifiedbitumen mix. The least percentage value of voids is 2.66%.
3. Void in the nominal mix is mostly filled by bitumen where as in modified bitumen; the voids are filled by bitumen and fiber. Hence, the bitumen percentage for filling voids is less when compared to nominal mix.
4. The flow value is more for the nominal mix whereas the modified bitumen is having much lesser flow value. In nominal mix, the maximum flow value is 4.1mm where as in modified bitumen mix flow value is 3.1mm.
5. The stability value is higher for the Mix-3 proportion than others. The maximum stability value for modified Bitumen is 16.79 kN,whereas for nominal mix the maximum value is 12.07kN.

7. REFERENCES

[1] VangariManikanta, GopiSaiReddy, "*Construction of Permeable Asphalt Pavement using Recron Fiber*", International Journal of Engineering Technology Science &Research (IJETSR), Vol. 05, January 2018, pp. 1124-1127.

[2] G. Pradeep Reddy, G. Tarun Krishna, K. Hemantha Raja, S. Shahabas, "*Designof Polyester (Recron-3s) Fiber Reinforced Flexible Pavement*", International Journal of Civil Engineering and Technology (IJCIET), Vol. 08, April 2017, pp. 1231-1237.

[3] P. Rajendhra Kumar, P. Archana, "*Effect of CBR of Black Cotton Soil Reinforced with Recron Fiber*", Imperial Journal of Interdisciplinary Research (IJIR), Vol. 03, November 2017, pp. 1093-1097.

[4] G. Jenitha, M. Shenbhagvalli, "*Experimental Study on Behavior of Recron Fiber Reinforced Concrete*", InternationalJournal of Science & Engineering Research, Vol. 07, December 2016, pp. 1137-1140.

[5] Remadevi, M. Anjali, G. Pillai, Elizabeth Baby George, Priya Narayan, Sophiya Sunny, "*Study of Fiber Reinforced Bituminous Concrete*", International Journalof Engineering Research and Development (IJERD), Vol. 10, March 2014, pp. 49-56.

[6] Muhammad Nawazish Husain, PraveenAgarwal, "*Application of Recron-3s Fiber in Improving Silty SubgradeBehavior*", IOSR Journal of Mechanical and Civil Engineering, Vol. 12, November, 2015, pp. 51-55.

[7] R.Aishwarya, G. Venkatesan, R. Regupathi, R. G. Jenith, "*Effect of Copper Slag and Recycled Aggregate in the Behavior of Concrete Composite*", International Journal of Applied Engineering Research, Vol. 10, July 2015, pp. 117-121.

[8] T. Sandeep, "*Recron Medium Strength Fiber Reinforced Concrete*", International Journal in IT and Engineering, Vol. 03, April 2015, pp. 131-134.

[9] FarhadZafari, Mohammed Rahi, "*The Improvement of Bitumen Properties by Adding Nano Silica*", Study of Civil Engineering &Architecture (SCEA), Vol. 03, September 2014, pp. 62-69.

[10] Kishan Khunt, "*Use of Additive Materials in Sub Grade for Road Construction*", International Journal of Engineering Trends &Technology (IJETT), Vol. 04, September 2013, pp. 3790-3795.

[11] P. Pradhan, R. Karand, A. Naik, "*Effectof Random Inclusion of Polypropylene Fibers on Strength Characteristics of Cohesive Soil*", International Journal of Applied Engineering Research, Vol. 30, March 2012, pp. 15-25.

[12] Satish Chandra, M.N. Viladkar, Prasanth P. Naglare, "*Mechanist Approach for Fiber Reinforced Flexible Pavements*", Journal of Transportation Engineering, ASCE / January, 2008.

IOP Conf. Series: Materials Science and Engineering 1112 (2021) 012022 doi:10.1088/1757-899X/1112/1/012022

CASE STUDIES ON CONDITION ASSESSMENT AND REPAIRS FOR FIRE DAMAGED REINFORCED CEMENT CONCRETE STRUCTURES

Raja Veerendra Yenigalla[1], Satish Sajja[2]
[1]Assistant Professor, Civil Engineering Department, VR Siddhartha Engineering College, A.P., India. e-mail: raja.yenigalla@vrsiddhartha.ac.in
[2]Assistant Professor, Civil Engineering Department, VR Siddhartha Engineering College, A.P., India. e-mail: ssatish@vrsiddhartha.ac.in
*Corresponding Author: raja.yenigalla@vrsiddhartha.ac.in

Abstract. In India, the shift in the mode of construction of structures from load bearing walled to framed, started in the early eighties. Concrete is highly durable and has very good compressive strength, but is expensive compared to the materials used for load bearing walled structures. Inspite of the cost, since the durability is more, most of the structures being constructed are RCC structures. However, concrete structures are damaged due to the action of fire. Though concrete has good resistance to fire, it is not completely fireproof. This study presents two case studies on structures damaged by fire. The condition assessment, Non Destructive Testing for quality evaluation of the concrete and the suggested repair techniques are discussed
Keywords. Repair, Rehabilitation, Non-Destructive Testing, Rebound Hammer, Ultrasonic Pulse Velocity

1. Introduction

Reinforced cement concrete constructions are highly durable. However, these structures are distressed either due to intrinsic causes like improper design, insufficient cover thickness, high water cement ratio used for concrete, improper centering and shuttering, poorly graded aggregates, inadequate compaction, inadequate curing, chemical attack due to chemicals in construction water etc or due to extrinsic causes like temperature changes, weathering, sudden external loads, chemical attack due to chemicals in atmosphere etc. In addition to the above, structures are also damaged due to fire and other accidents. RCC structures damaged due to one or more causes discussed above, deteriorate with time and hence the design life of the structure is decreased. The deterioration of structure typically starts with cracks, spalling of concrete leading to delamination and decrease in the strength of the structure. This deterioration, if not identified and properly repaired, will lead to demolition of the whole structure. Since the cost of construction of RCC structures is high, it is necessary to repair the damages to the structure before it leads to loss of the whole structure. For any repair technique, it is important to recognize the cause and extent of the damage to the structure in order to suggest suitable repair methods to rehabilitate the structures. In order to assess the extent of damage to the structure and the loss in strength of the structure due to various reasons, non-destructive testing can be used, since it is in-situ, relatively fast and doesn't disturb the structure.

This paper focuses on the assessment of damage to RCC structures by using two non-destructive testing techniques: Rebound Hammer Test and Ultrasonic Pulse Velocity test. Two case studies were selected and the test results, interpretation of the results and the suggested repair methodologies are also presented.

Non-Destructive testing, as the name suggests, is a no damage test on the structure to assess its quality. These tests can be performed on both old and new structures: on new structures for quality assurance of concrete and for damage assessment in old structures. These tests are very helpful to assess the extent of damage to the structure since they can be successful used at high number of locations compared to other destructive tests like concrete core cutting since core cutting disturbs the structure and hence cannot be used at more locations. In this study, two non-destructive tests were used: Rebound Hammer Test for assessing the hardness of concrete and Ultrasonic pulse velocity test for assessing the uniformity of concrete.

The case studies selected in this study are damaged due to fire. Though concrete has good fire resistance, it is not completely fire proof. Concrete possess high resistance to fire damage and prevents reduction in mechanical strength against all other building materials due to its higher specific heat capacity and low thermal conductivity [1,2]. The conductivity decreases with increase in temperature and increase in occurrence of micro cracks [3]. The thermal conductivity of concrete with respect to Temperature is shown in Figure 1.

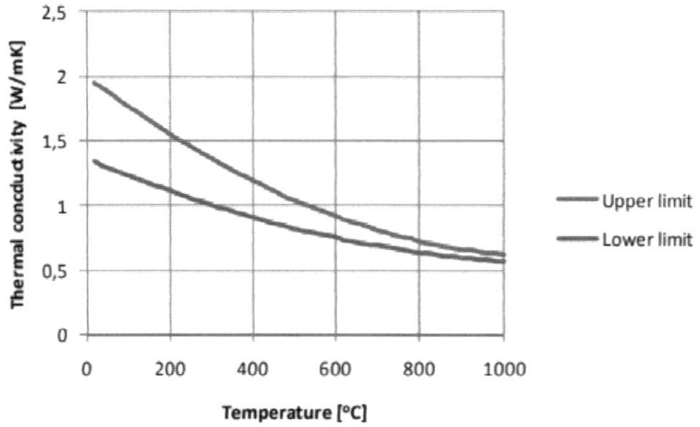

Figure1: Temperature dependant thermal conductivity [3]

Activity of fire on RCC structures causes various types of damages ranging from minor color change, severe cracking and spalling of concrete depending on the intensity of fire. Assessment of structure, evaluation and repair has to be carried out to the damaged structure [23]. The physical effects of the action of fire on Concrete structures are shown in Table 1.

Table 1: Physical effects of temperature on concrete [23]

Temperature	Color Change	Changes in Physical Appearance and Benchmark Temperatures	Concrete Condition
0 to 550 °F (0 to 290 °C)	None	Unaffected	Unaffected
550 to 1100 °F (290 to 590 °C)	Pink to red	Surface crazing: 570 °F (300 °C); Deep cracking: 1020 °F (550 °C); Popouts over chert or quartz aggregate: 1070 °F (575 °C)	Sound but strength significantly reduced
1100 to 1740 °F (590 to 950 °C)	Whitish Grey	Spalling, exposing not more than 25% of reinforcing bar surface: 1470 °F (800 °C); Powdered, light colored, dehydrated paste: 1650 °F (575 °C)	Weak and friable
1740+ °F (950+ °C)	Buff	Extensive spalling	Weak and friable

Demolition and rebuilding of a fire damaged structures is not a viable option, instead they can be repaired and rehabilitated [4 & 5]. Visual inspection, Non destructive testing has to be carried out on damaged structure which plays important role in condition assessment. This gives an understanding about the level of distress and proper restoration technique to be adopted to enhance the life of structure [19]. Selection of appropriate material and method to repair fire damaged RC members is important so that the RCC member can be restored to the original capacity economically and must be done with expertise using various techniques [6 & 15].

The condition assessment of the structure can be done in a process as provided by researcher [22]

The step by step process is shown below: [22]

1. Verify the structure is safe to enter
2. Perform preliminary site visit
3. Perform detailed evaluation
4. Analysis
5. Design Repairs

The Ultrasonic Pulse Velocity and Rebound hammer measurements in combination provides results with high accuracy and error below 10% in determination of modulus of elasticity and strength of concrete [17]. Researcher has studied rebound hammer, UPV technique and using synthetic simulation approach proposed a methodology in improving strength parameters. The reduction of compressive strength of RCC columns damaged due to fire will vary based on temperature conditions [18]. Repairing fire damaged structures is very important and it should be understood that not all repair techniques are suitable in all cases. The repair techniques vary based on the scenario. Shotcrete is more feasible to repair fire damaged structural elements and has various applications like tunnel support, slope support, underground structures. It gives high early strength and short setting time [7]. The shotcreting technique was successfully adopted to repair RC slab, joist of a stadium and RC joist of a school building [6, 8]. Column jacketing is another repair technique which gives more stiffness and improves load carrying capacity [14, 20, and 21]. Blanketing using CFRP and GFRP is also highly popular to repair defective structures and also they have more advantages including high strength to weight ratio, corrosion resistivity, resistance to chemical attack, ease of application and higher tensile strength compared to steel reinforcement [9-13]. In the present paper, two case studies are selected which are effected by fire. The condition assessment, Non Destructive Testing and the suggested repair methodologies are discussed.

2. Case Studies

The paper presents 2 case studies of structures which are damaged by fire. The process for repairing of the structure is detailed below.

2.1 Case Study 1

The project is a privately owned RCC framed structure in Turlapati Kutumba Rao Street, Labbipet, Vijayawada, Andhra Pradesh. The structure is a Stilt+G+3 structure with the first two floors being used as a Paper storage room for an educational institution and the top two floors are used as offices. The structure was constructed in 2011. The structure was damaged due to a fire in the first floor, caused by short circuit at around 1:30AM. The fire was active for a period of 3 hours before being subdued. This has caused major damage to the structural elements of the building in the first and second floors. To propose suitable repair techniques for the damaged structure, condition assessment was done. Since the structure is to be repaired, non destructive testing techniques were used to assess the quality of the structure. The plan of the structure is shown in Figure 2

2.1.1 Visual Inspection

The observations during the inspection of the structure are listed below:

- Columns provided only on the boundaries. Lateral and longitudinal beams provided connecting the columns.
- The fire was caused by short circuit. Due to the presence of paper stacks, the intensity of the fire has increased considerably
- The fire caused major damage to the columns and beams in the first floor
- The effect of fire was also observed in the second floor. The effect is low compared to the effect in the first floor
- The windows in the first floor, made of steel, were completely disfigured due to the fire
- Due to the intensity of the fire, cement has become parched causing loss of bonding

- Cracks were observed in the columns and beams of the structure

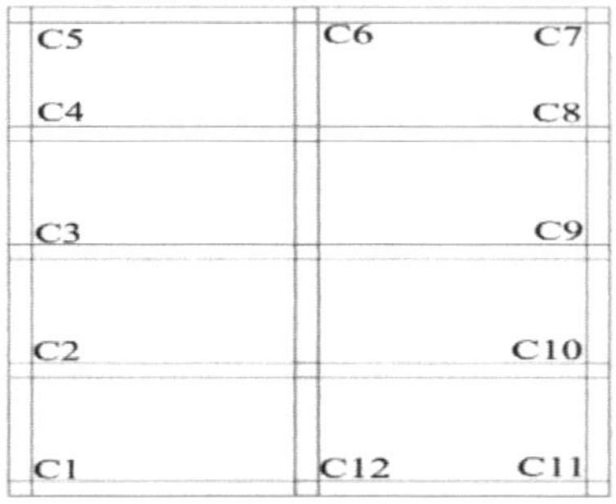

Figure 2: Figure showing the plan of structure with column numbering

2.1.2 Rebound Hammer Test

Rebound Hammer test involves finding the rebound number of a spring controlled mass attached to a plunger. The free end of the plunger is placed on the surface of concrete and is pushed so that the mass attached to it is pressed. At the maximum compressed position of the plunger, the mass snaps and rebounds off the concrete surface. The rebound value can be read from a graduated scale attached to the side. The harder the surface, the higher will be the rebound value. According to IS: 13311: Part 2, the accuracy of the rebound values obtained for concrete are in the range of +/- 25%. The Rebound Hammer Test was conducted using a Digital Schmidt Rebound Hammer – ST Type N of Proceq make. The results of the Rebound Hammer Test performed on the structure are presented in Table 2.

2.1.3 Ultrasonic Pulse Velocity Test

Ultrasonic pulse velocity test involves determination of the denseness of concrete which confirms the presence of voids. The test is conducted by sending and receiving an ultrasonic beam through concrete via two probes placed on either side of the concrete sample. The time of travel of the beam is noted and from the dimensions of the sample, the velocity of the beam is determined. The velocity of the beam is related to the denseness of the concrete and gives the presence of voids. Table 2 of IS: 13311: Part 1 gives the velocity criteria for concrete grading. The Ultrasonic Pulse Velocity test was performed using a Pundit Lab Ultrasonic Pulse Velocity Tester. The results of the UPV Test performed on the structure are also presented in Table 2.

Table 2. Rebound Hammer and UPV Test Results for Structure in Case Study 1

Sl.No	Location	Point Identification	Compressive Strength N/mm^2	Pulse Velocity m/sec
1	First Floor	Column 1	28.00	----
2		Column 2	21.00	1941
3		Column 3	20.00	2641
4		Column 4	19.00	2896
5		Column 5	13.50	2658
6		Column 6	15.00	758/1319

7		Column 7	19.00	1884
8		Column 8	15.00	3064
9		Column 9	18.00	2801
10		Column 10	15.50	2545
11		Column 11	22.00	2980
12		Column 12	19.50	3184
13		Beam adjacent to Column 12	22.00	2946
14		C2-C10 Beam	23.50	2475
15		C4-C8 Beam	15.50	1956/1282
16		C6-C12 Beam	15.00	1984
17		Slab Near 4	28.00	3229
18		Beam Column Joint at Column 4	18.50	3061
19		Beam Column Joint at Column 2	19.50	3453
20		Slab Near Column 11	41.0	----
21		C3-C9 Beam	----	3071
22	Second Floor	Column 2	20.00	2137
23		Column 3	17.00	3344
24		Column 10	29.0	4127
25		Column 7	18.00	3685

2.1.4 Repair and Restoration Measures

Based on the test results of Rebound Hammer and Ultrasonic Pulse Velocity, it was ascertained that the effect of fire was severe in the columns and beams in the first floor. Also the UPV test results indicate the presence of voids in almost all structural elements in the first floor. Since the damage is in one of the bottom floors, retrofitting is mandatory to make the structure safe and durable. The following repair techniques were suggested for the retrofitting of structure:

- Since the UPV test results indicate the presence of voids in all structural elements, it was suggested to grout the voids. Epoxy grouting was suggested because of the higher bonding
- The beams of the structure were also observed to have low strength and were hence retrofitted. Steel C-sections were attached to the beams in the lateral directions, to increase the strength. For the center longitudinal beam, a vertical support column in the form of a steel I-section was provided with shear connectors joining into the reinforcement of the roof and bottom slabs

2.2 Case Study 2

The project is a privately owned RCC framed structure in Plot No. 169&170, 5Th Road, Autonagar, Vijayawada, Andhra Pradesh. The structure is G+1 structure which is an industrial building used in the manufacturing of lubricants. The structure was constructed in 2007. The structure was damaged due to a fire in the first floor. The fire started around 7:00 AM and was active for a period of 4 hours before being subdued. To propose suitable repair techniques for the damaged structure, condition assessment was done. Since the structure is to be repaired, non destructive testing techniques were used to assess the quality of the structure. The floor plan of the structure along with the column numbering is shown in Figure 3.

2.2.1 Visual Inspection

The observations during the inspection of the structure are listed below:

- The intensity of the fire was high due to the chemicals placed in the structure
- The fire caused major damage to the columns slab in the first floor
- The effect of fire was also observed in the ground floor, especially in the slab. The effect is low compared to the effect in the first floor
- Due to the intensity of the fire, cement has become parched causing loss of bonding
- Cracks were observed in the slab, columns and beams of the structure

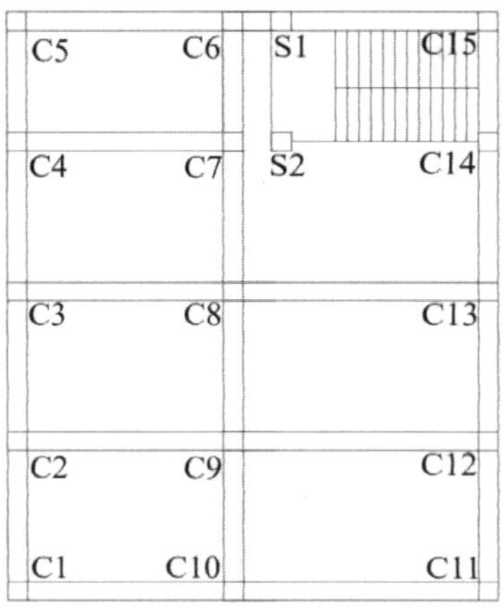

Figure 3: Figure showing the plan of the structure with column numbering

2.2.2 Rebound Hammer Test

The results of the Rebound Hammer Test performed on the structure are presented in Table 3.

2.2.3 Ultrasonic Pulse Velocity Test

The results of the UPV Test performed on the structure are also presented in Table 3.

IOP Conf. Series: Materials Science and Engineering 1112 (2021) 012022 doi:10.1088/1757-899X/1112/1/012022

Table 3. Rebound Hammer and UPV Test Results for Structure in Case Study 2

Sl.No	Location	Point Identification	Compressive Strength N/Mm^2	Pulse Velocity M/Sec
1	Ground Floor	C12 Column	18.00	2956
2		C13 Column	17.00	2854
3		C4-C7 Beam	16.00	3106
4		C14 Column	17.00	3251
5		Bay C7-C8-C13-C14 Slab	19.00	2961
6		C14-C15 Beam	20.50	3248
7		Bay C6-C7-C14-C15 Slab	15.00	2874
8		Bay C4-C5-C6-C7 Slab	18.50	2796
9		C7-C8 Beam	19.00	3029
10	Ground Floor	Ground Floor Roof Slab	24.50	2938
11		Ground Floor Roof Slab	20.00	2864
12	First Floor	C11 Column	19.00	3425
13		C10 Column	21.00	3569
14		S1 Staircase Column	14.00	3152
15		S2 Staircase Column	7.50	2243
16		C3 Column	13.50	2686
17		C4 Column	16.00	2732
18		C6 Column	17.50	2894
19		C7 Column	19.00	3018
20		C14 Column	7.00	1953

21		C13 Column	6.50	2068
22		C12 Column	5.00	1818
23		C9-C12 Beam	6.00	2056
24		Bay C9-C10-C11-C12 Slab	9.50	2148
25		C8-C13 Beam	7.50	2063
26		Bay C8-C9-C12-C13 Slab	13.00	2258
27		C3-C8 Beam	7.00	2186
28		Bay C2-C3-C8-C9 Slab	9.00	1965
29		C4-C7 Beam	11.00	1849
30		Bay C3-C4-C7-C8 Slab	12.00	2348
31		C5-C6 Beam	7.00	1927
32		C7-C14 Beam	7.00	2435
33		Bay C7-C8-C13-C14 Slab	9.00	1792
34		Staircase Slab	5.50	2481

2.2.4 Repair and Restoration Measures

Based on the test results of Rebound Hammer and Ultrasonic Pulse Velocity, it was ascertained that the effect of fire was severe in the columns and beams in the first floor. Also the UPV test results indicate the presence of voids in almost all structural elements in the first floor and also the first floor roof slab. The following repair techniques were suggested for the retrofitting of structure:

- To fill the voids caused due to the fire, epoxy grouting was recommended for all the columns and beams
- RCC jacketing with shear connectors was recommended for the columns with low strength as identified from the rebound hammer test. New reinforcement bars were anchored, from the foundation and continued till the first floor. Shear connectors were used to connect the new reinforcement to the old column reinforcement. Micro concrete was suggested for filling the column.
- For the first floor roof slab, on the interior side, plastering was removed and the surface was cleaned. Shotcreting was done by placing a wire mesh as reinforcement. Bonding chemical was applied on the surface before shotcreting. The total thickness of the shotcreting was around 25-30mm

3. Conclusions

The paper discusses case studies of two RCC structures damaged by fire. The condition assessment and the repair techniques adopted were discussed. The following conclusions can be drawn from the case studies:

1. Visual inspection should be carried out to understand the present state of the structure. Based on the visual inspection, suitability of the tests required to be performed can be assessed
2. Non Destructive testing can be used for condition assessment since it doesn't cause further damage to the structure. However, the results obtained from the Non-Destructive tests are only indicative of the concrete quality and cannot be taken as certain. Also, not all Non-destructive tests give the required quality data.
3. The same repair strategy cannot be used for all the structures. Depending on the loading and in-situ conditions, the repair strategy will vary

References:

[1] EN1994-1-2. European Committee for Standardization. (2005)

[2] EN1993-1-2. European Committee for Standardization. (2005)

[3] EN1992-1-2. European Committee for Standardization. (2004)

[4] M. Yaqub, C.G. Bailey, Repair of fire damaged circular reinforced concrete columns with FRP composites, Construction and Building Materials, Volume 25, Issue 1, Pages 359-370 (2011)

[5] M. Yaqub, C.G. Bailey, Cross sectional shape effects on the performance of post-heated Reinforced concrete columns wrapped with FRP composites, Composite Structures, Volume 93, Issue 3, Pages 1103-1117 (2011)

[6] N. K. Gosain, R.F. Drexler, D. Chouduri, "Evaluation and Repair of Fire-Damaged Buildings" Struct. Eng. Mag. Struct. Forensic. 18 (2008)

[7] Kai Xiang, Guo-Hui Wang, Calculation of Flexural Strengthening of Fire-Damaged Reinforced Concrete Beams with CFRP Sheets, Procedia Engineering, Volume 52, Pages 446-452 (2013)

[8] Prugar J.F., Osowski S.M., and Brighton J., Strengthening of Fire Damaged Concrete Joists at High School. Proceeding of 4th Congress on Forensic Engineering (American Society of Civil Engineering, USA, pp. 502-515 (2006)

[9] Al-Nimry, H., Jawarneh, M. Effects of increased slenderness in short heat-damaged RC columns confined with FRP composites. Mater Struct., 50, 95 (2017)

[10] Rajai Z. Al-Rousan, Rami H. Haddad, Alaa O. Swesi, Repair of shear-deficient normal weight concrete beams damaged by thermal shock using advanced composite materials, Composites Part B: Engineering, Volume 70, Pages 20-34 (2015)

[11] Kai Xiang, Guo-Hui Wang, Calculation of Flexural Strengthening of Fire-Damaged Reinforced Concrete Beams with CFRP Sheets, Procedia Engineering, Volume 52, Pages 446-452 (2013)

[12] L.A. Bisby, J.F. Chen, S.Q. Li, T.J. Stratford, N. Cueva, K. Crossling, Strengthening fire-damaged concrete by confinement with fibre-reinforced polymer wraps, Engineering Structures, Volume 33, Issue 12, Pages 3381-3391 (2011)

[13] Houssam A. Toutanji, William Gómez, Durability characteristics of concrete beams externally bonded with FRP composite sheets, Cement and Concrete Composites, Volume 19, Issue 4, Pages 351-358 (1997)

[14] A.P. Lampropoulos, S.A. Paschalis, O.T. Tsioulou, S.E. Dritsos, Strengthening of reinforced concrete beams using ultra high performance fibre reinforced concrete (UHPFRC), Engineering Structures, Volume 106, Pages 370-384 (2016)

[15] MZA Mohd Zahid, BH Abu Bakar, FM Nazri, MM Ahmad and K Muhamad, "Review of Repair Materials for Fire-Damaged Reinforced Concrete Structures", IOP Conference Series: Materials Science and Engineering, Volume 318, Malaysian Technical Universities Conference on Engineering and Technology 2017 (MUCET 2017) 6–7 December 2017, Penang, Malaysia

[16] Ourania Tsioulou, Andreas Lampropoulos, Spyridon Paschalis, Combined Non-Destructive Testing (NDT) method for the evaluation of the mechanical characteristics of Ultra High Performance Fibre Reinforced Concrete (UHPFRC), Construction and Building Materials, Volume 131, 30 January 2017, Pages 66-77, ISSN 0950-0618.

[17] M. Yaqub, C.G. Bailey, Non-destructive evaluation of residual compressive strength of post-heated reinforced concrete columns, Construction and Building Materials, Volume 120, 1 September 2016, Pages 482 493, ISSN 0950 0618.

[18] Maitham Alwash, Denys Breysse, Zoubir Mehdi Sbartaï, "Non-destructive strength evaluation of concrete: Analysis of some key factors using synthetic simulations", Construction and Building Materials, Volume 99, 30 November 2015, Pages 235-245, ISSN 0950-0618.

[19] Preethi Venkatesh, Mallika Alapati, "Condition Assessment of Existing Concrete Building Using NonDestructive Testing Methods for Effective Repair and Restoration-A Case Study", Civil Engineering Journal Vol. 3, No. 10, October, 2017, Pages 841-855, ISSN 2476-3055.

[20] B.J. Bett, R.E. Klingner, J.O. Jirsa, Lateral load response of strengthened and repaired reinforced concrete columns, ACI Struct. J. 85 (5) (1988) 499–508.

[21] U. Ersoy, A.T. Tankut, R. Suleiman, Behavior of jacketed columns, ACI Struct. J. 90 (1993) 288–293.

[22] Narendra K. Gosain, Ray F. Drexler, Dilip Choudhuri, "Evaluation and Repair of Fire-Damaged Buildings", Structural Forensics Sep, 2008, Pages 18-22

[23] Nabi Yüzer, Fevziye Aköz, Leyla Dokuzer Öztürk, "Compressive strength–color change relation in mortars at high temperature", Cement and Concrete Research, Volume 34, Issue 10, 2004, Pages 1803-1807

IOP Conf. Series: Materials Science and Engineering 1112 (2021) 012023 doi:10.1088/1757-899X/1112/1/012023

STABILIZATION OF EXPANSIVE SOIL BY USING LIME AND REINFORCEMENT WITH GEO-TEXTILE

Malathi N[1], Komala D.N[2], Sk Shabeena[3], Jaya Saahithya V. C[4]

[1]Department of civil engineering, VR Siddhartha Engineering college, Andhra Pradesh, India
[2]Project Manager, Dallas county planning and Development, Dallas, Texas, U.S.A.
[3]Department of civil engineering, VR Siddhartha Engineering college, Andhra Pradesh, India
[4]Department of civil engineering, VR Siddhartha Engineering college, Andhra Pradesh, India
*Corresponding Author: malathin76@gmail.com

Abstract. Expansive soils are those soil which have high rate of swelling. Hence the improvement of the soil properties has to be done. In this study, we are using of lime and geo synthetic material for stabilization of soil. Lime is an efficacious product to be mixed with fine grained soils with high plasticity and it improves certain properties of soil due to its chemical action. Stabilization of expansive soil with geo-textile will increase the stiffness and load bearing capacity of the soil through frictional interaction between the soil and geo-textile material. The lime is mixed in different percentages and reinforced with geo-textile by placing at certain depth with in sample height of four layers. In conclusion it shows an improvement of strength is in terms of CBR and peak load.

1. Introduction

Soil has been using as the construction material for roads, buildings, irrigation structures, etc. due to some of the weakness in its strength and mechanical properties it has to be improved (according to the requirement). In Andhra Pradesh, black cotton soil (BC soil) is covered in large area. BC soil is very hard in dry condition and lose its strength in wet condition. It has the properties to absorb more water and high in swelling. The BC soil is not suitable for the construction works because of its volumetric changes when they get in contact with water. Therefore, the BC soil stabilization is been done in the paper by using lime and geo-textile.

Soil stabilization is a method used to improve the engineering properties of a soil to make it useful for the intended purpose. It is done by blending and mixing material into soil to achieve a desired gradation or the mixing of commercially available additives that may alter the gradation, texture or plasticity, or act as a binder for cementation of the soil. The commonly used soil stabilizations are,

a) Cement stabilization
b) Lime stabilization
c) Bituminous stabilization
d) Chemical stabilization

Lime stabilization is performed by mixing lime to the soil and useful the clayey soils. The basic types of lime are

a) High calcium, quick lime, CaO

b) Hydrated, high calcium lime, Ca(OH)2

c) Dolomitic lime (CaO + MgO)

d) Normal, hydrated dolomitic lime {Ca(OH)2 + MgO}

e) Pressure, hydrated dolomitic lime {Ca(OH)2 + MgO2}

The main benefits of using lime stabilization is it improves the workability, strength, and stability. Workability is improved due to the flocculation makes the clay more friable, this assists combination for effective mixing and compaction. Lime increases the optimum water content for compaction, which is an advantage when dealing with wet soil. The compaction curve for lime treated clay is generally flatter, which makes moisture control less critical and reduces the variability of the density produced.

The geosynthetics are geo-textiles(woven & nonwoven), geogrids, geomembranes, erosion control blankets and materials, geosynthetic clay liners, geo-composite drainage materials and geonets. The major functions of geosynthetic materials are separation, filtration, reinforcement, drainage, and acting as a liquid barrier. The effectiveness of use of nonwoven geo-textiles as reinforcement material for stabilization of soil for different engineering works. Nonwoven Geo-textiles are planar polymeric materials that have been extensively used for separation, filtration, drainage and reinforcement. The effect of Lime and geo-textile on the properties of the soil. Their main objectives was to determine the properties of the soil before and after the addition of lime and geo-textile to it. The different tests they conducted were natural moisture content determination, specific gravity, Atterberg's limits, Compaction test, unconfined Compressive Strength test and CBR test. After studying and conducting the entire above test, the optimum lime content was found and they concluded that there was a substantial increase in the CBR and peak load.

2.Materials and Experimental Work

Soil: The disturbed expansive soil was collected from mandhadam near velagapudi region at a depth of 1.5m. The collected soil is dried and pulverized for conducting physical properties tests as per the required sieve sizes of a test. These tests are conducted as per the ASTM; Physical properties are shown in table 1

Table 1. Basic properties of collected soil sample

S.no	properties	observations
1	IS classification	CH
2	Free Swell index (%)	60
3	pH	7.5
4	Liquid limit (%)	69
5	Plastic limit (%)	43
6	Plasticity index (%)	26
7	Optimum moisture content (%)	28
8	Dry unit weight(gm/cc)	1.42
9	Specific gravity	2.56

Lime: Limeis from saraswathi chemicals Ltd. Guntur. It is manufactured by heating lime stone, sea shella etc. Which are mainly using in the method of soil stabilization. It is mainly increase strength and reduces the swell or shrink property. But further additions may decreases the plasticity of soil and excessive lime treatment contributes to brittle failure characteristic of soils that lead to rapid and great loss in strength when failure occurs.

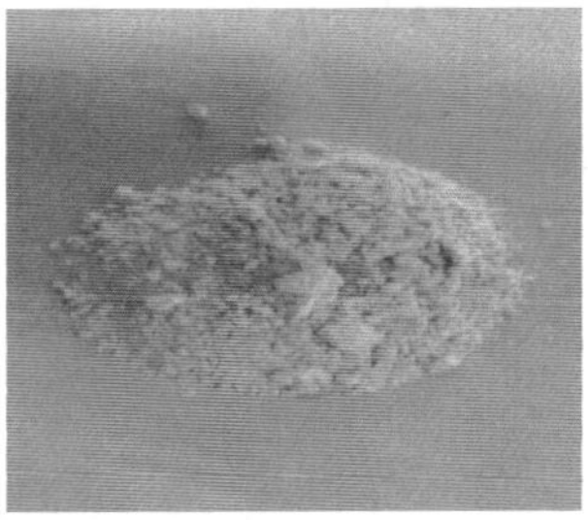

Properties of lime are listed in table 2.

Fig. 1 Lime

Table 2. Basic properties of lime

S.no	Properties	Observations
1	Physical appearance	Dry powder
2	color	white
3	Specific gravity	2.44

Geo-textile

The geo-textile used in this study was a medium-weight, nonwoven, spun-bond, and needle punched geotextile. The nonwoven geo-textile was procured from a local market and is available in form of rolls as shown in fig 2. The properties of geo-textile were evaluated in a laboratory and are presented in table 3

Fig: 2 nonwoven geo-textile

Table 3: Properties of geo-textile

S.no	properties	observation
1	Type	Nonwoven
2	Color	white
3	Thickness(mm)	1.6

4	Mass per unit area(gm/cm2)	0.0195
5	Tensile strength(kN/m)	38

Influence of lime on the geotechnical characteristics of expansive soil was investigating standard proctor tests, California bearing ratio test and unconfined compression tests. In this study evaluate the effectiveness of different percentages of lime. The results are presented as graphical plots and tables.

Table 4. *compaction, CBR and UCS values at different percentages of lime*

Soil sample % + lime %	OMC (%)	MDD (%)	CBR (%)	UCS (%)
100% soil sample	28	1.415	2.02	129.55
96% soil sample + 4% lime	27.5	1.445	7.278	301.13
94% soil sample + 6% lime	19.54	1.50	12.5	602.43
92% soil sample + 8% lime	21.42	1.44	20.2	740.07
90% soil sample + 10% lime	23.809	1.449	9.30	418.18

Compaction characteristics: The tests were performed as per Indian standard specifications for standard proctor compaction tests. Light compaction tests were carried out on the soil-lime mixtures. The compaction tests were performed for various combinations of lime. OMC is decreases andmaximum dry density increases with increasing of percentage of lime content in soil. Variation of OMC and MDD shown in graph 1&2

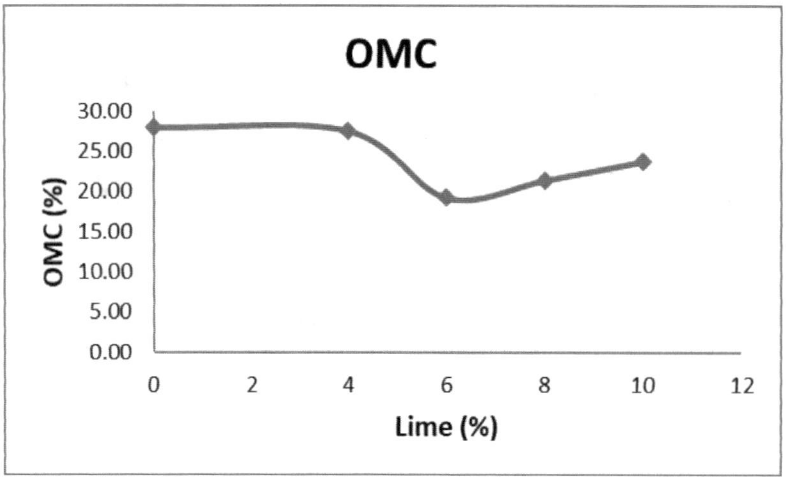

Graph: 1varying of optimum moisture content with various percentage of lime

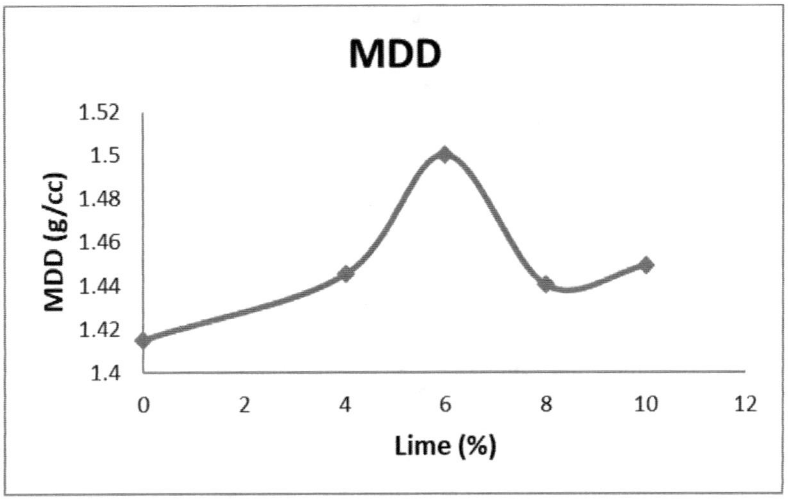

Graph:* 2*varying of dry unit weight with various percentage of lime

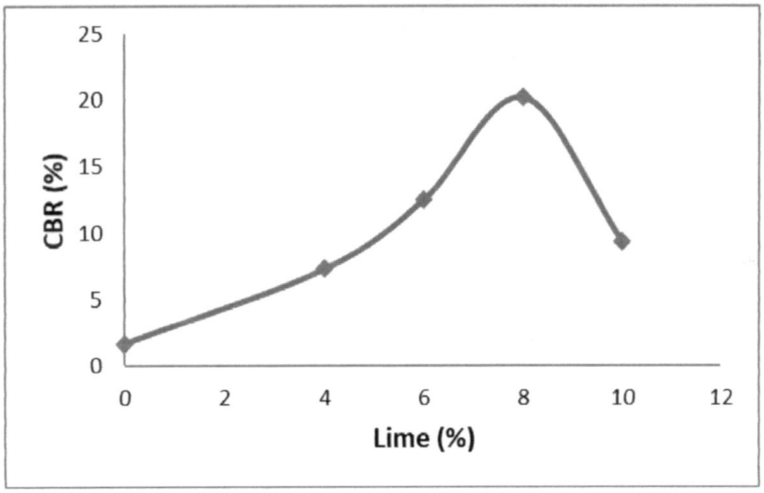

Graph:* 3 *Varying of CBR values at different % of lime

2.3: **California bearing ratio test**

CBR test conducted as per the IS2720 part16, for lime blended soil specimens with different percentages. Test specimen is prepared by mixing of lime in different percentages by the dry weight of soil and lime was mixed thoroughly with equivalent dry unit weight and optimum moisture content which is obtained from proctor test. Mixed soil sample is placed in five layers in CBR mould having dimensions of 150mm diameter and 175mm height and compacted each layer with a hammer of weight 25.5N in weight falling from a height of 310mm in 56 blows. The prepared soil sample is soaked for four days. After soaking condition conducting CBR test. Variation of CBR values at different percentages of lime as shown in fig 3

Fig .3 *. Experimental set up for conducting California bearing ratio test*

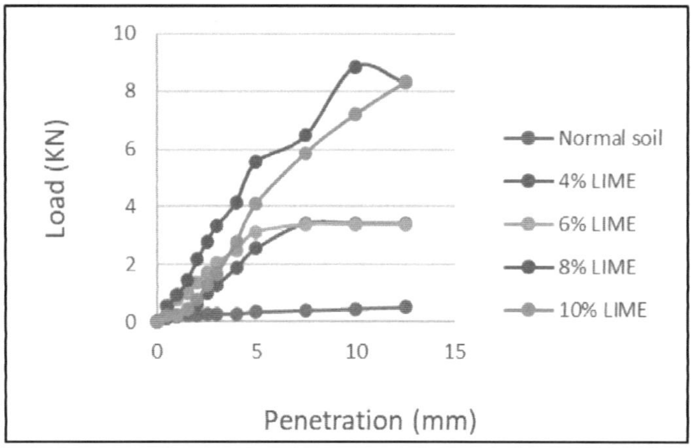

Graph: 4 Comparison of load penetration curves for lime blended soil specimen.

CBR test conducting on single and multiple layer reinforcement

CBR test conducted with 8% lime blended soil specimens then by placing geo-textile at single and multiple layers. In single layer reinforcement placed at 3cm depth from top of the mould. The load influence is up to 3cm only. In multiple layer reinforcement four layers (each 3cm) of geo-textile will be placed up to 12cm in the mould.

California Bearing Ratio (CBR) on Black cotton soil with 8% Lime and Geotextile @ single layer

IOP Conf. Series: Materials Science and Engineering 1112 (2021) 012023 doi:10.1088/1757-899X/1112/1/012023

Table.5 CBR values at single layer reinforcement

S.NO	Soil sample + lime (%)	Depth (cm)	CBR(%)
1	92%soil sample + 8% lime	0	20.2
2	92%soil sample + 8% lime	2.5	16.175
3	92%soil sample + 8% lime	3	22.645
4	92%soil sample + 8% lime	6	17.38

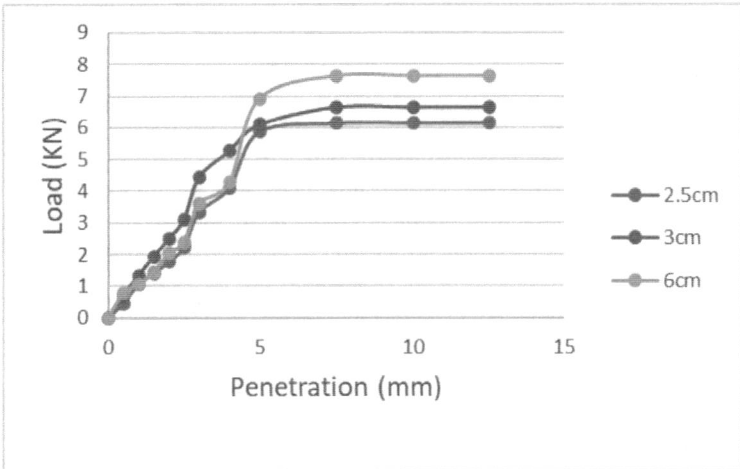

Grap.: 5 Comparison of load penetration curves for 8% lime and geo-textile placed at single layer

California Bearing Ratio (CBR) on Black cotton soil with 8% Lime and Geotextile @ multiple layer.

Table. 6 CBR values at multiple layer reinforcement

S.NO	NO.OF LAYERS OF GEOTEX TILE	% SOIL SAMPLE+% LIM E	DEPTH (cm)	CBR%
1	0	92% soil sample+8% lime	0	20.2
2	1	92% soil sample+8% lime	3	22.645
3	2	92% soil sample+8% lime	3& 6	25.475
4	3	92% soil sample+8% lime	3, 6& 9	30.32
5	4	92% soil sample+8% lime	3, 6, 9& 12	32.35

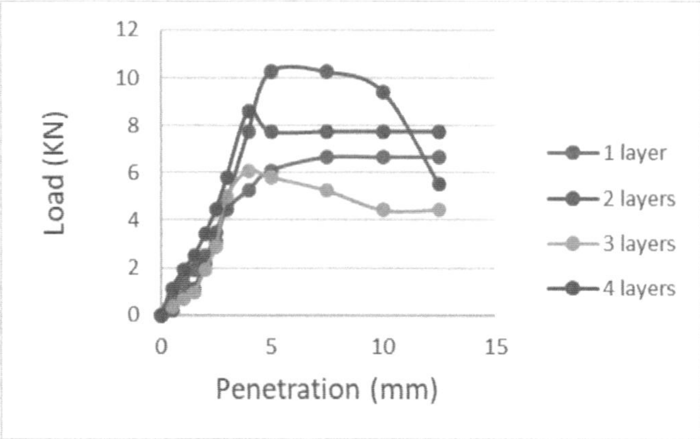

Graph: 6Comparison of load penetration curves for 8% lime and geo-textile placed at multiple layers

Peak load values at various depths in multiple layers: CBR test conducted with 8% lime blended soil specimens then by placing geo-textile at multiple layers. In multiple layer reinforcement four layers (each 3cm) of geo-textile will be placed up to 12cm in the mould. Peak load will be calculated by using CBR test.

Table.7 Peak load values at multiple layer reinforcement

s.no	No. of layers of geotextile	% soil sample+% lime	depth (cm)	peak load (kpa)
1	0	92% soil sample+8% lime	0	129.55
2	1	92% soil sample+8% lime	3	5531.87
3	2	92% soil sample+8% lime	3,6	6499.94
4	3	92% soil sample+8% lime	3, 6& 9	6555.26
5	4	92% soil sample+8% lime	3,6,9& 12	6776.54

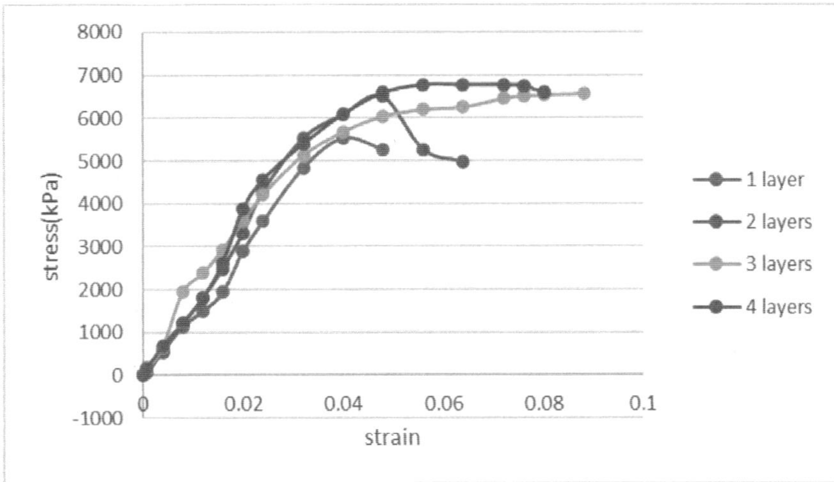

Graph. 7Comparison of stress-strain curves for 8% lime and geo-textile placed at multiple layers

2.4. Unconfined compression strength test

Unconfined compression tests is performed the comparison of soil improvement by adding of different percentages of lime. As per the IS code 2720 part 6 unconfined compression strength test have been conducted. At maximum dry density and optimum moisture content soil specimens were prepared with dimensions of 36mm diameter,76mm height and the specimens those are prepared by soil-lime samples for curing period 3days in a tightly covered plastic bags then testing is conducted.

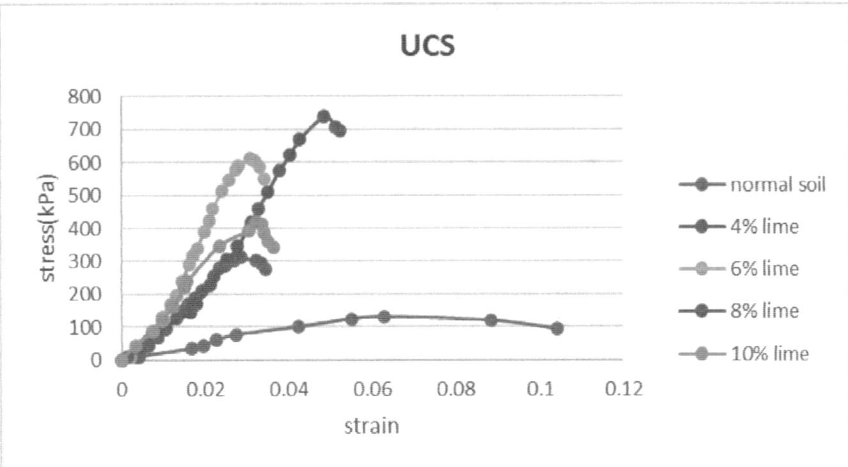

Graph: 8 Stress-Strain graphs for soil specimen prepared with various percentages of lime

3. Results and Discussions

CBR values increased 2.02% to 20.2% and unconfined compression strength values increased 129.55kpa to 740.07kpa with only lime up to 8%. At 10% lime CBR and UCS value decreases. But with lime and reinforced nonwoven geo-textile specimens results in much greater improvement in soil strength. In 8% lime content, non-woven geo-textile placed as single and multiple layers. Single layer reinforcement at 3cm depth from top of the mould gives highest CBR value22.645%. Four layers of 3, 6, 9 &12cm depth gives the highest CBR value 32.35% and peak load value6776.54%.

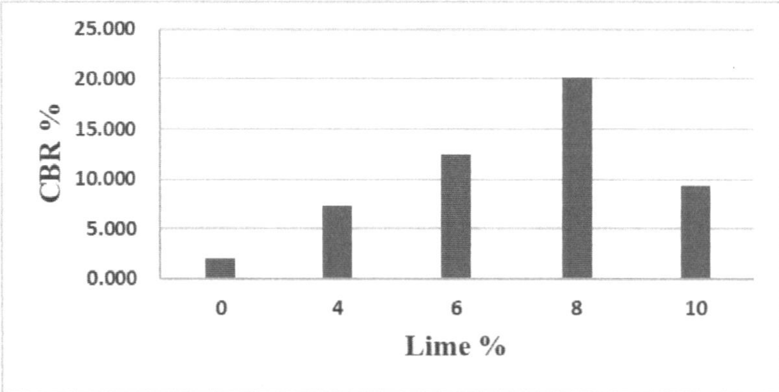

Graph: 9 CBR values at different percentages of lime

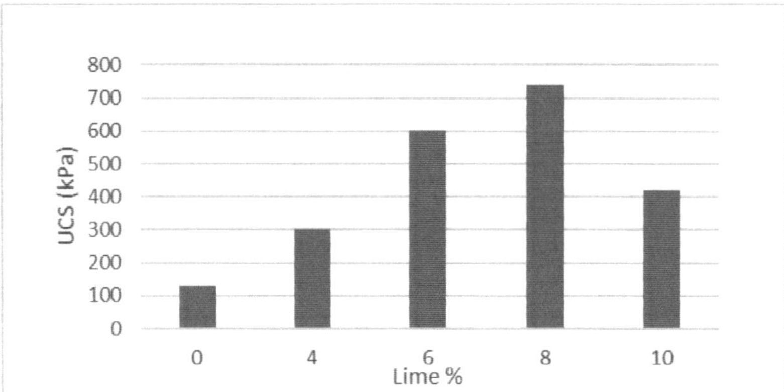

Graph: 10 UCS values at different percentages of lime

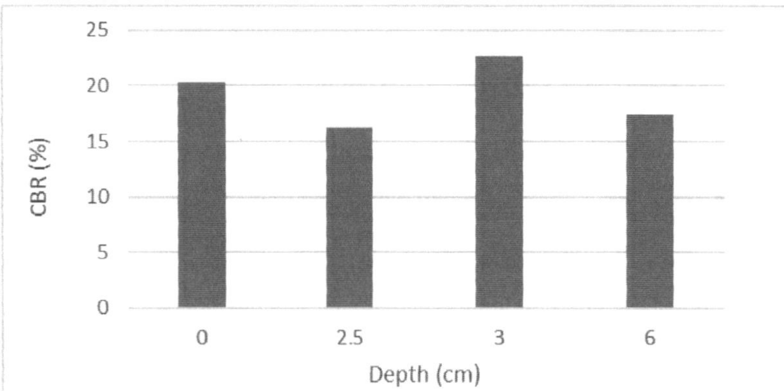

Graph. 11CBR values at various depths in single layer

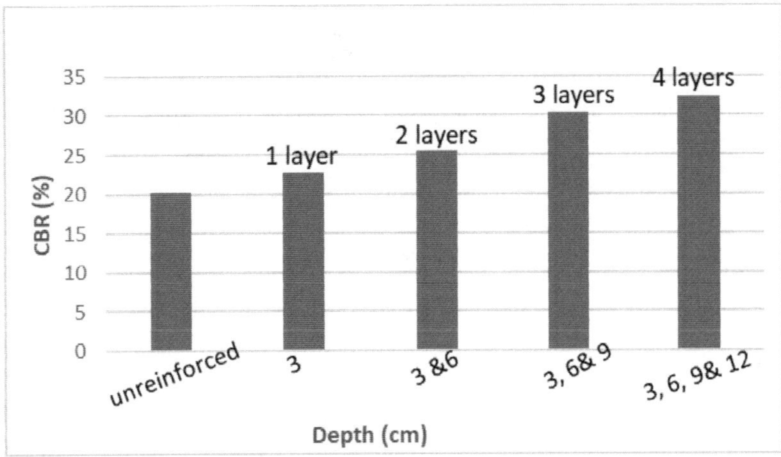

Graph. 12 CBR values at various depths in multiple layers

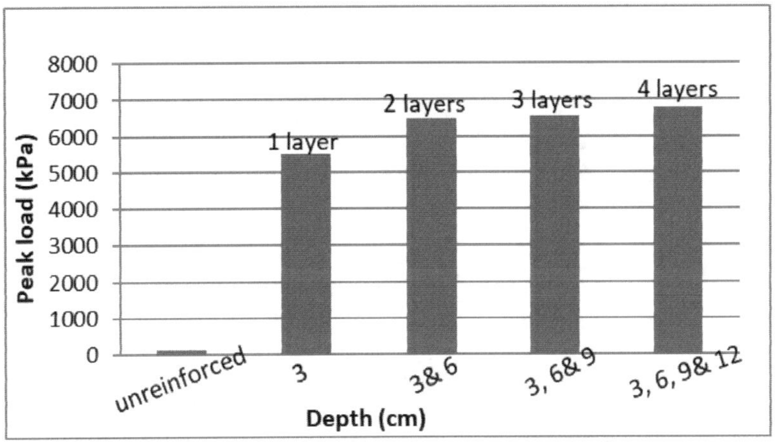

Graph. 13 Peak load values at various depths in multiple layers

Conclusions and Future Scope:

In this project work, it has been found that

1. The amount of lime increases there is apparent increase in maximum dry density and in optimum moisture content.
2. The optimum lime content is 8%.
3. There is significant increase in UCS values.
4. At 8% lime, number of layers of geotextile increased to increase the CBR value and peak load. There is substantial scope for carrying our further work in this area as the various types of geo-synthetic materials use.

4. REFERENCES:

[1] Ghosh, B., Ramesh, V. and Rajarajeswari, B.V., 2014. "Improvement of soil characteristics using jute geo-textile" International Journal of Science, Engineering and Technology Research (IJSETR), 3(7).

[2] Tailor, R.M., Desai, M.D. and Shah, N.C., 2011." Geotextile as reinforcement in flexible pavement for swelling subgrade". In Proceedings of Indian Geotechnical Conference (pp. 15-17).
[3] Sridharan, A., Soosan, T.G., Jose, B.T. and Abraham, B.M., 2006. "Shear strength studies on soil-quarry dust mixtures". Geotechnical & Geological Engineering, 24(5), pp.1163-1179.
[4] Al-Qadi, I.L., Dessouky, S.H., Kwon, J. and Tutumluer, E., 2012. "Geogrid-reinforced low-volume flexible pavements: pavement response and geogrid optimal location". Journal of Transportation Engineering, 138(9), pp.1083-1090.
[5] Dutta, R. and Sarda, V., 2008. "A comparative study of compaction and CBR behavior of stone dust reinforced with waste plastic strips". International Journal of Geotechnical Engineering, 2(3), pp.255-264. DOI: 10.3328/IJGE.2008.02.03.255-264
[6] Sharma, R.S., Kumar, B.P. and Nagendra, G., 2004. "Compressive load response of granular piles reinforced with geogrids". Canadian Geotechnical Journal, 41(1), pp.187-192.
[7] Viswanadham, B.V.S., Phanikumar, B.R. and Mukherjee, R.V., 2009. "Swelling behavior of a geofiber-reinforced expansive soil". Geotextiles and Geomembranes, 27(1), pp.73-76.
[8] Fourie, A.B. and Fabian, K.J., 1987. "Laboratory determination of clay-geotextile interaction". Geotextiles and Geomembranes, 6(4), pp.275-294.
[9] Nadgouda, K.A. and Hegde, R.A., 2010, December. "The effect of lime stabilization on properties of black cotton soil. In Indian Geotechnical Conference (pp. 511-514).
[10] Kumar, A., Walia, B.S. and Bajaj, A., 2007. "Influence of fly ash, lime, and polyester fibers on compaction and strength properties of expansive soil". Journal of materials in civil engineering, 19(3), pp.242-248.
[11] Twinkle, S. and Sayida, M.K., 2011. "Effect of Polypropylene Fiber and Lime Admixture on Engineering Properties of expansive soil". In Proceedings of Indian Geotechnical Conference December 15-17, 2011, Kochi (Paper No. H-085).
[12] Pai, R.R. and Patel, S., 2019." Effect of GGBS and Lime on the Strength Characteristics of Black Cotton Soil. In Ground Improvement Techniques and Geosynthetics (pp. 319-328). Springer, Singapore.
[13] Reddy, N.G., Tahasildar, J. and Rao, B.H., 2015. "Evaluating the influence of additives on swelling characteristics of expansive soils". International Journal of Geosynthetics and Ground Engineering, 1(1), p.7.
[14] Ghosh, A. and Subbarao, C., 2007. "Strength characteristics of class F fly ash modified with lime and gypsum". Journal of geotechnical and geoenvironmental engineering, 133(7), pp.757-766.
[15] Aggarwal, P. and Sharma, B., 2010, September." Application of jute fiber in the improvement of subgrade characteristics". In Proc of int conf on adva in civ eng, Trabzon, Turkey (pp. 27-30).
[16] ASTM Committee D-18 on Soil and Rock, 2017. Standard Practice for Classification of Soils for Engineering Purposes (Unified Soil Classification System) 1. ASTM international.
[17] IS: 2720 (1992) (Part 16) California Bearing Ratio Test, Bureau of Indian Standards, New Delhi, India.
[18] ASTM D5199 (2012) Standard Test Method for Measuring the Nominal Thickness of Geosynthetics, American Society for Testing of Materials, PA, USA.
[19] ASTM D5261 (2009) Standard Test Method for Measuring Mass per Unit Area of Geotextiles, American Society for Testing of Materials, PA.
[20] ASTM D6241 (2014) "Standard Test Method for Static Puncture Strength of Geotextiles and Geotextile-Related Products Using a 50-mm Probe", American Society for Testing of Materials, PA, USA.

IOP Conf. Series: Materials Science and Engineering 1112 (2021) 012024 doi:10.1088/1757-899X/1112/1/012024

FLOOD PLAIN MODELLING OF KRISHNA LOWER BASIN USING ARCGIS, HEC- GEORAS AND HEC-RAS

B Jagadeesh[1] K Krishna Veni [2]
[1]Department of Civil Engineering, MVGR College of Engineering, Vijayanagaram, 535005, India
[1]Department of Civil Engineering, Vasireddy Venkatadri Institute of Technology, Guntur-522508, India
**Corresponding Author:* [1]anil.jagadeesh2@gmail.com, [2]krishnakommanaboyina@gmail.com

Abstract. A flood is an overflow of water that submerges land which is usually dry. Floods can also occur in rivers when the flow rate exceeds the capacity of the river channel, particularly at bends or meanders in the waterway. This results in causing damage to human and property if they are in the natural flood plains of rivers or settled on the banks of rivers. To facilitate the appropriate measures for effective flood mitigation in advance, there is a need to model the flood plain which facilitates to locate the flood plain and its extent for effective flood mitigation measures. By understanding the extent of flooding and floodwater inundation, decision makers are able to make choices about how to best allocate resources to prepare for emergencies and to generally improve the quality of life. This research presents a straightforward approach for processing output of the HEC-RAS hydraulic model, to enable two and three dimensional floodplain mapping and analysis in the ArcView geographic information system. The methodology is applied to a stretch of river Krishna from downstream of Nagarjuna Sagar project to Kummaripalem which is located on upstream of Prakasam Barrage. The flood plain maps are developed for the flows corresponding to 2, 10, 25, 50 and 100 year return periods. The resulting surface model provides a good representation of the general landscape and contains additional detail within the stream channel. Overall, the results of the research indicate that GIS is an effective environment for floodplain mapping and analysis.

1. Introduction

A flood is a flow of high magnitude during which the river overflows its banks and inundates the adjoining area. Floods often cause damage to homes and commercial areas if they are located in the natural flood plains of rivers. Protective measures such as detention basins, levees, bunds, reservoirs and weirs are used to prevent waterway from overflowing their banks. Flood control structures such as dams can be built and maintained over time, to reduce the occurrence and severity of floods as well. By understanding the extent of flooding and floodwater inundation, decision makers are able to make choices about how to best allocate resources to prepare for emergencies and to generally improve the quality of life.

Depending upon the nature of work involved, flood protection and flood management measures may be broadly classified as structural and non- structural measures. The structural measures for flood control involve structurally intensive activities which bring relief to the flood prone area by reducing flood flows and thereby reducing the flood levels but these are cost intensive both in initial cost and maintenance cost. Non-structural measures involve the stakeholders in flood mitigation activities like creating awareness, preparation of the flood inundation maps, prevention of people occupying low lying and flood prone area and developing strategies for evacuating affected people during flood season. These measures provide a cost effective and reliable solution to flood management problems.

Floodplain management and mapping is a new applied method for the river engineering and is essential for prediction of flood hazards. The stretch of river Krishna between Nagarjuna Sagar Project and Prakasam barrage regularly gets affected due to flooding almost every year during the monsoon season, mainly due to lack of proper flood protection and limited resources. To understand the dynamics of flow of flood water in the river and the flood plain for the event, it is necessary to model the event using an appropriate model or combination of models. The results of the study can be used for hydraulic simulation by integrating with GIS analysis and could be effective for river training works and flood mitigation planning.

2. Literature review

Many studies of flood plain modeling have been carried out across the world. For the past few years, application of Remote sensing and GIS in terrain modeling and floodplain mapping was relatively limited. With the drastic advances in GIS, its usage has been increased in various fields and it began to be used to represent the flooded area on the land surface. Most of the prior works dealt with the digital elevation models (DEMs), analysis of square grids of regularly spaced elevation data for hydrologic applications. Callaghan et al.(1988) implemented methods for filling DEM depressions, creating flow direction and flow accumulation grids using DEM as the single input. These advances allowed for the delineation of watershed and drainage network. But, the use of DEM surface is generally not suitable for representation of large-scale terrain which is necessary for the hydraulic analysis of river. The triangular irregular network (TIN) model is preferred for hydraulic modelling of river channels,

Djokic et al. (1990) made use of TINs for storm drainage modelling in urban areas. He found that the TIN surface was very effective for the determination of parameters for design flow calculations. The flood extent for different return periods of 50, 100, and 200 years were estimated in the study. Hernandez et al. (2007) used GIS raster and feature data sets, to automate the pre-processing of input data and HEC-GeoHMS and HEC-GeoRAS for post-processing of HEC-RAS results. These tools are successfully linked with HEC-RAS to expedite the production of floodplain mapping and obtain accurate hydraulic results. HEC-RAS/HEC-GeoRAS have been used for steady flow simulation along 4km end of Zaremrood River to derive the extent of flood hazard by Karim et al (2009). The results indicated that by integrating HEC-RAS model, GIS can be used as an effective tool for various kinds of floodplain modelling and different scenarios of river training practices.

Alaghmand et al. (2010) used HEC-RAS and HEC-HMS models for river flood hazard mapping for Sungai Kayu Ara river basin, located in the west part of the Kuala Lumpur in Malaysia. Using HEC-GeoRAS, stream geometry data was extracted from the digital elevation model (DEM). The hydrologic data, which was generated with HEC-IIMS as initial flow and boundary conditions, was integrated with HEC-RAS steady flow data. HEC-RAS model results were post processed into ArcGIS using HEC-GeoRAS. The generated river flood hazard was based on water depth and flow velocity maps were prepared according to hydraulic model results in GIS environment.

For the development of a terrain model, Tate et al. (2010) used a GIS based approach, based on stream channel representation of the HEC-RAS hydraulic model. The approach required a digital elevation model DEM, GIS representation of the stream geometry and a completed HEC-RAS simulation as input. The channel data was exported from HEC-RAS to GIS, hydraulic model coordinates were converted to geographic coordinates. A digital terrain model was subsequently synthesized by merging HEC-RAS data of the stream channel with comparatively lower-resolution DEM data for the floodplain. The resultant surface model provides a good representation of the general landscape and contains additional detail within the stream channel.

A study on the flood zone of Karkheh River using hydraulic model HEC-RAS and GIS was carried out by Kamanbedastel et al. (2011). Topographical map, digital elevation model and hydrological data were used for the study. The results indicated that hydraulic simulation by integrating HEC-RAS model, GIS can be used as an effective tool for various kinds of floodplain managements and different scenarios of river training practices.

3. Study Area and Methodology

3.1 Study Area

The stretch of river Krishna from downstream of Nagarjuna Sagar Project (chainage 0.0 km), in Nalgonda District of Telangana state to Kummaripalem in Krishna district of Andhra Pradesh which is located on the upstream of Prakasam barrage (chainage 192 km) is considered for the present study. It is located between 79° 31' E and 80° 60 ' E and 16° 50' N and 16° 57 ' N which is shown in Figure 3.1.

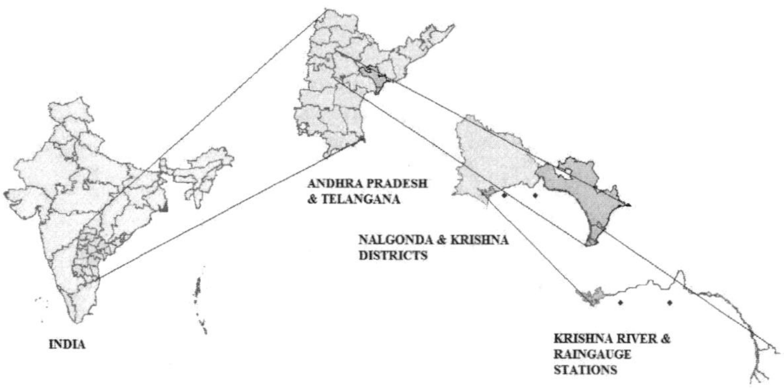

Figure 3.1 Location Map of Study Area Using GIS

The important soil types found in the basin are black soils (regur), red soils, laterite and lateritic soils, alluvium, mixed soils (red and black, red and yellow, etc.) and saline and alkaline soils.

3.2 Data Required:

The main objective of the HEC-RAS program is to compute water surface elevations at all locations of interest for given flow values. The data needed to perform these computations are divided into geometric data and steady flow data.

3.2.1 Geometric Data

The basic geometric data consists of the connectivity of the river system (schematic diagram of the river system), cross sections data reach, lengths, energy loss coefficients (friction losses, contraction and expansion losses), and stream junction information. If hydraulic structures (bridges, culverts, etc.) are present in the river system, the details and characteristics of those are also needed.

3.2.2 Steady Flow Data

Steady flow data, needed to perform a steady water surface profile calculation, consist of (i) Flow

IOP Conf. Series: Materials Science and Engineering 1112 (2021) 012024 doi:10.1088/1757-899X/1112/1/012024

regime, (ii) Boundary conditions, and (iii) Peak discharge information.

3.3 Data Availability

The data available for the study to carry out the simulation and their sources are presented in Table 3.1.

Table 3.1 Data Collected and their Sources

Data Collected	Source
Digital Elevation Model (30m resolution)	Earthexplorer.usgs.gov.in
Streamflow data	http://www.india-wris.nrsc.gov.in/eris.html

3.4 Methodology

In this study the topography of the river was obtained by using ArcGIS and HEC Geo-RAS. This GIS data was imported to HEC-RAS for further steps. Flood frequency analysis is carried out to find discharges corresponding to 2, 10, 25, 50 and 100 year return periods. For these discharges corresponding to 2, 10, 25, 50 and 100 year chance events HEC-RAS analysis is carried out to generate water surface profiles and other hydraulic details.

3.4.1 Watershed delineation of the study area

Watershed delineation process is carried out for identifying the stream location in Digital Elevation Model. Various tools available in Arc Toolbox, like Hydrology, Map Algebra, Spatial Analysis and Raster tools are used for doing all the steps involved in watershed delineation process. These steps consist of computing the Fill, Flow accumulation, Flow direction, Basin, Rater to Polygon, Raster Calculation and Raster to Polyline. The watershed delineation of the DEM is shown in Figure 3.2.

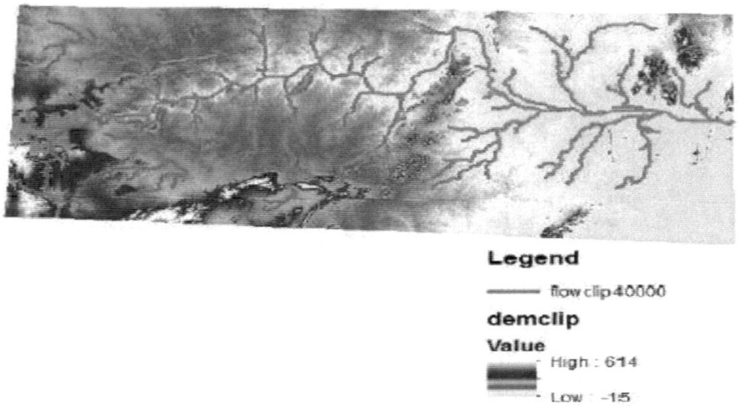

Figure 3.2. Flow Accumulation Grid of the DEM

3.4.2 Generation of Triangulated irregular network (TIN)

Triangulated Irregular Network (TIN) is a digital data structure used in GIS for the representation of a surface. A TIN is a vector based representation of the physical land surface or sea bottom, made up of irregularly distributed nodes and lines with three dimensional coordinates (x, y, and z) that are arranged in a network of non-overlapping triangles. TINs are often derived from the elevation data of a rasterised DEM. The first step in the pre-processing stage is to create a DEM of the river system in a TIN format. The TIN must be constructed with a special care in order to provide accurate analyses. Elevation data for each cross section is extracted from the TIN. The TIN also serves for determining floodplain boundaries and calculation of inundation depths. The TIN was constructed using 3D Analysis tools in ArcMap. The TIN is loaded and HECGeo-RAS extension is added for extracting the topography of the river and import to HEC-RAS. The pre-processing is done using the HEC-GeoRAS for creating physical attributes in GIS, and then exporting them to the HEC-RAS geometry file. For this, RAS layers in ArcMap, which are used to create geometric data sets will be modelled in HEC-RAS. Only two RAS layers are needed to map a floodplain (Stream Center line and Cross Section Cut lines). All other layers are optional. The RAS layers can be created in one step and will be stored collectively in a GeoDatabase, which HEC-GeoRAS creates automatically. By default, this GeoDatabase will be saved in the same location as the ArcMap project with which the users are working and will be given the same name.

3.4.3 Digitization of Stream centerline

The Stream centerline layer is very important, because it represents the river network for HEC-RAS. Digitizing of the stream centerline starts with selecting the sketch tool from the Editor Toolbar, and digitization proceeds in the direction of river flow. That is, digitizing must begin at the most upstream end of the stream, and proceed towards downstream ending at the outlet or beginning of the next downstream reach. After digitizing the Stream centerline layer, the next task is to assign River code. Each river in HEC-RAS, as well as each reach within a river, is assigned a unique name. This is accomplished by the selection of Assain River Code/Reach Code menu item and assigning appropriate name. Digitizing of river center line is shown in Figure 3.3.

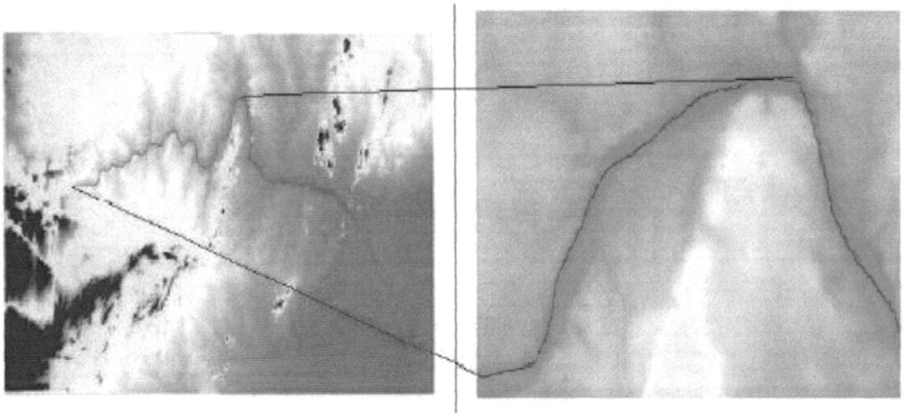

Figure 3.3 Digitizing of River Centerline Using DEM

3.4.4 Cross sectional cut lines

Cross-sectional cut lines are one of the most important inputs to HEC-RAS. They are used to extract the elevation data from the terrain and to create a ground profile across the flow. The intersection of cut lines with other RAS layers such as centre line and flow path lines are used to compute HEC-RAS attributes such as bank stations (locations that separate main channel from the floodplain), and downstream reach lengths (distance between cross-sections). Each cross-sectional cutline represents each cross-section of the river and they must be drawn perpendicular to the flow direction of river by intersecting stream center line, bank lines and flow path lines. Digitising of cross sectional cut lines is shown in Figure 3.4.

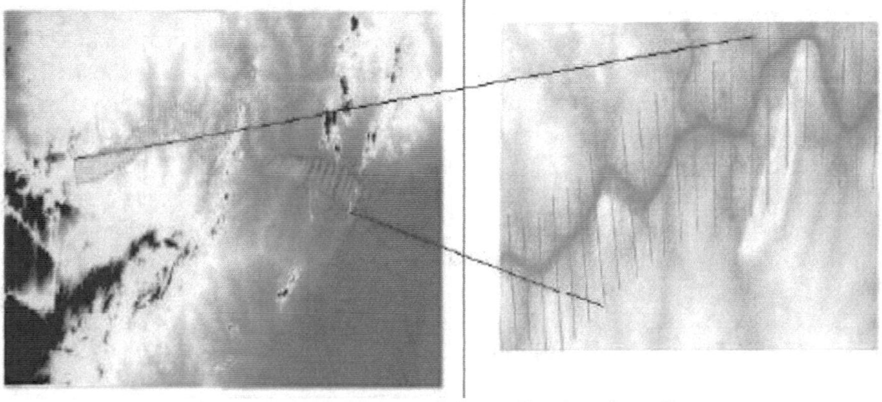

Figure 3.4 Digitizing of Cross Sectional cut lines

3.4.5 Exporting GIS data to HEC-RAS

The last step is to create a GIS import file for HEC-RAS so that it could import the GIS data to create the geometry file. Firstly, it is necessary to define which layers would be exported to HEC-RAS. Using RAS Geometry feature of HECGeo-RAS the GIS Data can be imported to HEC-RAS. Theimported geometry data of river cross- sections is shown in Figure 3.5

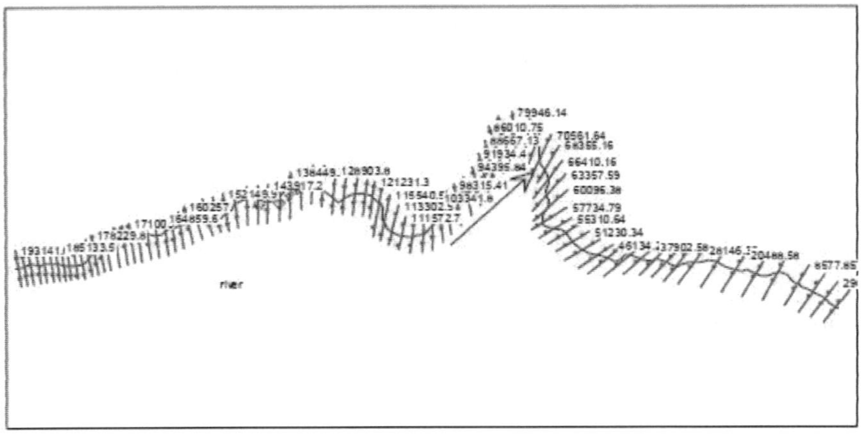

Figure 3.5 Geometric Data imported to HEC-RAS

3.4.6 Flood frequency analysis for the flow data

From the available daily flow data at all the hydro meteorological stations, annual maximum flows are found and flood frequency analysis is carried out for the annual maximum flow data. For fitting the suitable distribution to the data, Chi-Square goodness of fit test is performed at all the hydro meteorological stations of the study area. Chi-Square test is one of the most commonly used tests and can be used for testing the goodness of fit of any distribution. The test makes a comparison between the actual number of observations and the expected number of observations.

4. Results and analysis

The HEC-RAS model is calibrated with observed daily discharge data of monsoon season (June to September) of the year 1978 for different Manning's n i.e. 'n' starting from 0.03 up to 0.05. Though the simulation has been carried out for Manning's 'n' from 0.03 up to 0.05 with increment of 0.0025 (say 0.03, 0.0325, 0.035, 0.0375, 0.04 etc.), and it is found that at a value of Manning's 'n' 0.03, the model is giving better results by comparing observed and simulated flows as shown in Figure 4.1. Flood frequency analysis is carried for available flow data. HEC-RAS analysis is carried out for flows corresponding to 2, 10, 25, 50 and 100 year return periods. Floodplain maps are generated in ArcGIS by importing HEC-RAS outputs to ArcGIS again with the HEC Geo-RAS. To obtain floodplain map 'Base map' is added by loading 'GOOGLE EARTH' into ArcGIS. The generated water level boundary file of flow corresponding to each return period is superimposed on to the 'GOOGLE EARTH' and floodplain map is obtained. For sample, the water surface profile and flood plain map corresponding to flow of 100 year return period are shown in Figure 4.2 and Figure 4.3.

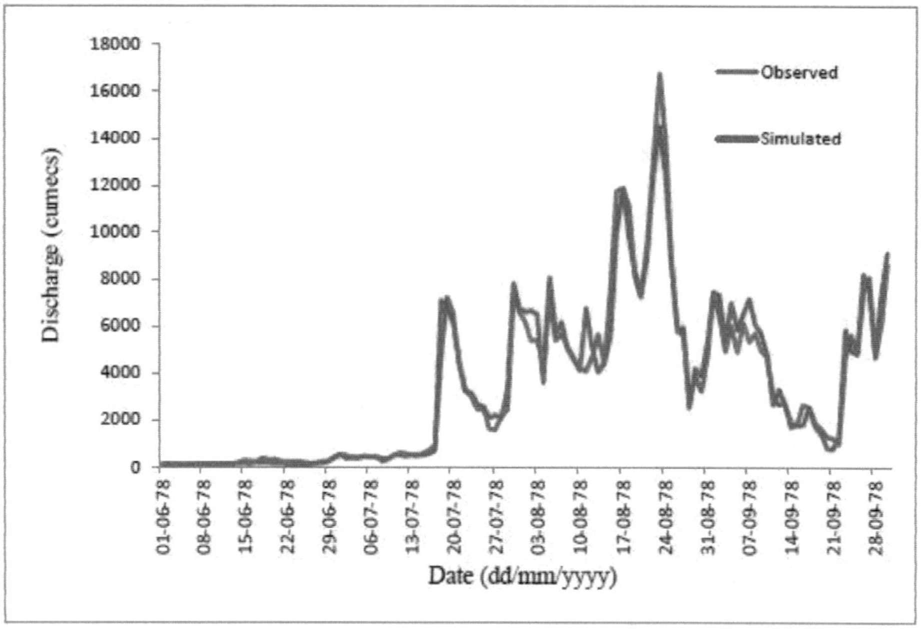

Figure 4.1 Comparison of Observed and Simulated Flows of Monsoon Season of Year 1978 at Wadenapally (RS 70561.64) for n=0.03

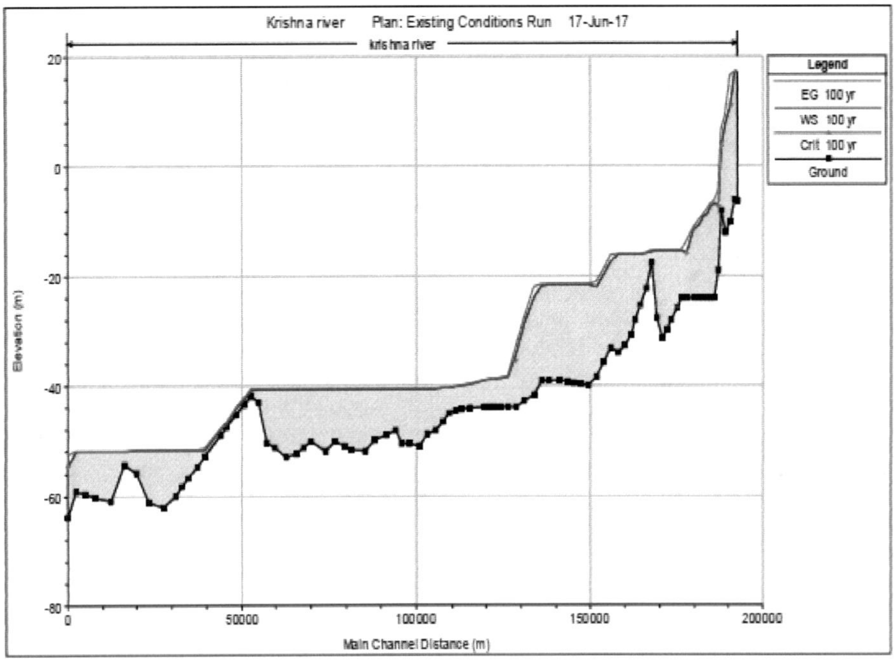

Figure 4.2 Water Surface Profile Plot of River Krishna for a Flow of 100 Year Return Period

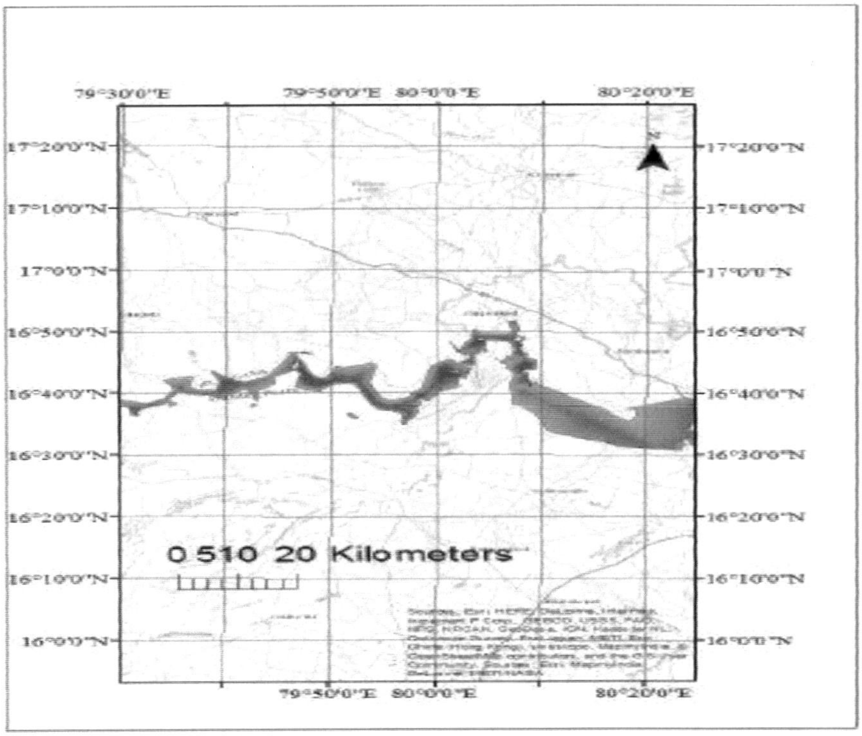

Figure 4.3 Flood Plain Map for a Flow of 100 Year Return Period

5. CONCLUSIONS

The following are the important conclusions drawn from the present study:

1. In the present study, the Manning's 'n' value for the river reach has been determined as 0.03 based on the calibration of the model. The model performed reasonably well during calibration using daily discharge data of monsoon season (June to September) of 1978 at Wadenapally with NSE as 0.963, deviation in peak as 2233.939 m3/s and percentage deviation in peak as 13.32% lower than observed peak flow.
2. Visualising the results in longitudinal profile (Water Surface Profile Plot) helps in understanding the variation in water levels at all cross sections at the same time. In the present study, the water surface profile plots and flood plain maps are generated for flows corresponding to 2, 10, 25, 50 and 100 year return period.
3. It can be observed, from the generated floodplain maps, that in the downstream side of the river there is more flood inundation extent because of the effect of Prakasam barrage.
4. From chainage of 140 km to 160 km the slope is too steep, with lesser Froude's number and narrower cross section, which leads to high velocity which may cause erosion.
5. The results obtained in this study provide essential information for administrative to analyse and manage flood hazards and also formulate remedial strategy.

6. REFERENCES

[1] Ackerman, C. T., Jensen, M. R., & Brunner, G. W. (2009). New floodplain delineation capabilities in HEC-RAS. In *World Environmental and Water Resources Congress 2009: Great Rivers* (pp. 1-7).

[2] Ahn, J., Cho, W., Kim, T., Shin, H., & Heo, J. H. (2014). Flood frequency analysis for the annual peak flows simulated by an event-based rainfall-runoff model in an urban drainage basin. *Water*, *6*(12), 3841-3863.

[3] Alaghmand, S., Bin Abdullah, R., Abustan, I., & Vosoogh, B. (2010). GIS-based river flood hazard mapping in urban area (a case study in Kayu Ara River Basin, Malaysia). *International Journal of Engineering and Technology*, *2*(6), 488-500.

[4] Cesur, D. (2007). GIS as an information technology framework for water modeling. *Journal of Hydroinformatics*, *9*(2), 123-134.

[5] Cook, A., & Merwade, V. (2009). Effect of topographic data, geometric configuration and modeling approach on flood inundation mapping. *Journal of Hydrology*, *377*(1), 131-142.

[6] Hernandez, T., & Zhang, B. (2007). Floodplain analysis using computational tools. In *World Environmental and Water Resources Congress 2007: Restoring Our Natural Habitat* (pp. 1-13).

[7] Heydari, M., Sadeghian, M. S., & Moharrampour, M. (2013). Flood Zoning Simulation byHEC-RAS Model (Case Study: Johor River-Kota Tinggi Region).

[8] Hu, H. H., & Walton, R. (2008). Advanced guidance on use of steady HEC-RAS. In *World Environmental and Water Resources Congress 2008: Ahupua'A* (pp. 1-10).

[9] Khattak, M. S., Anwar, F., Saeed, T. U., Sharif, M., Sheraz, K., & Ahmed, A. (2016). Floodplain mapping using HEC-RAS and ArcGIS: a case study of Kabul River. *Arabian Journal for Science and Engineering*, *41*(4), 1375-1390.

[10] Knebl, M. R., Yang, Z. L., Hutchison, K., & Maidment, D. R. (2005). Regional scale flood modeling using NEXRAD rainfall, GIS, and HEC-HMS/RAS: a case study for the San Antonio River Basin Summer 2002 storm event. *Journal of Environmental Management*, *75*(4), 325-336.

[11] Krause, P., Boyle, D. P., & Bäse, F. (2005). Comparison of different efficiency criteria for hydrological model assessment. *Advances in Geosciences*, *5*, 89-97.

[12] Merwade, V., Cook, A., & Coonrod, J. (2008). GIS techniques for creating river terrain

models for hydrodynamic modeling and flood inundation mapping. *Environmental Modelling & Software*, *23*(10), 1300-1311.

[13] Meyer, S., & Olivera, F. (2007). Flood plain mapping and hydraulic analysis with HEC-GeoRAS 4.1. 1 and ArcGIS 9. *Report by Texas A&M University, College Station, TX, USA*.

[14] Napradean, I., & Chira, R. (2006). The hydrological modeling of the Usturoi valley-using two modeling programs-WetSpa and HecRas. *Carpathian Journal of Earth and Environmental Sciences*, *1*(1), 53-62.

[15] NUT, N., & PLERMKAMON, V. (2015). Floodplain mapping using HEC-RAS and GIS in Nam Phong River Basin, Thailand. *International Journal of Environmental and Rural Development*, *6*(1), 153-158.

[16] Paz, A. R., Bravo, J. M., Allasia, D., Collischonn, W., & Tucci, C. E. M. (2009). Large-scale hydrodynamic modeling of a complex river network and floodplains. *Journal of hydrologic engineering*, *15*(2), 152-165.

[17] Pistocchi, A., & Mazzoli, P. (2002). Use of HEC-RAS and HEC-HMS models with ArcView for hydrologic risk management. In *IEMS 2002-Proc. of the Int. Environmental Modelling and Software Society Conference, Lugano, Switzerland* (pp. 305-310).

[18] Qi, H., & Altinakar, M. S. (2011). A GIS-based decision support system for integrated flood management under uncertainty with two dimensional numerical simulations. *Environmental Modelling & Software*, *26*(6), 817-821.

[19] Rivera, S., Hernandez, A. J., Ramsey, R. D., & Suarez, G. (2007). Predicting flood hazard areas: a swat and HEC-RAS simulations conducted in Aguan river basin of Honduras, central America. In *ASPRS 2007 Annual Conference* (pp. 7-11).

IOP Conf. Series: Materials Science and Engineering 1112 (2021) 012025 doi:10.1088/1757-899X/1112/1/012025

A CASE STUDY ON EVALUATION AND DESIGN OF ON-STREET PARKING IN NARASARAOPET TOWN IN ANDHRA PRADESH

G Leela Suresh[1], Dr R Ratna Prasad[2] ,T Venkateswara Rao[3]

[1]Department of Civil Engineering, VVIT, Guntur, Andhra Pradesh, India
[2]Department of Civil Engineering, VVIT, Guntur, Andhra Pradesh, India
[3]Department of Civil Engineering, VVIT, Guntur, Andhra Pradesh, India
*Corresponding Author: Gsuresh@vvit.net

Abstract. One of the most significant transport issues has been the lack of parking policies. Parking problem in the central business districts and major cities across the globe is trending. When we look at Narasaraopet, market growth and training contribute to a rise in transport facilities and thus parking problems. Lack of off-road space and the propensity to park vehicles near the destination contributes to high demands for parking. Further reasons for increased demand are car ownership and low efficiency of the transit system. These factors reduce the main wagon route width, reduce the flow speed, boost crash rates and generate excessive traffic congestion. There is therefore a need to provide parking on the lane. So by means of in-out or sampling of a fixed time or license plate scheme, we have to assess the current on-street parking conditions in Narasaraopet and see whether it's appropriate or not.If they're not enough, we need to prepare Narasaraopet on-street parking conditions. In this research, proposed techniques should be implemented by the transportation department from an effective point ofviewof transportation system.

Key words. on street parking, parking volume, parking requirements, parking bays

1. Introduction

Parking is a place where a car can be stored. Parking areas in cities where the availability of parking is a primary concern of car users are considered to be among the basic land uses. The parking problem is one of the burning issues in the CBD of metropolises due to increasing population concentration, rapid urbanization and certain economic activities. Parking issues in all sorts of places can be found. Owing to large volumes of traffic in comparatively small spaces, City centre is impacted by its inability to have sufficient areas for parking.A major concern for the transport engineers has been the rise in demand for parking loads as a result of the rapid growth in private vehicle ownership in India. Parking demands are typically made by on- and/or off-road parking provisions.

One of the main topics of movement control is the behavioral pattern of road car parking. The city's road system's capacity management is essential to ensure that street parking is properly administered on high demand roads. On-street parking slows down, assists and interrupts traffic flow, particularly if vehicles are maneuvering in and from the curb; but if on-street parking is properly planned, modified and properly included in the calculation of traffic capacity, using an equation model, the disruption of on-street parking to traffic flow can be reduced There are unique problems associated with mixed traffic in urban streets in increasing international locations.

The same street area is shared by peasants, bicycles, taxis, motorcycles/ scooters, auto rickshaws, bike-rickshaws and different modes of transport, which generate inefficient conditions under which the economic potential of cities is stolen in developing world locations. These problems are yet to be overcome due to various causes. They are becoming more complicated because of the inability to settle these matters. The scenario is more disturbing if motor cars are kept in rising foreign locations. Road control and road regulation, as well as road installations, require a number of road studies in order to plan, build and operate them.

The city's road system's capacity management is essential to ensure that street parking is properly administered on high demand roads. On-street parking slows down, assists and interrupts traffic flow, particularly if vehicles are maneuvering in and from the curb; but if on-street parking is properly planned, modified and properly included in the calculation of traffic capacity, using an equation model, the disruption of on-street parking to traffic flow can be reduced There are unique problems associated with mixed traffic in urban streets in increasing international locations.

The same area is shared by pedestrians, cyclists, taxis, engines, motorcycles/scooter, car rickshaws, cycling rickshaws and different transport modes which are inefficient types of mobility that are stolen from urban economies in the growth of international locations. These problems are yet to be overcome due to various causes. They are becoming more complicated because of the inability to settle these matters. The scenario is more disturbing if motor cars are kept in rising foreign locations. Road control and road regulation, as well as road installations, require a number of road studies in order to plan, build and operate them. Their characteristics become tougher and time consuming with various aspects of traffic flow in the region. Furthermore, it is not possible to draw on experience in this area with a large range of traffic and transmission due to practical difficulties.

This study has been aimed to evaluate the traffic parameters of on street parking system by proposing the new methods which include the collection of data and analysis under recommended analysis techniques and to compare the results with original transportation properties.

2. Methodology and data collection

In order to collect the aforesaid parking data, parking surveys are carried out. In-out surveys, fixed-period sampling and license plate methods were the most common surveys performed in parking. Parking Characteristics: The research aims at understanding the existing parking patterns in the region under study and the general parking issues. Parking on the road or kerb parking, the parking pattern, the smooth flow of traffic caused by parking and un-parking activities and the incidents involved during parking operations also have to be investigated. If the car park is offstreet, minimum, maximum and average walking distance values should be identified to reach various destinations after the vehicle has been parked.

Parking space inventory: There is a complete survey of the area under research and a map of every place where kerb parking can be provided and off-street parking facilities to satisfy the parking demand.

Parking demand: The accumulation term parking indicates the cumulative number of vehicles parked in a given location over a defined period of time. Depending on the timing of the day and the venue, demand for parking increases from the beginning to a maximum in the afternoon and decreases later in the night. The time a car spends on a parking spot is the time of parking. The number of parking spaces in a parking lot is a percentage of total available parking spaces. The number of times the parking space was used during the day is generally measured as parking turnover.

3. Results and Discussions

Design of parking at Mallamma center:

Parking length available	= 220m
Number of bays available for parking	= 205
Angle of parking at present	= 60°
Total width of road	= 16.77 m
Parking width	= 2.17 m
Travel width available	= 14.6 m
Length occupied by street hawkers and parking	= 3.12 m
Number of bays required for parking from the collect data is	= 398
Land acquired for construction of road	=16.77m

From the above collected data we should understand there is no place for the extension of present existing road and at a time we have large parking volume. Present parking angle is 60° which is insufficient for present parking demand. From that vehicles are parked sides of the carriageway which leads to decreasing of the travel width up to 21.3% and increasing of the journey time. for overcame of all this problems change the angle of parking from60° to 45°

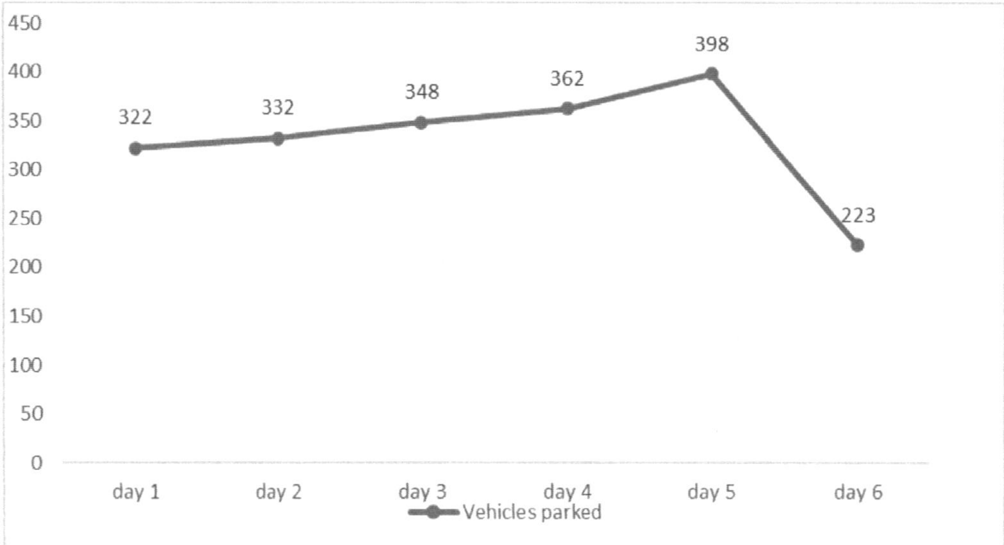

Fig 1. Graphical Representation of the Peak Parking Demand of 6 Days

Design of parking at kota center:

Length of the road	= 128.6m
Parking length available	= 257.2m
Number of bays available for parking	= 0
Present parking system	= not existing
Total width of road	= 5.55m
Travel width available	= 3 m
Length occupied by street hawkers and parking	= 2.55m
Number of bays required for parking from the collect data is	= 316
Land acquired for construction of road in FMB	=28.67m

IOP Conf. Series: Materials Science and Engineering 1112 (2021) 012025 doi:10.1088/1757-899X/1112/1/012025

Width available for extension of present road =23.12m
Width taken for providing of parking system =4.5

From the above collected data we should understand there is sufficient place for the extension of present existing road. But now all shops occupy the total road width only 5.55m is available. Out of the available width both parking and vehicle moment has occur so the number traffic jams are observed and the travelling time also increases this lead to unfavorable conditions to the road users. So necessary to improve the road width and providing of sufficient parking system to avoid the traffic jams and for creating of free flow conditions. The available road width is only used for the vehicle moment and side parking is provided by extension of shoulders 2.25m both side of the carriage way.

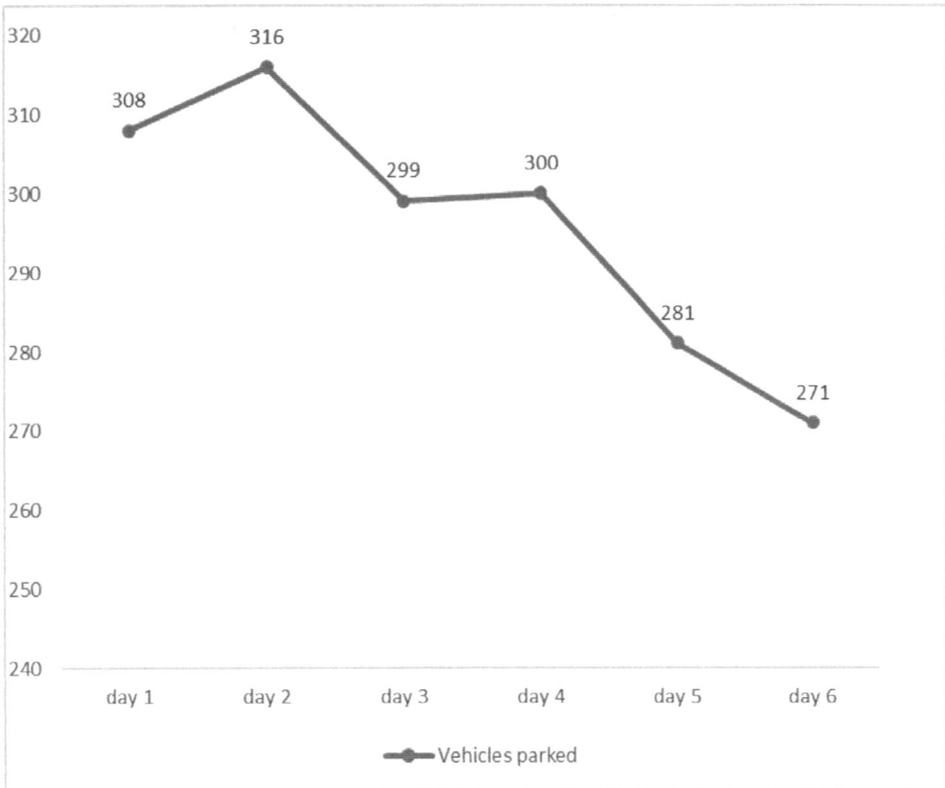

Fig 2. Graphical Representation the Peak Parking Demand Of 6 Days

Parking length we provide = 257.2m
Present parking system = not existing
Number of bays required for parking from the collect data is =316
Designed dimensions of 60° parking =2.2m × 1.07m
No of bays designed for parking =321
For future purpose we provide extra bays = 5

Design of parking at more center:

Parking length available	= 542.72m
Number of bays available for parking	= 0
Present parking system	= not existing
Total width of road	= 10.9m
Travel width available	= 4m
Length occupied by street hawkers and parking	= 6.9m
Number of bays required for parking from the collect data	=920
Land acquired for construction of road in FMB	=13.725m
Width available for extension of present road	=2.825m

From the above collected data we should understand there is sufficient place for the extension of present existing road. But street hawkers and road side parking leads to reduction of carriage way width. Which leads to decreasing of the travel width 7 and increasing of the journey time. Out of the available width both parking and vehicle moment has occur so the number traffic jams are observed and the travelling time also increases this lead to unfavorable conditions to the road users. So necessary to improve the road width and providing of sufficient parking system. The available road width is only used for the vehicle moment and side parking is provided by extension of shoulders.

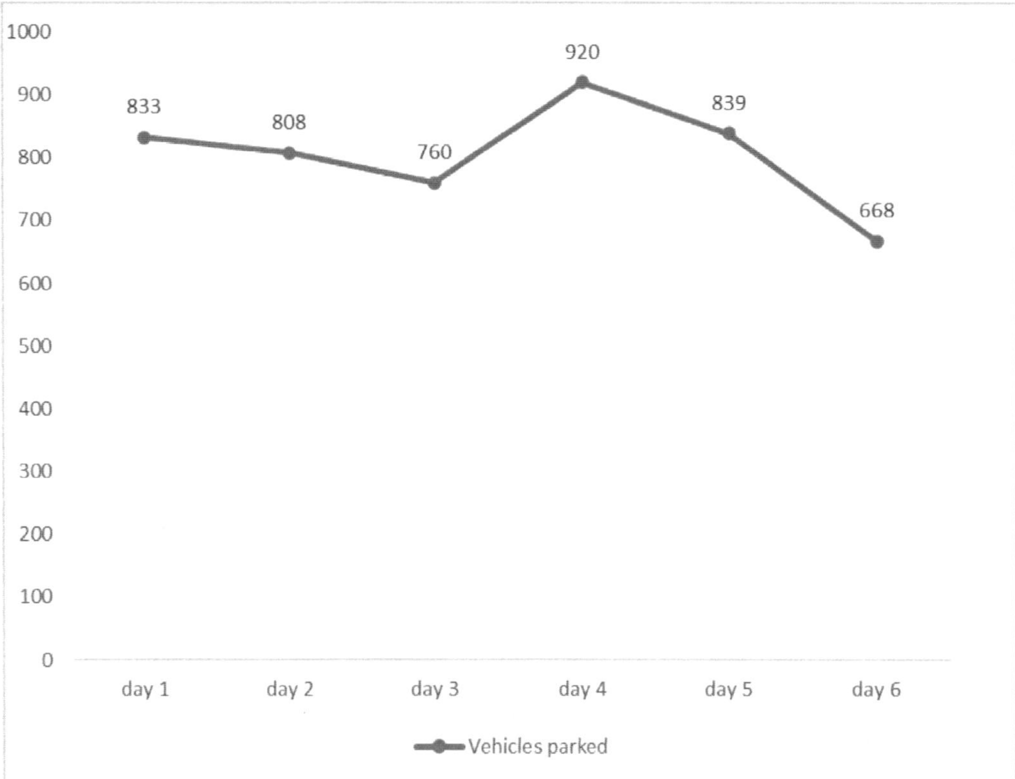

Fig 3 . Graphical Representation the Peak Parking Demand Of 6 Days

Parking length available = 542.72m
Present parking system = not existing
Number of bays required for parking from the collect data is =920
Designed dimensions of 60° parking =2.2m × 1.07m
No of bays designed for parking =678
No of bays designed for of street parking =242
We have no place for providing on street for 242 bays so park that vehicles on off street parking in and around of 1km there is no government land.

Design of parking at Palnadu center

Length of road available = 125m
Number of bays available for parking = 0
Present parking system = not existing
Total width of road = 15m
Travel width available = 7.6m
Length occupied by street hawkers and parking = 7.4m
Number of bays required for parking from the collect data =293
Land acquired for construction of road in FMB =21.73
Width available for extension of present road =2.825m

From the above collected data we should understand there is sufficient place for the extension of present existing road. But street hawkers and road side parking leads to reduction of carriage way width. Which leads to decreasing of the travel width and increasing of the journey time. Out of the available width both parking and vehicle moment has occurred so the number traffic jams are observed and the travelling time also increases this lead to unfavorable conditions to the road users. So necessary to improve the road width and providing of sufficient parking system. The available road width is only used for the vehicle moment and side parking is provided by extension of shoulders.

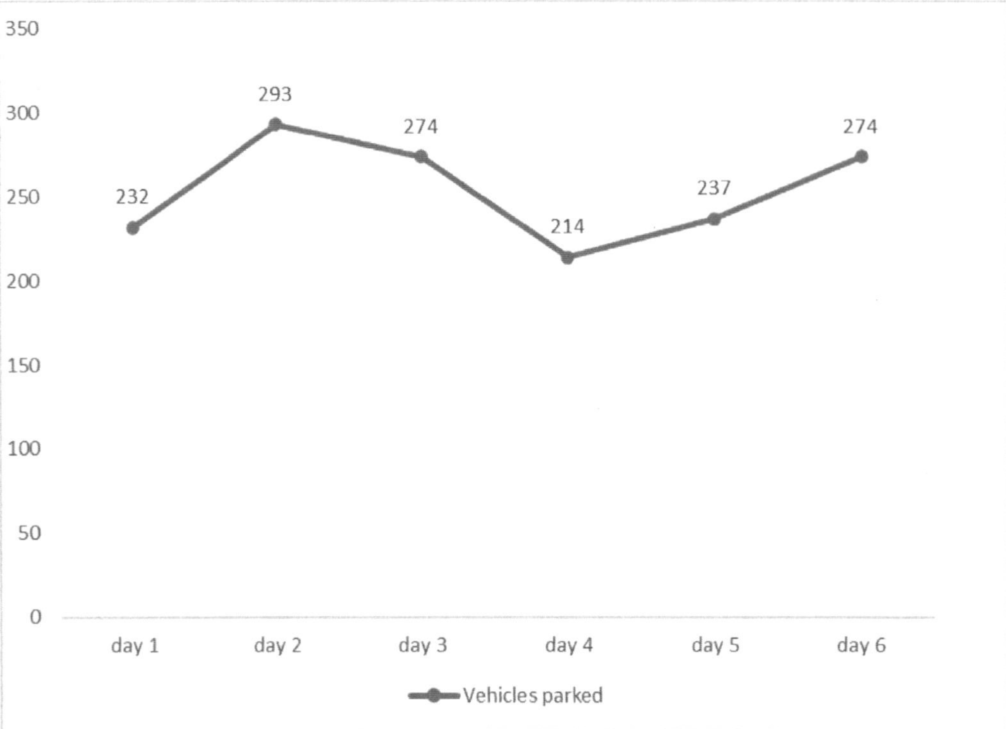

Fig 4 . Graphical Representation the Peak Parking Demand Of 6 Days

Parking length we provide	= 250m
Present parking system	= not existing
Number of bays required for parking from the collect data is	=293
Designed dimensions of 60° parking	=2.2m × 1.07m
No of bays designed for parking	=312
For future purpose we provide extra bays	=19

4. Conclusion

a) Mallamma Center

Mallamma center is one of the center having more traffic in Narasaraopet. In this center a four road junction is present. Among four roads two roads contains on street parking at the middle of the road for bikes. On street parking with bay markings with an angle of 60^0. Marketing is more in this center so we proposed this center to evaluate existing on street parking. For evaluation of existing on street parking we did different activities at this center they are

- We done six days in and out survey with 15 minutes interval.
- By this survey we count the number of vehicles parked at the bays and side of the road.
- The road users parked their vehicles at the side of the road due to insufficient space at the middle of the road which is allotted for parking only.

b) KOTA CENTER

We should understand there is sufficient place for the extension of present existing road. But now all shops occupy the total road width only 5.55m is available. Out of the available width both parking and vehicle moment has occur so the number traffic jams are observed and the travelling time also increases this lead to unfavorable conditions to the road users. So necessary to improve the road width and providing of sufficient parking system to avoid the traffic jams and for creating of free flow conditions. The available road width is only used for the vehicle moment and side parking is provided by extension of shoulders 2.25m both side of the carriage way.

c) MORE CENTER

We should understand there is sufficient place for the extension of present existing road. But street hawkers and road side parking leads to reduction of carriage way width. Which leads to decreasing of the travel width and increasing of the journey time. Out of the available width both parking and vehicle moment has occur so the number traffic jams are observed and the travelling time also increases this lead to unfavorable conditions to the road users. So necessary to improve the road width and providing of sufficient parking system. The available road width is only used for the vehicle moment and side parking is provided by extension of shoulders.

d) PALNADUCENTER

This centre is busy with various activities; a number of commercial establishments, markets, government offices and the bus stand are situated here. Since most of these places are frequented by public and busy with floating population, the demand for parking is very high. On-street parking is observed on all the roads surrounding palnadu bus stand. Many cars and two wheelers are seen parked on sides of the roads. Both angular as well as parallel type of parking was noticed on almost all the stretches of the roads. This has reduced the capacity of the carriageway and endangering pedestrians and motorists alike. The frontage of almost all the roads in this area has been converted into commercial land use without taking into account the demand for parking of the vehicles. There is no planned parking space available. Based on collecting data at this center we proposed that:
The parking space is more in this center but there is no planned parking space.so if we provide 60^0parking to two wheeler motor vehicles the parking space will be reduced. Which results increase in travel width and decrease in congestion.

References

[1] AlirezaNaseria (2013)"Analysis of On-street parking Impact of Traffic Flow Velocity" Middle-East Journal of Scientific Research, pp.652-657.

[2] Arjun C A and M.S. Nagakumar (2014)"Studies on On-Street Parking Using License Plate Method in Basavangudi, Bangalore", International Journal of Emerging Technologies and Engineering.

[3] Charles Anum Adams, Richter Opoku-Boahen, M. Abdul MuhsinZambbang (2014)"Parking Management in Metropolitan Cities in West Africa Case Study of the Kumasi Paid Parking Scheme,Ghana", International Refereed Journal of Engineering and Science, Vol.3, pp.01-08.

[4] Chung & Cassidy (2007)"Relationship between traffic density and capacity drops at three freeway bottle neck" Journal of Transportation Research Part B; Methodology, 41:82-95..

[5] Marshall, W.E and N.W Garrick (2006)"Parking at mixed- use centres in small cities" Journal of the transportation Research board, No.1977, pp.164-171

[6] Russell and Anthony (1999)"A Parking search model"Journal of Transportation Research Part-B: Methodology, 41:82-95

[7] R.A. Asiyambola, A.A. Akinpelu (2012) "The Challenges of On-street Parking in Nigerian Cities' Transport Routes" International Journal of Development and Sustainability, , Vol.1, pp.476-489.

[8] L.R.Kadiyali,A Text book of "Traffic Engineering and Transport Planning", Khanna Publishers, 2013.

[9] IRC-106"Guidelines For Capacity Of Urban Roads In Plain Areas", The Indian Road Congress, 1990

[10] Sirous Heidar Reihani, Alireza naseri, Kasra Zehforoush (2013) "Modeling the Impact of On-street Parking on Main Parameters on Vehicular Traffic", Journal of American Sciences, pp.99(5s).

[11] IRC-SP-12"Tentative Recommendations On The Provision Of Parking Spaces For Urban Areas", Indian Road Congress Special Publication, 1988.

[12] Y. Yorozu, M. Hirano, K. Oka, and Y. Tagawa, "Electron spectroscopy studies on magneto-optical media and plastic substrate interface," IEEE Transl. J. Magn. Japan, vol. 2, pp. 740-741, August 1987 [Digests 9th Annual Conf. Magnetics Japan, p. 301, 1982].

A comparative study on seismic analysis of G+6 building with ribbed slab & conventional slab using SAP2000 software

Mirza Mahaboob Baig[1], Abdul Rashid[2], Y Pavan Sai Durga Reddy[3], T.G.N.C Vamsi Krishna[4]

[1] Assistant Professor, Department of Civil Engineering, Vasireddy Venkatadri Institute of Technology, Nambur, Guntur, Andhra Pradesh, India

[2] Assistant Professor, Department of Civil Engineering, Vasireddy Venkatadri Institute of Technology, Nambur, Guntur, Andhra Pradesh, India

[3] Assistant Professor, Department of Civil Engineering, Vasireddy Venkatadri Institute of Technology, Nambur, Guntur, Andhra Pradesh, India

[4] Assistant Professor, Department of Civil Engineering, Chalapathi Institute of Engineering and Technology, Lam, Guntur, Andhra Pradesh, India

E-mail: mahaboobbaig8@gmail.com, raforashid999@gmail.com, pavansai463@gmail.com, tgncvamsikrishna@gmail.com.

Abstract. The stability of structures mainly depends on the structural components like beams, columns and slab. Here in this study columns are reduced in the large open areas to access the whole room, which can be done by using the ribbed slab, grid slab and flat slab, which will help in minimizing the no. of columns which are required in the structure. A comparative study on seismic behaviour of G+6 school building is done with conventional slab and ribbed slab at seismic zone III and medium soil conditions by using SAP2000 software, it is found that in gravity analysis the maximum bending moments and shear force obtained for normal slab structure are 84.96 kN-m and 80.86 kN respectively, whereas for ribbed slab structure is 137.39 kN-m and 108.01 kN. The maximum storey stiffness of the ribbed slab structure is found to be 0.94 times of maximum storey stiffness of normal slab structure. It is found that the ribbed slab structure doesn't perform well in moderate to high seismic areas. The stability of the ribbed slab structure can be enhanced by retrofitting the existing structure.

1. Introduction

An Earthquake is Earth's Shaking or the arrival of energy because of the movement of tectonic plates. This can be dangerous to slaughter a great many individuals and bring gigantic financial misfortune. Buildings are exceptionally influenced by a quake: and sometimes they are broken to the earth level. At the point when the ground shaking happens underneath the structure's establishments, they will get vibrated. There has been an extending enthusiasm for the development of seismic tremor safe structures; the fundamental target of this investigation is to get the behaviour of a structure with the ribbed slab. Also, the slab is a fundamental portion of the RCC type of structure which is intended to withstand both vertical and horizontal burden coming about during quakes. A ribbed slab is a lighter and stiffer slab when compared to the flat slab. They work very excellently where slab vibration is an issue, ribbed slab sections comprised of wide beams with narrow ribs crossing in a perpendicular direction to each other and a thin slab over it. Advantages of Ribbed type of slabs are it is flexible, moderately light, economical for a longer range of spans, construction is speed, thin floor depths, robustness, excellent vibration control.

Ribbed slab consists of parallel reinforced T shaped concrete beams which are framed into RCC girders. The slab is considered as the flange of the beam and extended parts as ribs. There are generally two types of ribbed slabs.

1. One way ribbed slab
2. Two-way ribbed slab (Waffle slab)

2. Objective and Methodology

1. To identify the seismic performance of ribbed slab structures after the removal of obstructed columns in a hall or room of a building.
2. To analyze and design the structure as per IS code recommendations.
3. To analyze and compare the behaviour of structure with normal slab and ribbed slab.

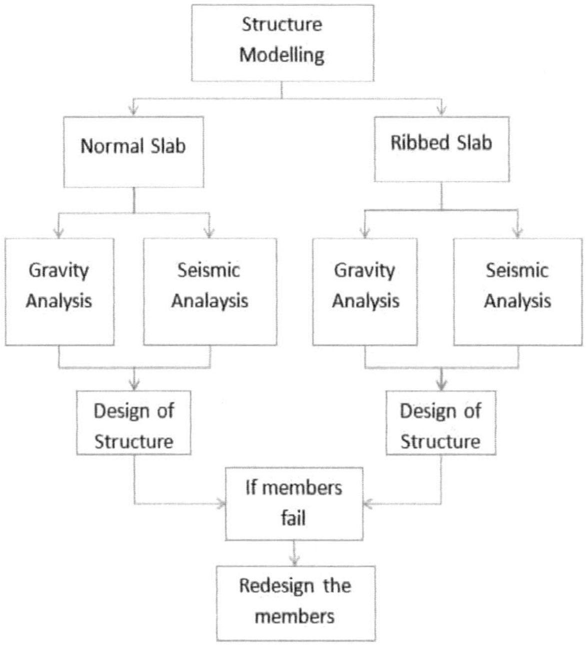

Fig 1: Flowchart of methodology

3. Structural modelling

3.1 General properties

A G+6 building is modelled in SAP 2000 Software, with storey height of 3m, length of the structure in one direction is 25m and in other direction, it is 39m and the member sizes will vary according to design requirements. The thickness of the slab is 15cm; the thickness of the external and internal wall is 23cm and 11.5cm respectively. The model is analyzed with two different Structural elements one is Normal Slab Structure and the other is Ribbed Slab Structure and designed according to IS 456-2000.

Table 1: Material properties

Name	Type	E MPa	Unit Weight kN/m^3	Design Strengths
Concrete	M30	27386	25	Fc=30 MPa
HYSD415	Rebar	200000	76.97	Fy=415 MPa, Fu=485 MPa
Mild250	Rebar	200000	76.97	Fy=250 MPa, Fu=410 MPa

IOP Conf. Series: Materials Science and Engineering 1112 (2021) 012026 doi:10.1088/1757-899X/1112/1/012026

3.2 Plan views of normal slab and ribbed slab structures

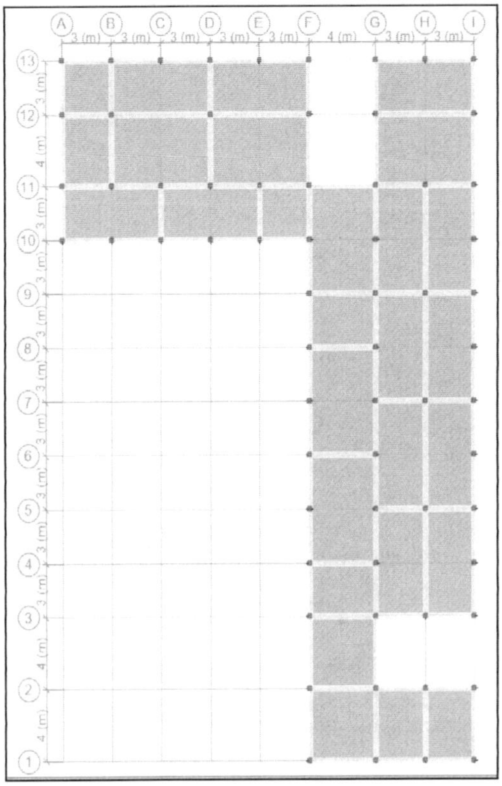

Fig 2: Plan view of normal slab structure

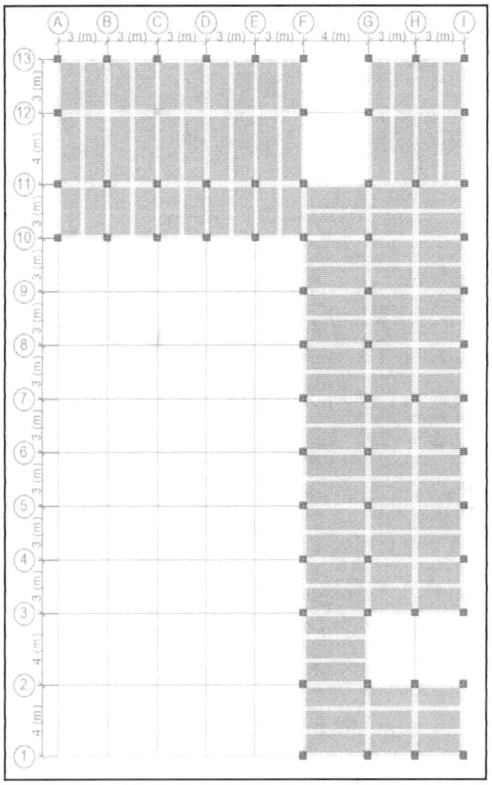

Fig 3: Plan view of ribbed slab structure

3.3 3D views of structures

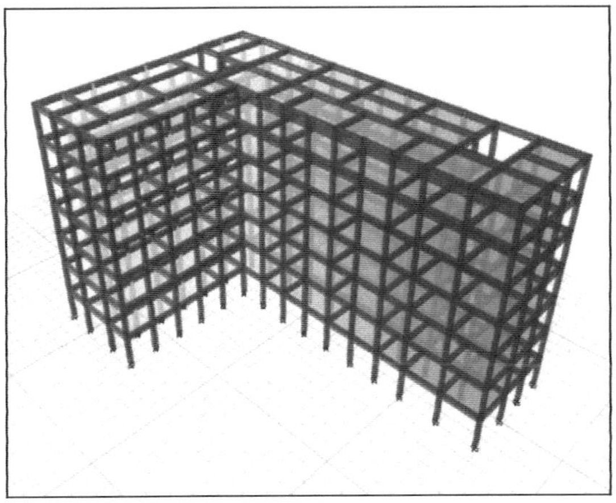

Fig 4: 3D model of ribbed slab structure

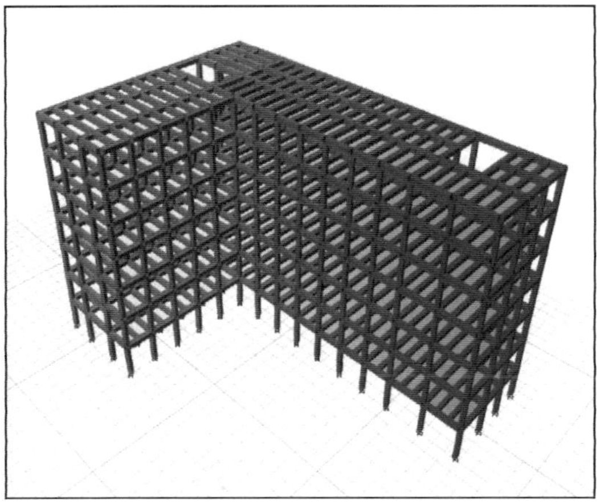

Fig 5: 3D model of normal slab structure

4. Loading on structure

Loads are taken from Indian standard codebooks for dead loads we have IS 875 part 1, for Live loads IS 875 part 2 and seismic analysis is done according to the IS 1893 part 1 2016.

Table 2: Loading data

Type of Load	Intensity of Load
Live load	3 kN/m^2 (IS 875-Part 2)
Floor finishing	1.1 kN/m^2 (IS 875-Part 1)
Wall loads	External = 14 kN/m Internal = 7 kN/m
Slab and Lift wall thickness	Slab = 150 mm Lift wall = 250mm
Seismic zone	III
Type of soil	Medium

5. Results And Discussion

5.1 Results of gravity analysis

In gravity analysis, the maximum bending moments and shear force obtained for Normal slab are 84.96 kN-m and 80.86 kN respectively, whereas for Ribbed slab the gravity analysis result shows maximum bending moment as 137.39 kN-m and maximum shear force as 108.01 kN. In the table given below N.S.S stands for Normal Slab Structure whereas R.S.S stands for Ribbed Slab Structure.

Table 3: Maximum values of gravity analysis

Type of member	Load combination	Axial forces (kN)	Shear force (kN)	Bending Moment (kN-m)	Reinforcement (mm^2)
N.S.S Beam (400 x 300)	1.5(DL+LL)	------	80.86	84.96	663
R.S.S Beam (400 x 300)	1.5(DL+LL)	------	108.0	137.39	1120
N.S.S Column (350 x 300)	1.5(DL+LL)	2217.51	34.17	82.33	4448
R.S.S Column (400 x 350)	1.5(DL+LL)	2260.90	58.56	268.5	7268

5.2 Results of seismic analysis

Seismic analysis is done by the linear static (equivalent static) method to get structural responses like lateral displacement, story drifts, overturning moment, story stiffness and base shear for both Normal Slab Structure and Ribbed Slab Structure.

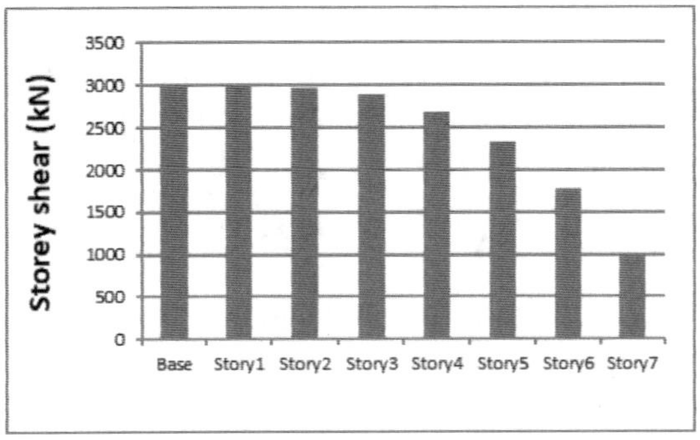

Fig 6: Storey shears of normal slab structure

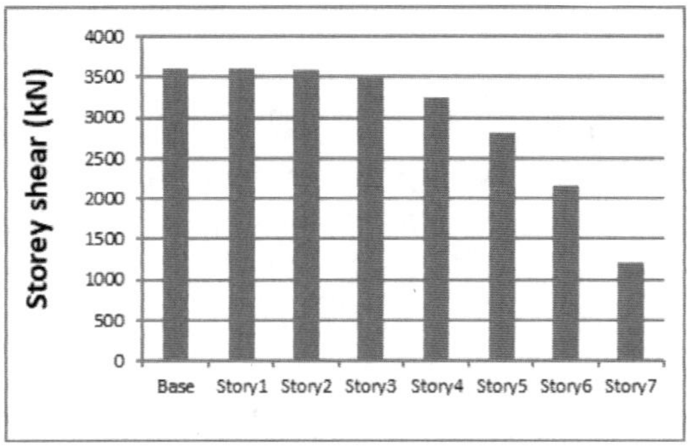

Fig 7: Storey shears of ribbed slab structure

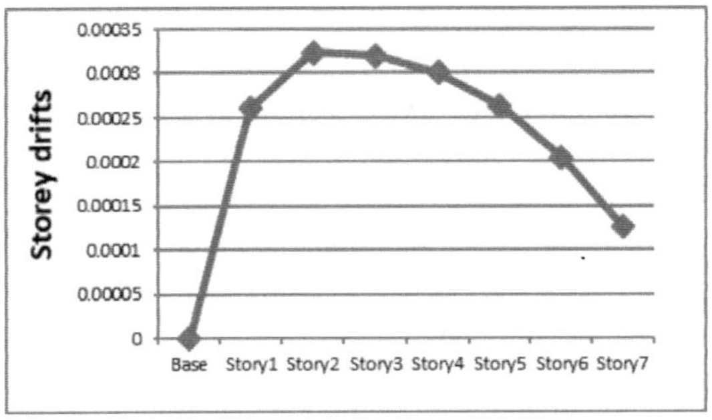

Fig 8: Storey drifts of normal slab structure

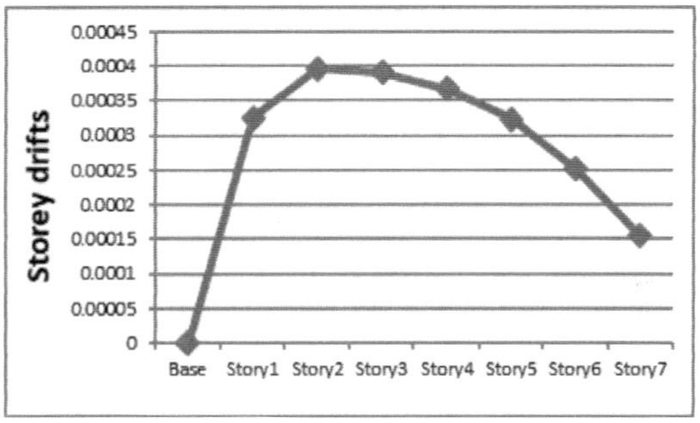

Fig 9; Storey drifts of ribbed slab structure

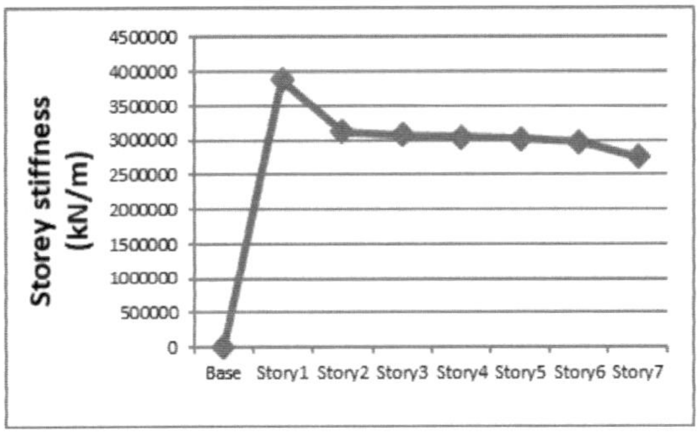

Fig 10: Storey stiffness of normal slab structure

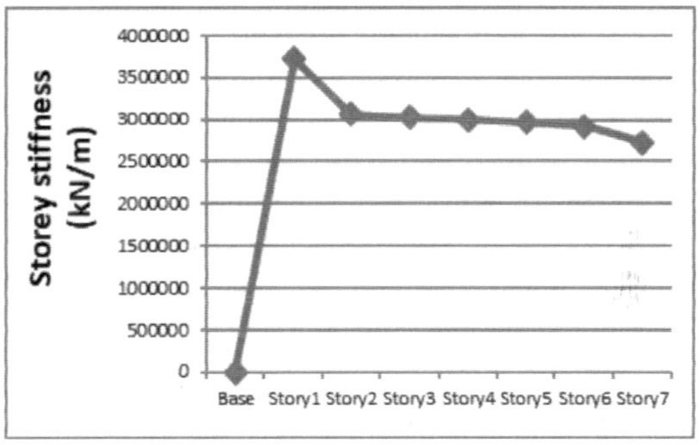

Fig 11: Storey stiffness of ribbed slab structure

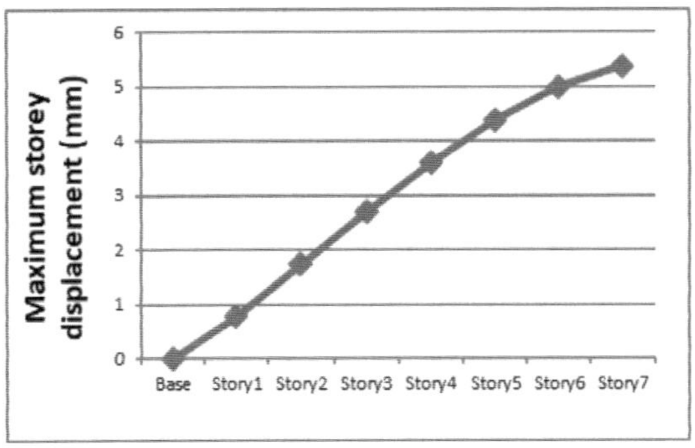

Fig 12: Storey displacements of normal slab structure

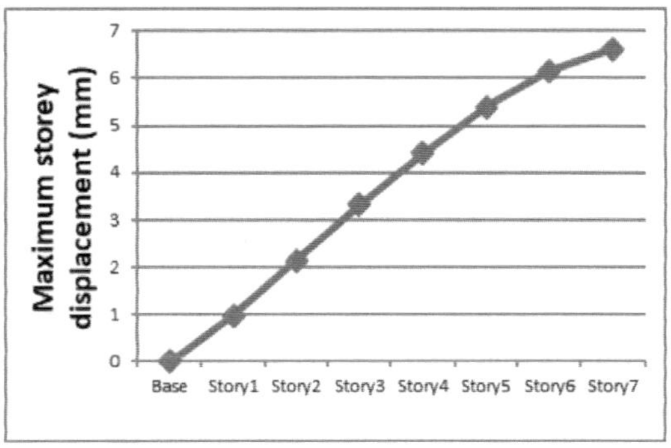

Fig 13: Storey displacements of ribbed slab structure

IOP Conf. Series: Materials Science and Engineering 1112 (2021) 012026 doi:10.1088/1757-899X/1112/1/012026

- From the above storey shear graphs, it is observed that lateral loads are more on ribbed slab structure when compared to normal slab structure, that means the ribbed slab structure has more self-weight.
- For both types of structures, it is observed that storey drifts are in permissible limits as recommended by the IS 1893 code. There is a little amount of variation for storey drifts when both the structures are compared.
- Storey stiffness is a bit more for normal slab structure as there are lesser columns in ribbed slab structure. In the software, storey stiffness will be calculated from storey1.
- Storey displacements were found to have much less variation when both of the structures are compared to each other.

6. Conclusions

It can be concluded that ribbed slab structures compared to the normal slab structure in high seismic zones does not give a good performance because of lesser columns, whereas its performance can be enhanced by retrofitting the structure.

1. Maximum storey shear for ribbed slab structure is increased by 22% of normal slab structure.
2. Maximum Lateral displacement for the ribbed slab structure increased by 23% of normal slab structure.
3. Lateral loads on a ribbed slab structure are more when compared to the normal slab structure due to the increase of self-weight.
4. Storey drifts for normal slab structure are 20% less than the ribbed slab structure.
5. The maximum overturning moment for ribbed slab structure is 20% more than the normal slab structure.
6. Maximum storey stiffness of ribbed slab structure is found to be 0.94 times of maximum storey stiffness of normal slab structure.

The maximum overturning moment for ribbed slab structure was found to be 60321.2 kN-m whereas for normal slab structure it is 49888.4 kN-m.

References

[1] A Benavent-Climent, X Cahís, A Catalán – Engineering. Structures, 2008 Elsevier. "Seismic behavior of interior connections in existing ribbed plates structure".

[2] A Kumar, S Pal - J. CivilEng. Environ. Technol, 2015 - krishisanskriti.org. "Parametric study and analysis of ribbed slab".

[3] MT BI, MR ME - 2018 - academia.edu. "A Study on the Behavior of Grid Slab Subjected to Seismic Loading".

[4] AS Samsudin, M Hashim, M Hisbany Scientific Research , 2018-ir.uitm.edu.my "Bending strength of steel fibre reinforced concrete ribbed slab panel".

[5] Duggal S K (2010), Fourth Edition, Oxford University Press, New Delhi. "Earthquake Resistance Design of Structure".

[6] Wilkinson S and Hiley R (2006), Computers and Structures, Vol. 84. " A Non- Linear Response History Model for the Seismic Analysis of High - Rise Framed Buildings".

[7] Cahís, X., Benavent-Climent (2008), A & Catalan, seismic evaluation of old RC waffle flat-plate systems. "The 14th World Conference on Earthquake Engineering (WCEE)".

[8] Pankaj Agarwal and ManishShirkhande. "Earthquake resistant design of structures".

[9] Www.csiamerica.com, "Structural and Earthquake Engineering Software, Computers and Structures.inc**".**

[10] Mahesh bakale **.**“ Seismic behavior of multi-story structure with different types of slabs**”.**

[11] Ying-Jie Jia, Quan Yuan, Qian-Feng Yao. **“**Research on Shear Resistant Capacity and Reliability of Multi-Ribbed Composite Wall High-rise Building rise Building rise**”.**

[12] Francesco Porco, Sergio Ruggieri, and Domenico Raffaele. **“**Influence of rigid floor assumption in seismicanalysis of rc existing buildings**”.**

[13] IS 456–2000, “*code of practice for reinforced and plain concrete* ",Bureau of Indian Standards, New Delhi, India, 2000.

[14] IS 1893-2016 (part 1),“*criteria for designing earthquake resistant* structures”.

[15] IS 875 (part 1 and part 2) “*Code of practice for building and structure design loads (other than earthquake)*”.

[16] Hancock JW. and Mackenzie AC. (1976). On the mechanisms of ductile failure in high strength steels subjected to multi-axial stress states. Journal of the Mechanics and Physics of Solids

[17] Kaufmann, E. J. (1997). Dynamic tension tests of simulated moment resisting frame weld joints. Chicago. Steel Tips,Structural Steel Education Council. American Institute of Steel Construction

[18] Shiravand, M. and Deylami, A. (2010). Application of Full Depth Side Plate to the Moment Connection of I-Beam to Double-I Column. Advances in Structural Engineering 13:6,1047-1062.

IOP Conf. Series: Materials Science and Engineering 1112 (2021) 012027 doi:10.1088/1757-899X/1112/1/012027

BLAST RESISTANCE OF G+10 COMMERCIAL BUILDING WITH AND WITHOUT SHEAR WALL SHAPE COMPARISON

Harish.Pendem[1], Sukesh Chandana[2]

[1]*Post Graduate In structuralEngineering, VVIT college, Guntur,522508, Andhra Pradesh, India*
[2]*Assistant professor, Dept of civil Engineering, VVIT college, Guntur,522508, Andhra Pradesh, India*
*Corresponding Author: [1]pharish0123@gmail.com , [2] chsukesh@vvit.net

Abstract. The effect of blast load on the structure was an essential aspect to be taken into account during the design process even though such kind of attacks are very scare in nature ,blast effect must be consider while designing a building just like wind ,seismic load and shear wall with and without shear wall load is be measured in this project mainly discuss about the shape comparison of the G+10 commercial building and that be with shear wall and without shear wall .when the blast load is applied to the structure then how much of deflection of the structure is appeared with and without shear wall shape comparison of the building so that we find the deflection of the structure by recording the values on the graph and so that how much of deflection is occurs and then that structure is safe with shear wall are not . applying loads on the each section of the building like beams and columns so that may get the how much of deformation the structure is appeared so may that the structure is safe without shear wall . mainly in this we find the displacement, acceleration, max displacement and max drift by taking that values and finding the deflection of the structure and that was the exact values of the deformed shape will be given. when subjected to blast load using ETABS software with prominence given to different standoff distances the blast load and incorporating various charge weight of TNT according to IS code 4991.

Key Words. ETABS software , Graphs, max Drift, max Displacement , Acceleration ,Displacement, TNT(Trinitrotoluene).

Introduction. The number of terrorist activities and threats has been exponentially increased in the recent few years across the globe and hence it has become a vital issue to consider blast load effects from the stage of designing a structure just like the consideration of Earthquake loads, Wind Loads etc. The terrorist actions has become an important issue and safe guard of the citizens against terrorist acts includes prediction, prevention & mitigation of such actions has gained importance.

The central intension of this examination is to reveal insight into the system of managing with the blast induced effects from the planning and design stage of the structure and to know the reaction of the structure when exposed blast loading with significance importance given on different Standoff separations of the impact and considering different charge loads of TNT as according to IS Code 4991 utilizing ETABS software.Impact idea ought to be considered at idea level since it assists with building a very much planned structure which can display improved blast resistance along with aesthetic appearance. For guaranteeing security of the current structures against impact, a strategy for assessment, investigation and retrofitting is required.

A clear understanding of the characteristics of blast, will help us to design the buildings to be blast resistant more competently. Basic methods for enhancing the ability of a structure to survive a blast should be with respect to both architectural & structural approach. The main endeavor of the study is to provide guidance for blast resistant design of buildings & using ETABS software to identify the response of a structure when subjected to blast load with importance given on a range of Standoff distances of the blast on the building and considering various charge weights of TNT according to IS Code 4991. Based on the needs & purpose of the building, the Dead loads, Partition Wall loads & Live loads are considered for the study of the structure according to the specifications of IS 875-1987

OBJECTIVES OF THE WORK

1) The purpose of the study is to through light on designing blast resistant buildings, which will enhance the building security against the effects of explosions.
2) To understand the explosion process and to have clear idea about the effects of explosion on buildings.
3) For assessing the probability of occurrence of an explosion and the impact that is to be included in the design based on the type of structure or importance of the structure.
4) To know the response of a structure when a building is subjected to blast loads using ETABS software in accordance with IS Code 4991.
5) To study the response of a building when imposed with blast load for various standoff distances and different charge weights.
6) To come to a conclusion of optimum percentage of safety that is to be included in the design for making a building economical.

Explosions and Blast occurrence

Explosions are grouped based on their type asnuclear explosion, physical explosion & chemical explosions. In case of nuclear blasts the energy will be released because of the high speed nuclear reaction which is a result of the formation of unlike atomic nuclei formed because of relocation of the neutrons & protons in an interacting nuclei. In the case of an explosion the energy will be released from the actions such as mixing of liquids, burning of explosive materials, bursting of gas cylinder etc. Rapid oxidation of fuel essentials mainly carbon & hydrogen atoms results in chemical explosion. Based on the physical state of the explosive materials they can be classified as solids, liquids & gases. The major type of explosives known are of solid state explosives. Based on the sensitivity of ignition the explosives are classified as primary & secondary explosions. Primary explosions are the one which gets easily detonated by easy ignition just with a spark of fire or when subjected to flame. For instance primary explosives such as Mercury fulminate, lead azide etc. Secondary explosions are the materials which generate blast waves which results in extensive damage to the nearby areas. When high explosive is detonated it will produce very hot gases with the generation of pressure up to a measure of 3×10^5 kg/cm^2& developing temperature in the range of 3000°C to 4000°C. The hot gases produced starts expanding in all the directions which results in the movement of the surrounding air outwards in compressed form.

IS-Code Books

The IS-code utilized for this undertaking was IS-4991:1968(CRITERIA FOR BLAST RESISTANT DESIGN OF STRUCTURES FOR EXPLOSIONS ABOVE GROUND) And for the safety check the earthquake resistance code is also usedIS-1893(Part1):2002(CRITEARIA FOR EARTHQUAKE RESISTANT DESIGN OF STRUCTURES) and for wind loads IS-875 Part-3:1987(WIND LOAD) And for shear wall IS-456:2000 according to clause - 32 and little data about in IS-13920-1993 statement 9.

Modelling of Structure

A RCC structure is fundamentally a get together of slabs ,beams ,columns and foundation inter connected to each other as a unit, we have embraced 3 cases by expecting various shapes for a similar structure .the structure are appearing below with shear wall and without shear walls. As explained below.

1. C-SHAPE PLAN
2. L-SHAPE PLAN
3. I- SHAPE PLAN

IOP Conf. Series: Materials Science and Engineering 1112 (2021) 012027 doi:10.1088/1757-899X/1112/1/012027

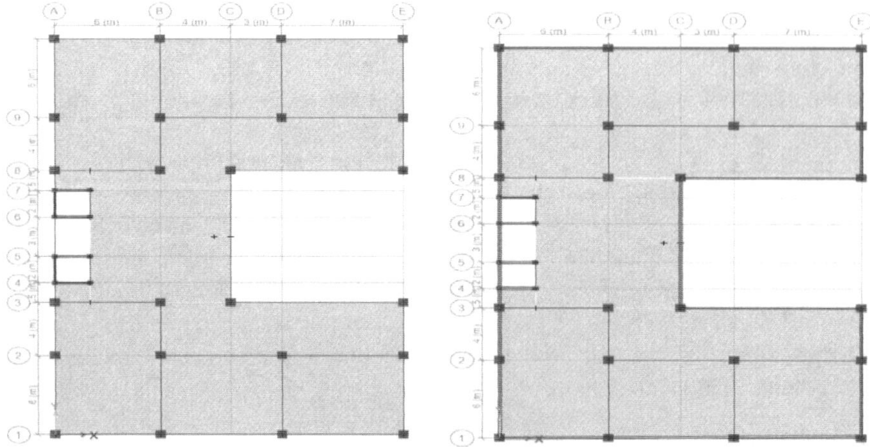

C-Shape Without Shearwall& with shearwall

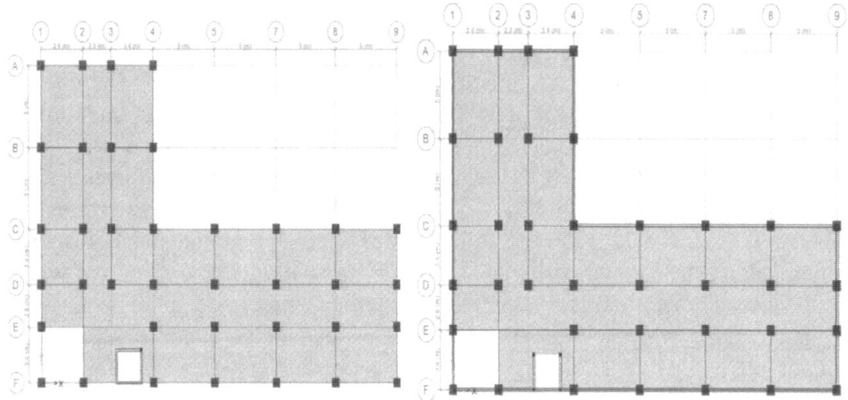

L- Shape Without Shearwall& with shearwall

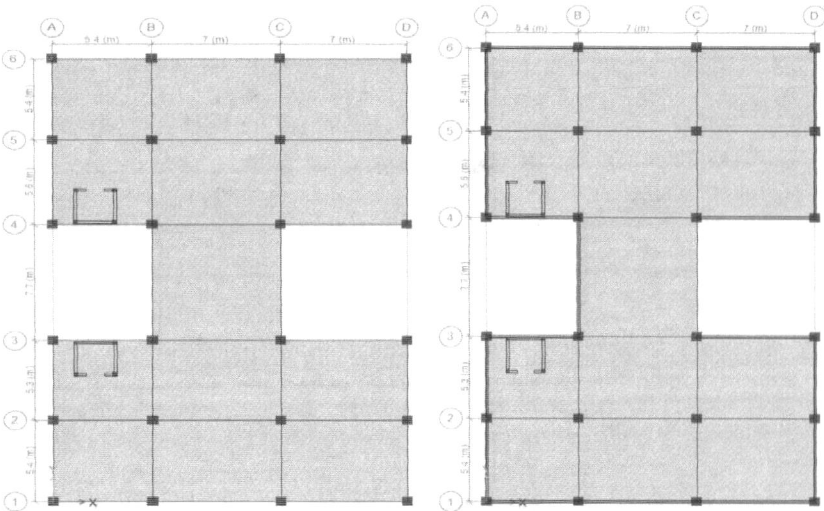

I-Shape Without Shearwall& with shearwall

TABLE-1
BUILDING DESCRIPTION:

Length x breadth	30m x20m
No of storeys	G+10
Storey height	3m
Beam dimensions	400mmx600mm
Column dimensions	600mmx600mm
Column dimensions	230mmx230mm
Shear wall thickness	150mm
Slab thickness	150mm

Load applying on Structure:-

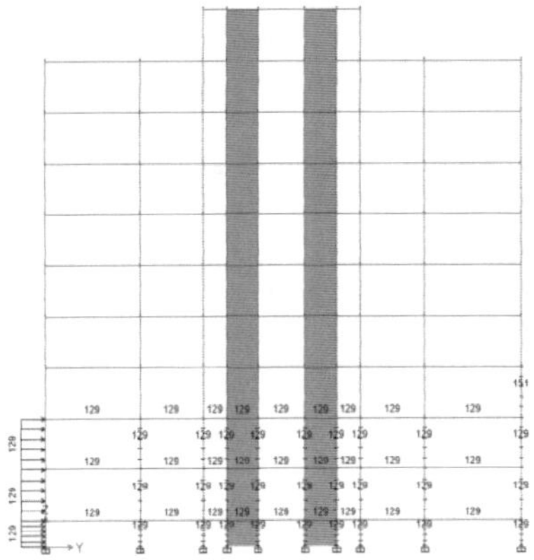

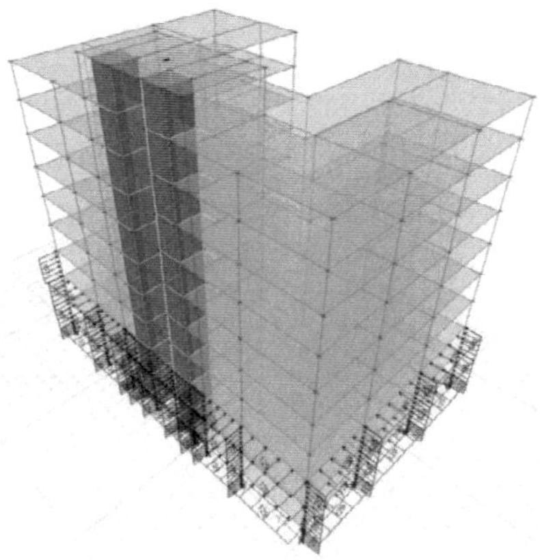

The applied load on the structure was calculated from the codebook:4991:1968. The load combinations of the structure was given below.

Blast Load Combinations

1.5xDL+1.5xLL
1.2xDL+1.2xLL+1xBL
1.5xDL+1xBL
0.9xDL+1xBL

Calculation of Blast Loads

Calculation of Blast Loads for various Charge weights & standoff distances are done below. Total of 12 cases are considered by varying charge weight & standoff distances.

Blast parameters obtained due to the denotation of a 0.150 tonne TNT (150kg's charge weight) explosive & Standoff distance of 10m are evaluated on an above ground rectangular structure for
Height of the structure: 31.5m
Length of the structure: 30m
Width of the Structure: 20m
Stand-Off distance is varied from 10m ,15m, 20m.
i)Characteristics of blast-

Scaled distance $x = 10/(0.150)^{1/3} = 18.82$m

From Table 1 assuming $P_a=1$ kg/cm^2 and for the scaled distance 20m, the pressures are directly obtained:
$P_{so}=5.46$kg/cm^2
$P_{ro}=19.88$ kg/cm^2
$q_o=4.506$ kg/cm^2

The scaled times t_o and t_d obtained from Table 1 for scaled distance 15 m are multiplied by $(0.150)^{1/3}$ to get the values of the respective quantities for the actual explosion of 0.150 tonne charge.

t_o=12.47 x $(0.150)^{1/3}$ = 6.625 milliseconds

t_d=7.76 x $(0.150)^{1/3}$= 4.123millisecods

U=shock front velocity= M x a

$$M = \sqrt{1+\frac{6Pso}{7Pa}} = \sqrt{1+\frac{6*5.46}{7*1}} = 2.383$$

M=Mach number

a=344m/s (Velocity of sound in air at mean sea level)

U=2.383x344=819.84m/millisecond.

ii) Pressure on the Building;

H=31.5m, B=20m, L=30m

S=H or B/2 whichever is less= 31.5 & 20/2=10 so S=10m

$$t_c = \frac{3S}{U} = \frac{3*10}{0.819} = 36.594 \text{ millisec}$$

$$t_t = \frac{L}{U} = \frac{30}{0.819} = 36.594 \text{ millisec}$$

$$t_r = \frac{4S}{U} = \frac{4*10}{0.819} = 48.792 \text{millisec}$$

$t_r > t_d$ So, no pressure on the Backface are considered.

For Side face C_d= -0.2, $P_{so}+C_dq_o$ =5.46 + (-0.2)x4.506= 4.55 kg/cm^2

Conversion from Kg/cm^{2}to KN/m^2

= 4.55Kg/cm^2= 4.55x 9.81 N/cm^2 = (44.6N) /(10^{-4}m^2)

=446 KN/m^2

Graphs & Results

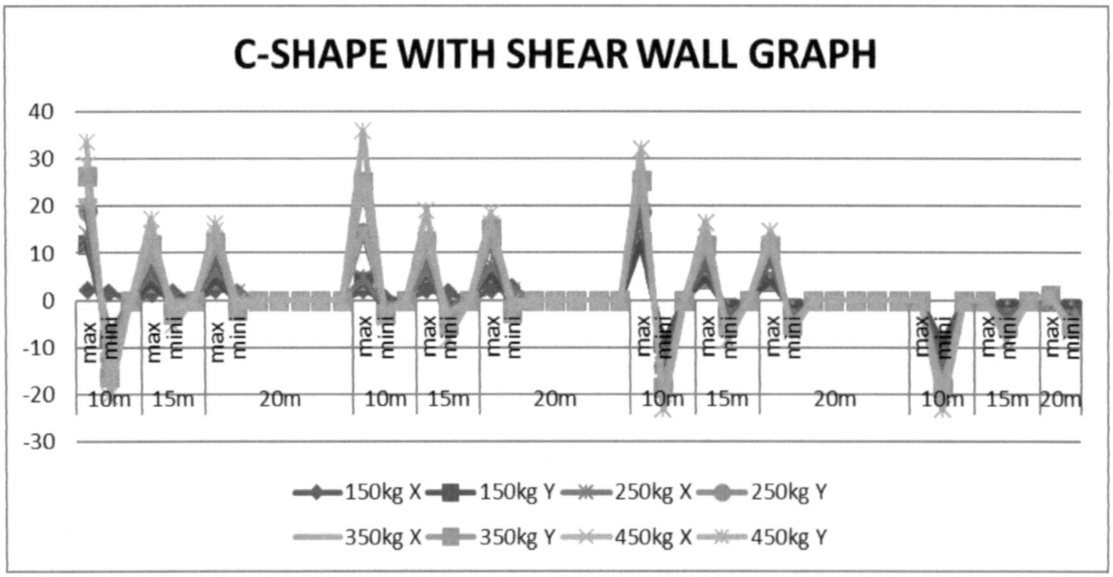

Chart-1 Overall C-Shape Comparison With Shear wall

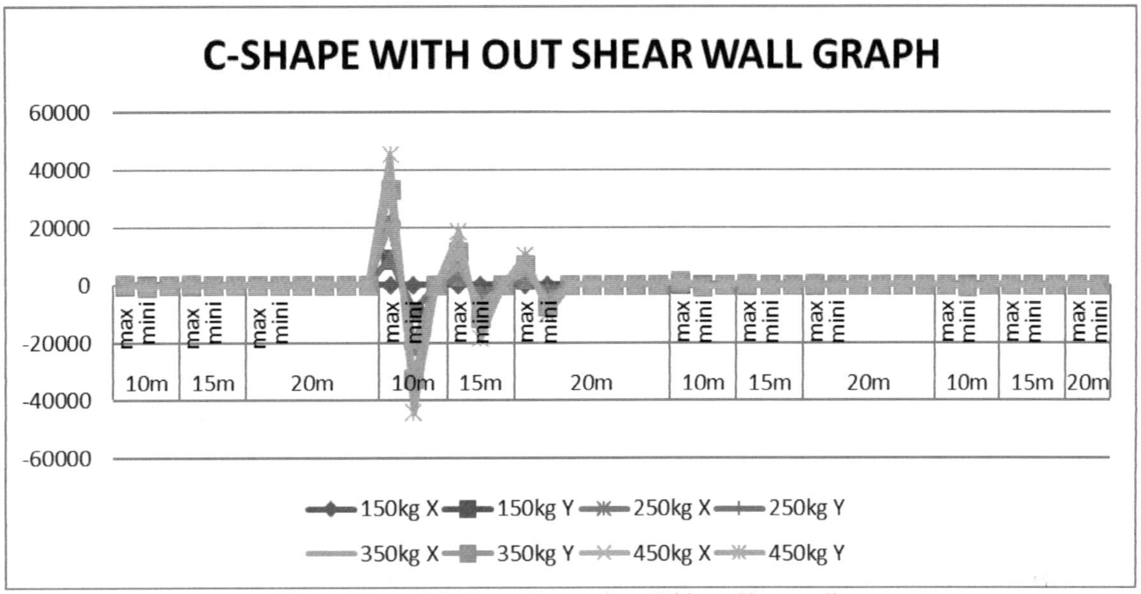

Chart-2 Overall C-Shape Comparison Without Shear wall

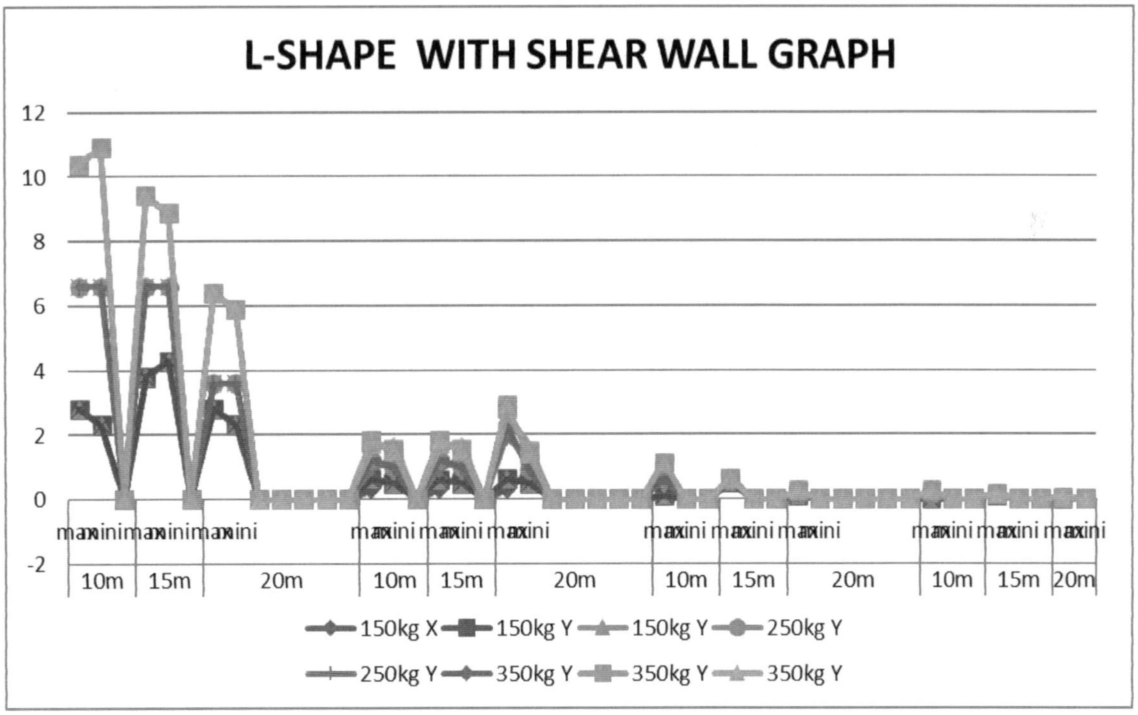

Chart-3 Overall L-Shape Comparison With Shear wall

IOP Conf. Series: Materials Science and Engineering 1112 (2021) 012027 doi:10.1088/1757-899X/1112/1/012027

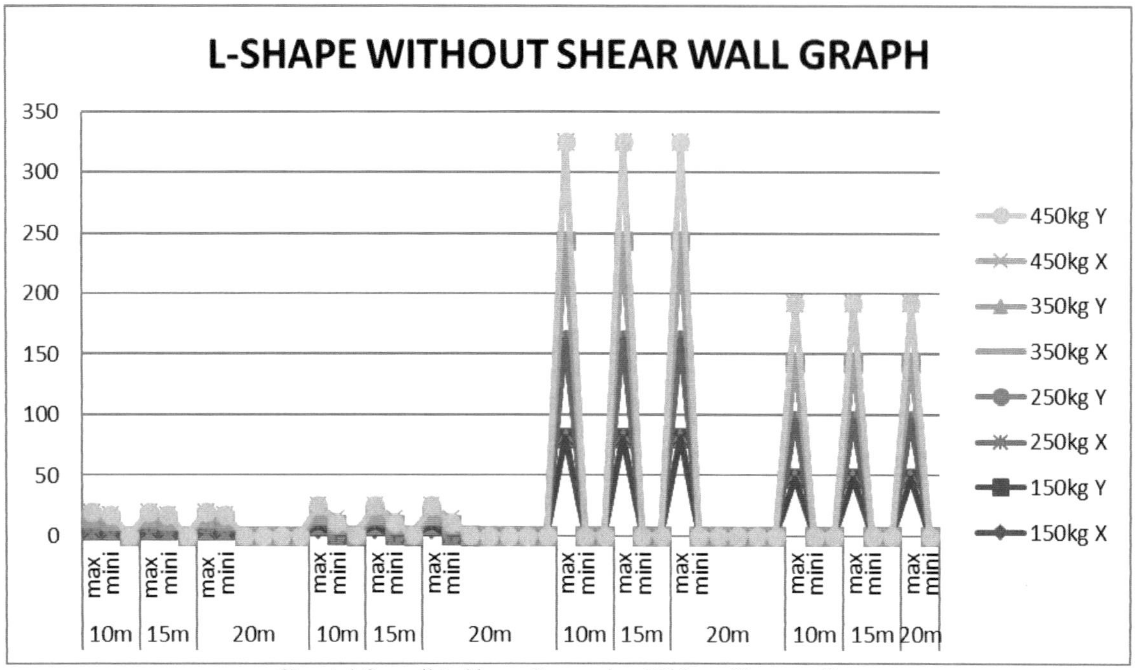

Chart-4 Overall L-Shape ComparisonWithout Shear wall

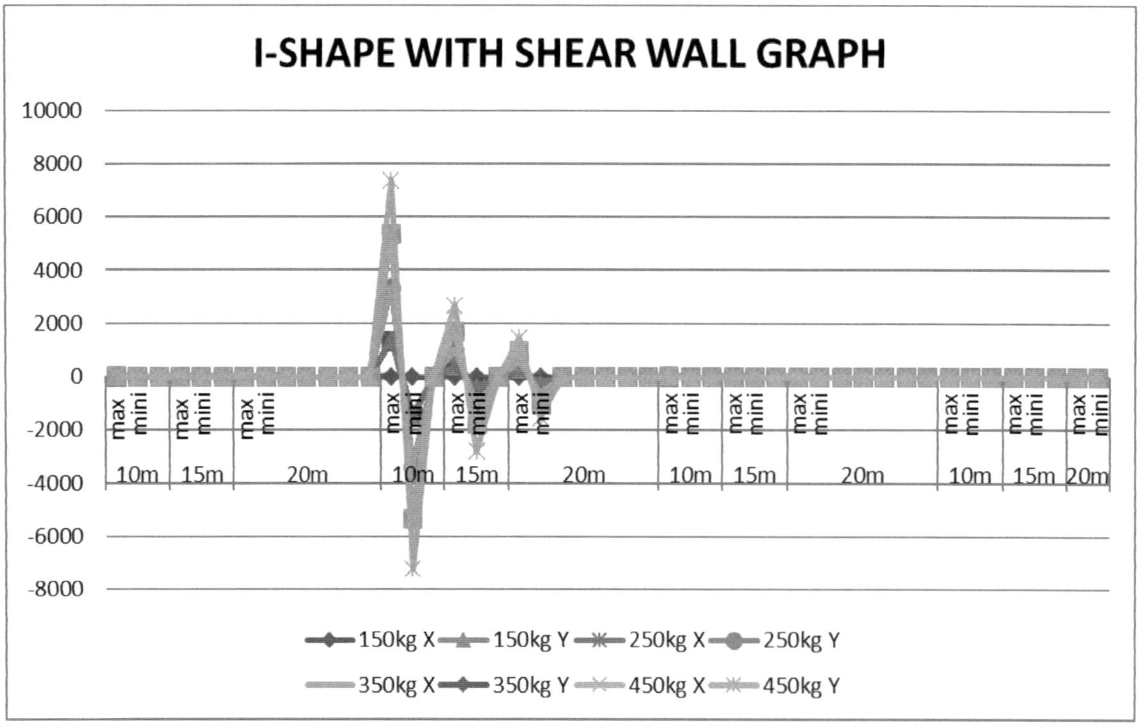

Chart-5 Overall I-Shape Comparison With Shear wall

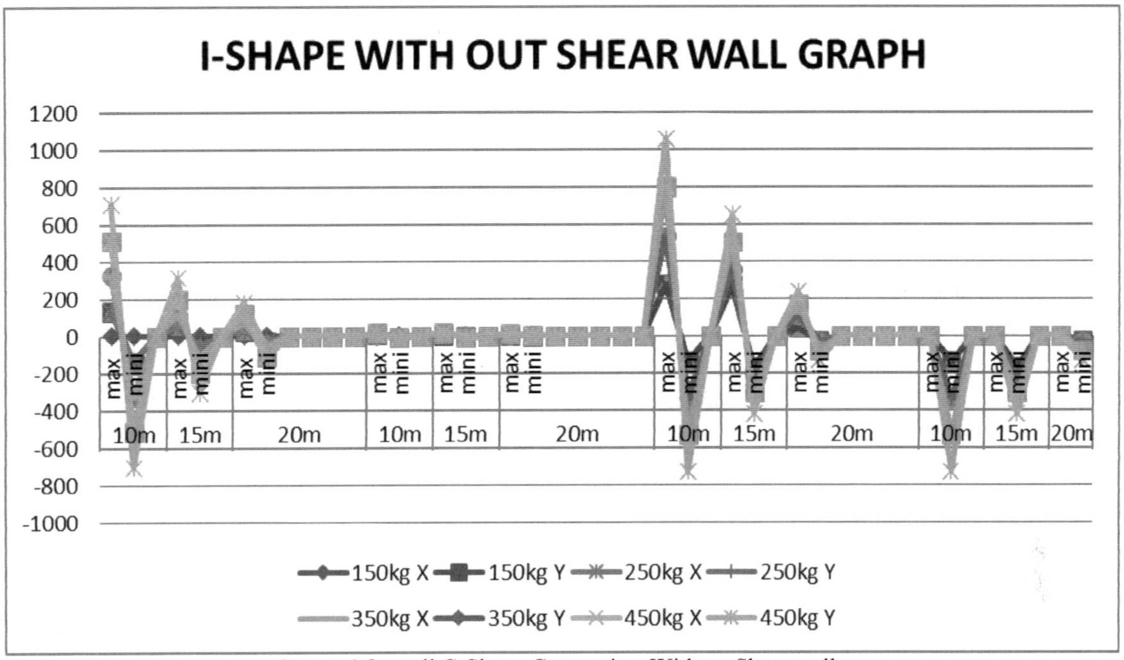

Chart-6 Overall C-Shape ComparisonWithout Shear wall

CONCLUSION

1. Now a days the terrorist attacks were huge so for that purpose the main aim of this project is to resist the blast pressure or blast load.
2. When the bomb were exploses the design of this structure is were according to the code book IS 4991 so that all measures are Taken from the code book the standoff distance and the TNT the as we take in this project 150kg,250kg,350kg,450kg this are the TNT and the standoff distances are 10m,15m,20m, and these are standoff distances.
3. When the load is applying on the structure the structure is resisting the load while applying the blast load.
4. Main aim of the project is shape finding with shear wall so shape deflection is occurred with the high amount of load is applied on the structure so 450 kg of TNT and the standoff distance 15meter load is applied.
5. The deflection is occurring but the building is resisting the load which was applied on the structure.
6. And without shear wall so shape deflection is occurred on large amount but the building is resisting the load because of the beams and the column but overstress is appearing and the structure is safe and by design check all beams and columns are passing by the check
7. So the structure is safe if otherwise the design check is showing failed then the structure is failed here the structure is safe.
8. The shapes are considered by the displacement, acceleration, maximum displacement, max drift so by the values of the above mention by these the shape comparison was done.
9. So the overall project was done by these values only the aim of the project is shape comparison with shear wall and without shear wall it has been done .

ACKNOWLEDGEMENT

I would like to thank my guide, Head of the department, principal, college, friends, family and all who have helped me in the completion of this dissertation work.

REFERNCES

[1] Gautham T N, Dr. M N Hegde, International Research Journal of Engineering and Technology (IRJET) Volume: 04 Issue: 09 | Sep -2017 .

[2] Aditya C. Bhatt , Snehal V. Mevada , Sumant B. Patel Comparative Study of Response of Structures Subjected To Blast and Earthquake Loading Aditya C. Bhatt.et al. Int. Journal of Engineering Research and Applications www.ijera.com ISSN: 2248-9622, Vol. 6, Issue 5, (Part - 2) May 2016, pp.62-66.

[3] Umesh Jamakhandi1, Dr. S. B. Vanakudre Design and Analysis International Research Journal of Engineering and Technology (IRJET) e-ISSN: 2395-0056 Volume: 02 Issue: 07 | Oct-2015

EFFECT OF NANO SILICA & OTHER INDUSTRIAL WASTES ON STRENGTH AND ECONOMY OF CONCRETE

D. Aditya Sai Ram[1] K Sai Ramya[2] D. Kamalakar[3]

[1] *Asst. Professor, Vasireddy Venkatadri Institute of Technology, Guntur-522508, India*

[2] *Asst. Professor, Vasireddy Venkatadri Institute of Technology, Guntur-522508, India*

[3] *PG Student, Vasireddy Venkatadri Institute of Technology, Guntur-522508, India*

Email: dkamalakar973@gmail.com, dasairam102@gmail.com, sairamya.kattula@gmail.com

Abstract. The application of Nanotechnology in concrete has added a new dimension to improve its properties. Nano-materials, by virtue of their very small particle size can affect the concrete properties by altering the microstructure. In this study an attempt is made to analyze the effect of Nano-Silica on the strength of control concrete and replaced cement concrete. In replaced cement concrete, cement is partially replaced with the dolomite powder and fine aggregate with granite fine in different percentages. Workability and Strength tests were conducted on both the control concrete and concrete with Nano-Silica, dolomite and granite powder.

The replace percentage of dolomite powder 15% and 20% of granite fine gave good results, more than this percentage of mix the strength of concrete decreased. Nano-silica of different percentages 0%, 0.5%, 1%, 1.5%, 2%, 2.5% and 3% added to this selected mix (15% DP and 20% GP) and control concrete. The test results reveal that the replacement of Nano-silica from the replacement percentage of 0.5, the strength of concrete significantly increases with increasing percentage of Nano-silica upto 2.5%. The Cost of the replaced concrete increased as well as the strength also with reference to control concrete. However this is valid upto certain point, later the strength decreases with increase in cost.

1. Introduction

Concrete is the most commonly using material in present as well as in future. The wide use of concrete in structures, from buildings to factories, from bridges to airports, makes it one of the most investigated materials of the 21st century. Due to the rapid population explosion and the technology boom to cater to these needs, there is an urgent need to develop the strength and durability of concrete. Out of the different materials utilized within the generation of concrete, cement plays a major part due to its measure and cement property. So, to deliver concrete with progressed properties, the component of cement hydration needs to be considered properly and superior substitutes to it ought to be proposed. Distinctive materials known as supplementary cementitious materials or SCMs are included to concrete move forward its properties. Of the different advances in utilize, Nano-technology looks to be a promising approach in making strides the properties of concrete.

Nano-Materials

Nano-materials are very small sized materials with particle size in Nanometers. These materials are extremely useful in altering the properties of concrete at the ultrafine level by the virtue of their extremely small size. These Nano-materials improve the strength and permeability of concrete by filling up the minute voids and pores in the microstructure.

Nano-Silica

Nano-silica is a breakthrough in Nano materials that has been applied in ultra high performance concrete (UHPC). In general, Nano-silica was produced from micro based silica. The positive reactions created by Nano-silica in ultra-high performance concrete are similar to silica fume or micro silica which is in terms of performance strength and durability enhancement studies. Concrete with addition of Nano-silica gained early strength as compared to that of silica fume, and also the addition of Nano-silica in concrete improved workability of concrete while the addition of super plasticizers is at a minimum dosage. About 20%–30% of cement content can be reduced by Nano-silica. However, the disadvantage of Nano-silica is its price and availability in certain countries. Some countries have to import Nano-silica to be used in concrete industry.

2. Literature Review

Upto now various admixtures are used in concrete to improve its strength and durability properties. In present investigation mineral admixtures such as Nano-silica, dolomite powder, and granite powder are used to study the strength properties of concrete. The following related information is base for this study.

Nadine Hani et.al. (2017) observed the effect of increasing water/binder ratio on the fresh and hardened properties of self-compacting concrete containing Nano-silica with different percentages. Selected 12 different mixes with different water binder ratios of 0.41, 0.45 and 0.5% and Nano-silica 0%0.25%, 0.50% and 0.75% replaced by weight of cement. Tests were conducted on fresh and hardened properties of self-compacting concrete. Test results concluded that mixes with 0.75% Nano-silica compressive strength improved by 26.9%, 32.7% and 48.8% for w/b of 0.41, 0.45, and 0.5 respectively.

Qinyong Ma et.al. (2017) experimentally investigated the microstructure and compressive and tensile properties of Nano-silica concrete containing basalt fibers. Compressive, split tensile tests were conducted on mix with the addition of different percentages of Nano-silica and basalt fibers, 0, 0.6, 1.2, 1.8 percentages by weight of cement and 0 kg/m^3, 2kg/m^3, 3kg/m^3, 4kg/m^3, 5kg/m^3 by volume of concrete. Results concluded by Adding an appropriate amount of Nano- SiO2 and basalt fiber to concrete improves both its compressive strength and splitting tensile strength. The strength enhancement is greatest when the Nano-SiO2 and basalt fiber dosages are 1.2% and 3 kg/m3, respectively, and the compressive strength and splitting tensile strength increased by 9.04% and 17.42%, respectively, compared with those of plain concrete.

Peng Zhang et.al. (2017) studied Application of Nano-SiO2 (NS) particles to reinforce high performance concrete (HPC). He mainly focused on Effect of Nano SiO2 on the fresh properties, microstructure, mechanical and durability properties of high performance concrete. The replacement of Nano-Silica in place of cement increases the compressive and flexural strength of concrete, setting time, durability properties of concrete. Finally, concluded replacement of Nano-Silica up to 5% increased fresh, mechanical and durability properties of HPC greatly.

Dr.G.PrinceArulraj et.al. (2013) experimentally investigated on the effect of granite powder partially replaced with sand by weight with different percentages 0, 5, 10, 15, 20 and 25%.on mix Compressive and split tensile tests were conducted. Tests results shown that is 15% replacement of granite powder gives a good results compared to other mixes.

Manasseh Joel et.al. (2010) Investigated the Suitability of Crushed granite fine (CGF) to replace river sand in concrete with different percentages 0, 10, 20, 30, 40, 50, 60, 70, 80, 90, 100%. Slump, compressive and split tensile strength tests were conducted on fresh and hardened concrete for 7, 14, 28 days'. compressive and split tensile strength results 40.70N/mm^2 and 2.30N/mm^{2}at 28 days of respectively was obtained, with the partial replacement of river sand with 20% crushed granite fine, as against values of 35.00N/mm^2 and 1.75N/mm^2, obtained with the use of river sand as fine aggregate. Based on economic analysis and results of tests, river sand replaced with 20% crushed granite fine is recommended for use in the production of concrete for use in rigid pavement.

3. Materials and Methodology

Materials

In this present investigation materials like Cement, Nano-Silica, dolomite powder, granite powder and Coarse aggregate are used. The description of properties for each of these materials wore evaluated in the following sections. The materials used in this experimental investigation are tested to obtain their properties as per the relevant Indian Standard Codes.

3.1 Cement

The cement used in this entire experimental investigation was commercially available Ordinary Portland cement (OPC) of 53 grade manufactured by Sagar cement company confirming to IS: 12269. Physical properties of cement tested are Normal consistency, fineness of cement, specific gravity, Initial and final setting times.

Results obtained for these tests are tabulated below:

Table 1 Properties of Cement

S.No	Characteristics	Experimental values	As per IS: 4031
1	Specific gravity	3.141	3.10 – 3.16
2	Fineness of cement	3.356 %	Less than 10%
3	Normal consistency	32 %	30 – 35%
4	Initial setting time	75 min	Not less than 30 min
5	Final setting time	550 min	Not more than 600 min

3.2 Nano-Silica

Nano-Silica is one of the newest technologies in Nano process, it consumes calcium hydroxide during the pozzolanic reaction which has been used as an alternative to silica fume. Nano-Silica has large surface area and acts as good filler material in concrete. Nano-Silica increases the strength and durability of concrete.

Table 2 Properties of Nano-silica

Test Items	Standard Requirements	Test Results
Specific surface area	200+20	202
PH value	3.7 – 4.5	4. 12
Loss on drying @ 105° (%)	< 1. 5	0. 47
Loss on Ignition @ 100° (%)	< 2.0	0.66
Sieve Residue (%)	$\leq$0. 04	0. 02
Tamped Density (g/l)	40 – 60	44
Sio2 Content (%)	$\geq$99. 8	99. 88
Carbon Content (%)	$\leq$0. 15	0. 06
Chloride Content (%)	$\leq$0. 0202	0. 009
AL_2O_3	$\leq$0. 03	0. 005
TiO_2	$\leq$0. 02	0. 004
Fe_2O_3	$\leq$0. 003	0. 001
Specific Gravity	2. 2 – 2. 4 (GENERALISED)	
Particle Size	17 NANO	

3.3 Dolomite Powder

Dolomite powder has some similar characteristics of cement. Usage of dolomite powder in concrete can reduce the cost of concrete and may increase the strength to some extent.

Table 3 Properties of dolomite powder

S.No	Property	Dolomite powder
1	Formula	$CaMg(CO_3)_2$
2	Specific gravity	2.75-2.85
3	Colour	White
4	Tenacity	Brittle
5	Moisture content	Nil
6	Crystal system	Trigonal
7	Sieve analysis	Zone III

3.4 River Sand

The sand used in this investigation for preparation of concrete samples is natural river sand conforming to grading zone-III as per **IS: 383-1970** with specific gravity 2.488 and having fineness modulus as 3.096. The testing of sand is done as per **IS: 2386 – 1963**.

3.5 Granite Powder

Granite powder belongs to igneous rock family, obtained from the polishing unit of granite factories. Density of granite powder is between 2.64 to 2.75 g/cm^3 and the compressive strength will be greater than 200Mpa. It is a waste material obtained during polishing process of granite rocks. This waste material is creating problems in the process of disposal and it creating environmental hazards.

Table 4 Properties of granite powder

S.No	Properties	Values	Permissible limits as per IS 456
1	Specific gravity	2.64	
2	Water absorption	23%-25%	
3	Fineness	88%	
4	Chloride content	0.0098%	less than 0.025%
5	Sulphate content	0.052%	less than 0.2%

3.6 Coarse Aggregate

Aggregates are the most important constituent in concrete. They aggregate to the concrete, reduce shrinkage and effort economy. The fact is that the aggregates occupy 70%-80% of the volume of concrete. Coarse aggregate available locally were used in casting of concrete cubes, cylinders. The aggregate used in this project mainly of gravel which comes under normal weight category 20 mm size aggregates were used for the entire work.

Table 5 Physical properties of coarse aggregates

S.No	Characteristics	Experimental values	As per IS 4031
1	Specific gravity	2.967	2.5 to 3
2	Water absorption	1.80 %	
3	Fineness modulus	9.494	Less than 10%

3.7 Water

Water should be free from acids, oils, alkalis, vegetables or other organic Impurities. Polluted water produces weaker concrete. Water has two functions in a concrete mix. Starting, it reacts chemically with the cement to form a cement paste in which the inert aggregates are mixed until the cement paste has hardened. Middle of the process, it serves as a vehicle or lubricant in the mixture of fine aggregates and cement.

3.8 Methodology

The test specimens were prepared and procedures adopted for the strength tests of concrete as per IS516-1959. Mix design for M30 has done as per IS10262-2009 for all the studies in this investigation.

3.8.1 Test Specimens for Compression and Tensile Test

The concrete cubes of size 150mm×150mm×150 mm were used as test specimens to determine the compressive strength. For split tensile strength cylinders of size 150mm diameter and 300mm height were used as test samples. The specimens were cast for M30 grade of concrete. The workability of fresh concrete was measured in terms of slump values by conducting slump cone test.

3.8.2 Mix Design

The Mix design was done by using IS 10262:2009 and was tabulated as follows:

Water	:	Cement	:	F.A	:	C.A
171	:	380	:	664	:	1242
0.45	:	1	:	1.74	:	3.2

4. Results and Discussions

4.1 Compressive Strength

Compressive strength values of the cube specimens were tested at 7, 14 and 28 days for all the mixes. It is observed that the compressive strength of concrete increased with a little percentage of Nano-Silica in mixes, Control Concrete and concrete with Dolomite and Granite powder. Upto 2.5% addition of Nano-Silica, compression strength values of concrete are increasing and beyond that the value gets decreased. At a percentage of dolomite powder (DP) 15% and granite powder (GP) 20%, the concrete gave good results.

Control Concrete:

Table 6. Compressive strength of Control Concrete

Type of concrete	Compressive strength of control concrete		
	7days	14days	28 days
Control concrete	19.50	28.45	35.77

Concrete with Dolomite powder and Granite powder

Table 7. 28 Days Compressive Strength of Concrete with Dolomite and Granite Powder

Granite powder percentage	Compressive strength of concrete (in N/mm^2)				
	5% DP	10% DP	15% DP	20% DP	25% DP
0%	30.9	32.8	35.9	31.3	29.3
10% GP	33.2	35.2	38.5	34	31.8
15% GP	35.2	37.2	39.8	34.3	33
20% GP	37.8	39.6	**41.8**	37.2	35.5
25% GP	33.4	35.8	37.3	33.9	31.6
30% GP	30.4	33.9	36.1	32.9	30.1

Figure 1. 28 Days Compressive Strength of Replaced Concrete

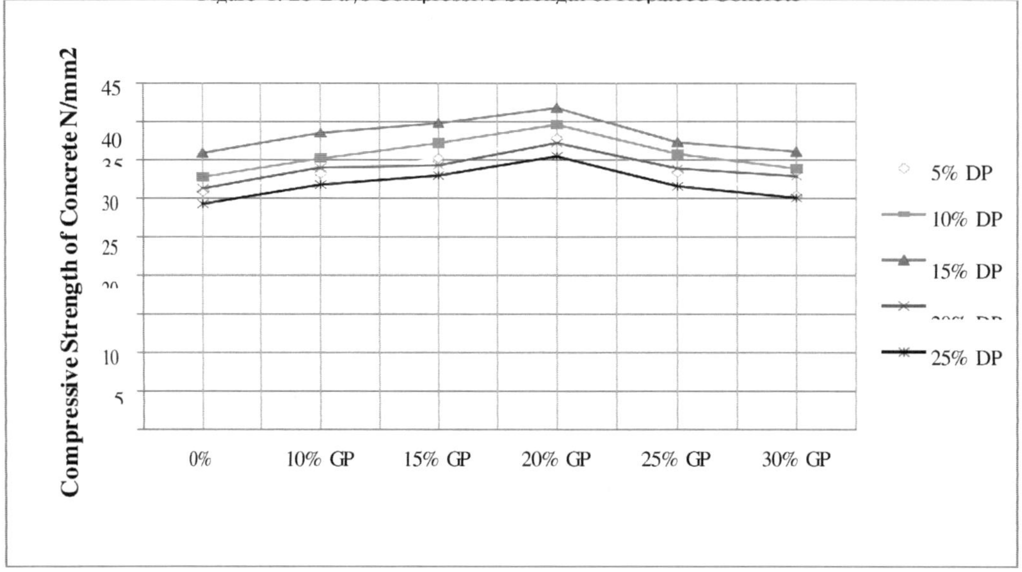

It was found from the above table and figure that at **20% GP and 15% DP** Concrete provided good results. These percentages were fixed for further investigations in concrete with and without Nano-silica.

Concrete with Nano-Silica

Table 8. 28 Days Compressive Strength of Concrete with & without Nano-silica

Nano-Silica	Compressive Strength of Concrete N/mm2	
	Control Concrete	Concrete with Dolomite and Granite Powder
0%	35.77	41.8
0.5%	37.78	43.9
1.0%	39.65	45.85
1.5%	41.85	46.92
2%	43.2	48.97
2.5%	**45.56**	**50.70**
3%	43.8	47.85

Figure 2. 28 Days Compressive Strength of Concrete with Nano-silica

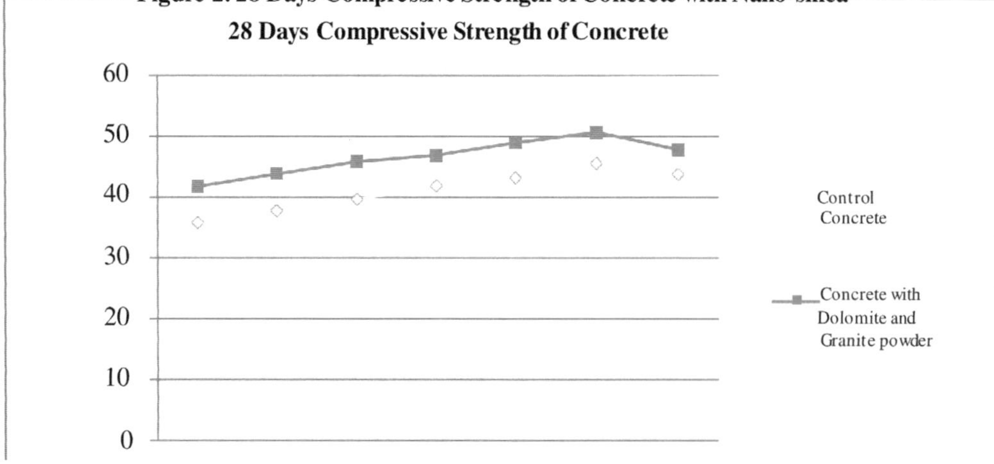

It was found from the above table that, with 20% GP, 15% DP and **2.5% Nano-Silica**, Concrete provided good results.

4.2 Split Tensile Strength

Split Tensile strength of the cylinder specimens were tested at 7 and 28 days for all concrete mixes with Dolomite and Granite powder. It is also observed that the tensile strength of concrete increased with increasing percentage of Nano-Silica in both Control Concrete and concrete with Dolomite and Granite powder upto 2.5%.

Control Concrete

Table 9. Tensile strength of Control Concrete

Type of concrete	Split Tensile Strength of Concrete (N/mm2)	
	7 days	28days
Control Concrete	2.30	3.85

Concrete with Dolomite powder and Granite powder

Table 10. 28 days split tensile strength of concrete with Dolomite and Granite powder

Granite Powder percentage	Split tensile strength (in N/mm²)				
	5% DP	10% DP	15% DP	20% DP	25% DP
0% GP	2.89	3.35	3.65	3.12	2.96
10% GP	3.20	3.52	3.87	3.30	3.12
15% GP	3.42	3.82	4.20	3.75	3.59
20% GP	3.76	4.25	**4.59**	3.97	3.69
25% GP	3.15	3.60	3.89	3.42	3.19
30% GP	2.81	3.29	3.62	3.16	2.87

Figure 3. 28 Days Split Tensile Strength of Concrete with Nano-silica

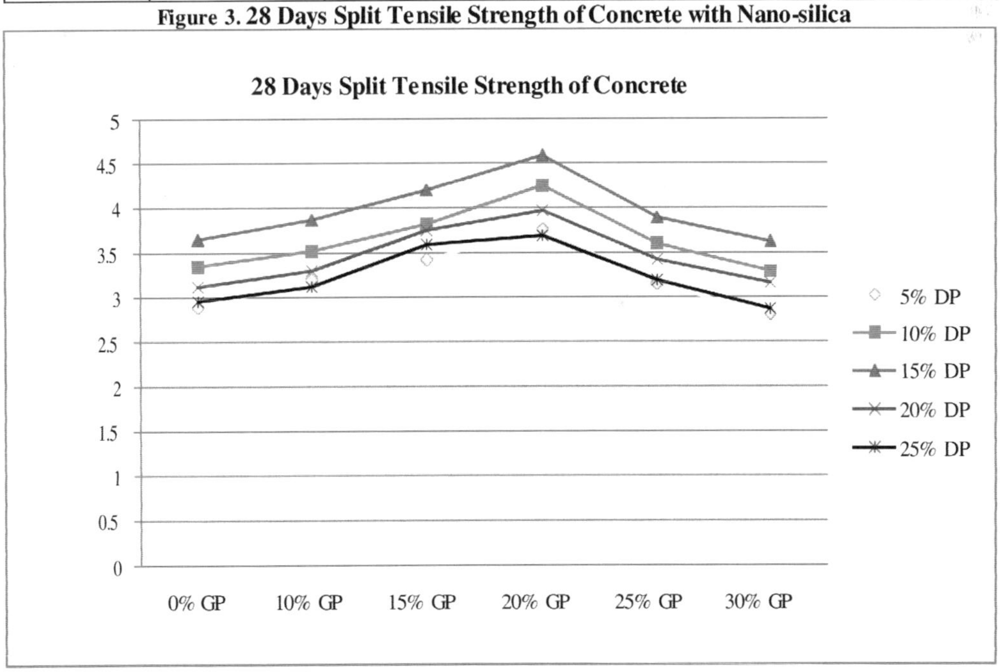

It was found from the above table that, at **20% GP and 15% DP** Concrete provided good results with respect to Split tensile strength also.

Concrete with Nano-Silica

Table 11. 28 Days Split Tensile Strength of Concrete with Nano-Silica

Nano-Silica Percentage	Control Concrete (N/mm2)	Concrete with Dolomite and Granite powder (N/mm2)
0%	3.85	4.59
0.5%	3.97	4.7
1.0%	4.2	5.12
1.5	4.6	5.3
2%	4.95	5.57
2.5%	**5.6**	**5.92**
3%	5.1	5.36

Figure 4. 28 Days Split Tensile Strength of Concrete with Nano-Silica

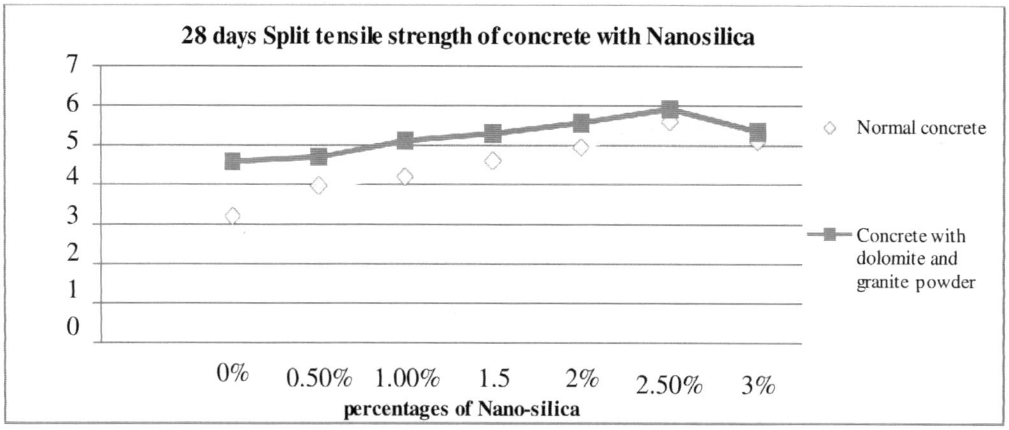

It was found from the above table that, with 20% GP, 15% DP and **2.5% Nano-Silica**, Concrete provided good results with respect to Split Tensile Strength also.

4.3 Economy
The variation of cost with variation of strength of the replaced concrete with reference to control concrete was presented below:

Table 12 Percentage of Strength Variation VS Cost Variation of Concrete with Nano-Silica with Reference to Control Concrete

Type of mix	Control concrete	0.5% N.S	1% N.S	1.5% N.S	2 % N.S	2.5% N.S	3 % N.S
Percentage of cost variation (increase) with control concrete	0	7.68	15.36	23.04	30.73	38.42	46.06
Percentage of strength variation (increase) with control concrete	0	5.6	10.84	16.99	20.77	27.36	22.4

Figure 5. Percentage of Strength Variation VS Cost Variation of Concrete with Nano-Silica with Reference to Control Concrete

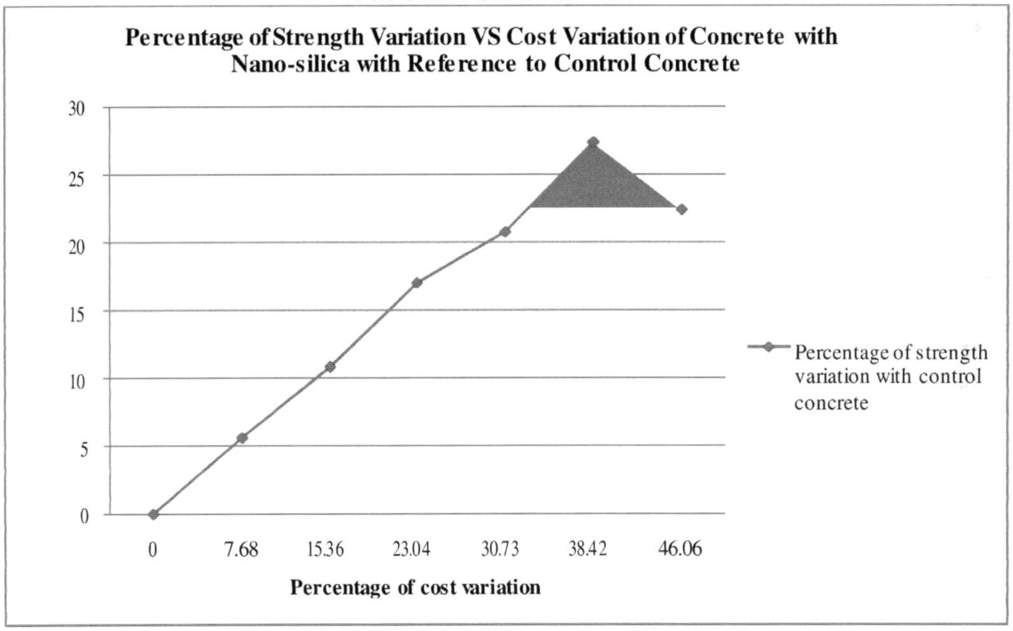

Table 13 Percentage of Strength Variation VS Cost Variation of Concrete with Nano-Silica, Dolomite and Granite Powder with Reference to Control Concrete

TYPE OF MIX	15%DP+ 20% GP	0.5%NS +15%DP +20% GP	1.0%NS +15%DP + 20% GP	1.5%NS +15%DP + 20% GP	2.0%NS +15%DP + 20% GP	2.5%N S+15% DP+20 %GP	3.0%NS +15%D P+20% GP
Percentage of cost variation (increase) with control concrete	-8.16	0.48	7.22	11.57	22.5	30.25	37.93
Percentage of strength variation (increase) with control concrete	16.85	22.7	28.18	31.17	36.9	41.73	33.77

Figure 6. **Percentage of Strength Variation VS Cost Variation of Concrete with Nano-Silica, Dolomite and Granite Powder with Reference to Control Concrete**

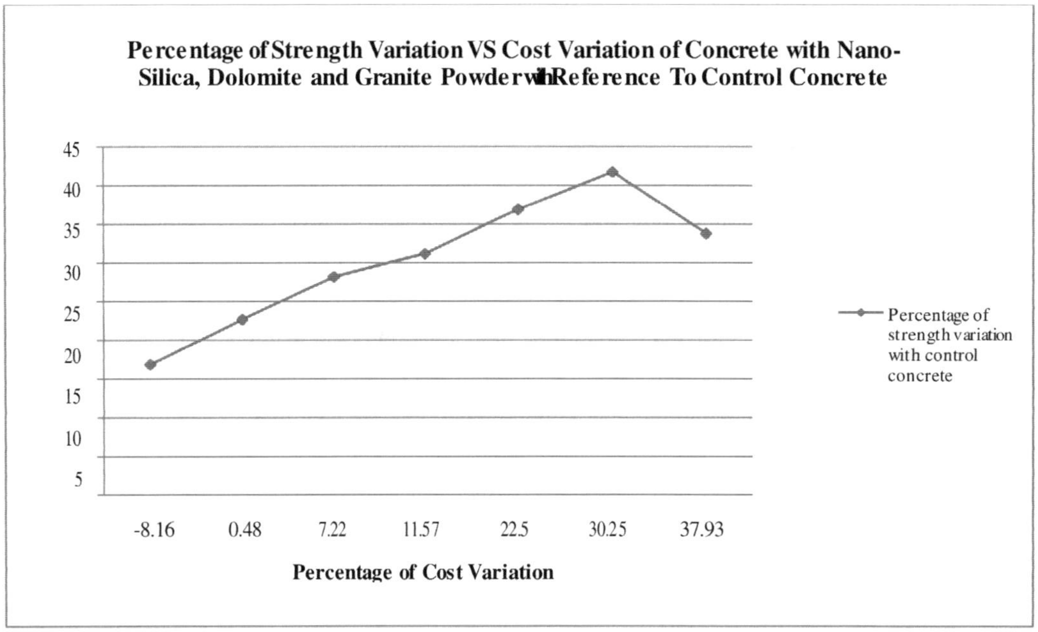

5. CONCLUSIONS

The following conclusions were made after carrying out test on four concrete mixes (Control Concrete, concrete with Dolomite and Granite powder, Control Concrete with Nano-Silica, Dolomite and Granite powder replaced concrete with Nano-Silica):

1. The test result of concrete with Dolomite and Granite powder with different percentages shows that, at replacement percentages of dolomite with 15% and granite with 20%, the compressive strength is increased by **16.85%.**

2. The compressive strength of concrete with Dolomite and Granite powder of 15%, 20% and 2.5% of Nano-silica, is increased by **11.28%.**

3. The test results of split tensile strength showed the same pattern as that of the compressive strength results i.e., at replacement percentages of dolomite with 15% and granite with 20%, the split tensile strength is increased by **19.22%.**

4. It is observed that the split tensile strength of concrete with Dolomite and Granite powder of 15%, 20% and 2.5% of Nano-silica, is increased by **5.71%.**

5. It is observed that the strength of concrete started increasing with just **0.5%** replacement of cement with Nano-silica in the concrete mix.

6. As a result, the cost was found to be increased by **30.25%** at **(2.5%NS+**

15%DP+20%GP) for reaching Maximum Strength, but can be utilized in lesser percentages also based on the Economy and applications of Concrete.

6. REFERENCES

[1] Nadine Hani, Omar Nawawy, Khaled S. Ragab, and MohamedKohail2018 "The effect of different water/binder ratio and Nano-silica dosage on the fresh and hardened properties of self- compacting concrete" published in Elsevier Construction and Building Materials 165 pp504–513

[2] Qinyong Ma, Ying Zhu 2017 "Experimental research on the microstructure and compressive and tensile properties of Nano-SiO2 concrete containing basalt fibers" published in science Direct Underground Space 2 pp 175–181

[3] PengZhang,Jinyi Wan, KejinWang,Qingfu, 2017 "Influence of Nano-SiO2 on properties of fresh and hardened high performance concrete: A state-of-the-art review" presented in Elsevier Construction and Building Materials 148 pp648–658.

[4] Saber Fallah, Mahdi Nematzadeh 2016 "Mechanical properties and durability of high strength concrete containing macro-polymeric and polypropylene fibers with Nano-silica and silica fume" published in Elsevier Construction and Building Materials 132 pp 170–187

[5] KsenijaJankovic, SrboljubStankovic, DraganBojovic, Marko Stojanovic, Lana Antic 2016 "The influence of Nano-silica and barite aggregate on properties of ultra-high performance concrete" presented in Elsevier Construction and Building Materials (An international journal dedicated to the investigation and innovative use of materials in construction and repair) 126 pp 147–156

[6] Min-Hong Zhang, Jahidul Islam 2011 "Use of Nano-silica to reduce setting time and increase early strength of concretes with high volumes of fly ash or slag" presented an article in Construction and Building Materials 29 pp 573–580Hongjian Du, Suhuan Du, Xuemei Liu 2014 "Durability performances of concrete with Nano-silica" presented in Elsevier science direct Construction and Building Materials 73 pp705–712.

[7] A.M. Said,M.S.Zeidan, M.T. Bassuoni , Y. Tian 2012 "Properties of concrete incorporating Nano-Silica" presented in Elsevier Construction and Building Materials 36 pp 838–844.

[8] AbhishekMathurPradeep K. Goyal2015 "Effect on strength and durability properties of concrete incorporating granite and marble residues as fillers" published by International Journal of Civil Engineering and Technology (IJCIET) Vol 6, Issue 11, pp. 173-183.

[9] Ragulraj,Vennila A, Venkatasubramani R, Sreevidya V 2017, " A Study on Self-Compacting Concrete by Replacing Fine Aggregate and Cement by Foundry Sand and Dolomite Powder" published in International Journal of Chem.Tech Research, volume number10(8), pp 119-126.

[10] Deepthi C, Shindon Baby, Ernakulum 2016 "Study on Compressive Strength of Concrete with Dolomite Powder and Crushed Tiles" published in International Journal of Innovative Research in Science, Engineering and Technology Vol. 5, Issue 9, pp16584-16590

IOP Conf. Series: Materials Science and Engineering 1112 (2021) 012029 doi:10.1088/1757-899X/1112/1/012029

Performance of Fine Aggregate Replaced Pond Ash on strength of Concrete

A Srikanth[1], K Adithya Nandini[2], Y Anand Babu[3].

[1] Assistant Professor, Department of civil engineering, VVIT.

[2] Assistant Professor, Department of civil engineering, VVIT.

[3]Lecturer, School of Civil and Environmental Engineering, Jimma University, JIMMA, Ethiopia

Abstract: There is a great demand for the concrete, the second largest consumed material and concrete making materials, for constructions due to stimulated growth of infrastructure worldwide. This lead to the scarcity of natural resources, resulting in creating vast interest in the research area to look for alternative materials which could satisfy both strength and performance criteria of concrete constructions. Pond ash, a waste product of thermal power plants, is one such material, that can be adopted as a suitable material as a fine aggregate in concrete, replacing natural sand partially or fully. Encouraging the usage of such a waste material as constituent in concrete so as to address the issues related to its disposal , environmental and ecological problems, is a social responsibility of researchers, thus contributing to 3Rs-Reduce Reuse and Recycle, there by promoting sustainable construction . It is commonly thought that as inert filter, fine aggregate has a very little effect on the finished concrete properties. However fine aggregate (natural sand or alternative material) and its characteristics play a substantial role in controlling the workability, strength, and durability of concrete constructions. Thus its detailed characterization is essential to boost the confidence of user. So in this report behavior of pond ash is studied. Change in compressive strength and slump values are studied.

1. Thirty six concrete matrices were designed with 0, 10, 20, 30, 40 and 100 percent pond ash by replacing fine aggregate.
2. The concretes of M20 with 0 percent pond ash and 100 percent fine aggregate shows compressive strength at 28 days curing 89.51 percent higher than conventional comparison mixes respectively.
3. The slump values got on decreasing from 50mm for 0 percent pond ash to 0mm for 100 percent pond ash.

1. INTRODUCTION

POND ASH

The idea of using by-products to replace natural aggregates is another alternative solution to achieve environmental conservation as well as to obtain a reasonable concrete cost. After burning the coal, 70% extracted as fly ash and remaining 30% precipitated as bottom ash. The obtained bottom ash is usually combined with fly ash. This blended fly ash and bottom ash together is referred to as Pond ash. Approximately 30 percent of the coal ash is handled wet and disposed as Pond ash. With a growing content of pond ash, there has been are relatively greater increase of compressive strength, compared to normal concrete, and such a trend might be a consequence of decreased water/cement ratio induced by the absorption of mixing water.

NECCESITY OF CHARACTERIZATION

Pond ash is potentially useable as fine aggregate, but the properties of pond ash depend on various factors such as source of coal, and its type, design of coal fired boilers, power plant operating parameters, point of disposal of wet coal ash, the efficiency and type of coal grinding processes adapted at the power plant. More over because of differences in the unit weight of fly ash and bottom ash, the coarser bottom ash particle settle first and the finer fly ash remains in suspension longer. The ash deposited in about 100m of ash slurry discharge point in the pond is coarser as compared to the ash deposited away from 100m, in between these two areas is of medium particle size. Hence it is the necessary for the researchers to know the variation in properties of pond ash for a given plant by collecting pond ash at different locations around the discharge outlet, studying its properties as fine aggregate and comparing it with the properties of natural sand. Characterization of pond ash thus plays a very important role in assessing its sustainability as a material in the field of concrete construction.

LIMITATIONS OF STUDY

As all of us knew, every coin has two sides. Until this moment we tend to saw solely the straightforward approach of use of pond ash as replacement of natural sand however from another facet there are some limitations on such replacement. This study is work on the concept of partial and full replacement of sand (one of the important ingredient of concrete) by pond ash (waste byproduct of thermal power plant), with different proportions. The investigation of previous papers show that, if the replacement of sand exceeds some limit that affects the properties of concrete on large scale in practically.

- Skilled supervision will require while replacing the pond ash with fine aggregate in concrete.
- This replacement is suitable only for mass concreting projects as like construction of dam structures. This replacement is not economical if the concreting is on small scale.
- Also pond ash might not out there simply, availableness of pond ash depends upon the construction site distance and ash ponds.
- The CaO content is a smaller amount within the pond ash so plasticizer property of pond ash is small. Hence, we will observe the decrement in compressive strength.

2. MATERIALS AND METHODS

Samples of pond ash were collected as per IS 6491-1972 at 4 different pits (P1-P4) from Ash pond at Dr. Narlatatarao thermal power station, Ibrahimpatnam, selecting 2 pits around 200m away from each other and 2 pits 75 m from discharge outlet of Ash pond. These samples were oven dried at 100°C and detailed characterization was carried out as fine aggregate in concrete as per IS codes of practice. 53 grade OPC (Ordinary Portland Cement) and coarse aggregate with particle size 20mm was used for the preparation of concrete mix. Table 1 shows the physical characteristics of each material.

Table 1: Specific gravities of materials used

Material	Specific gravity
Cement	3.16
Fine aggregate	2.69
Coarse aggregate	2.98
Pond ash	2.22

MIX DESIGN FOR M20 GRADE OF CONCRETE

HOW MY PROJECT DIFFERS FROM OTHERS

From the all the journals and literature review I have observed one common point i.e., replacement of fine aggregate by pond ash is done in weight replacement method. Let us consider an example of 10 kg of fine aggregate in which 25% replacement of fine aggregate has to be done with pond ash. Weight of pond ash quantity replaced is equal to 2.5 kgs in weight replacement method.

Table 1: General weight replacement of Pond ash

Sl. no	Fine aggregate(kg)	% of replacement	Fine aggregate(kg)	Pond ash(kg)
1	10	25	7.5	2.5

From the Table1, we clearly observed that specific gravity of pond ash is 2.22 which is less than specific gravity of fine aggregate i.e., 2.69. As specific gravity is inversely proportional to volume, less the specific gravity more the volume occupied. Doing replacement in such type increase the volume of mix which is limited to kg/m^3.

In order to maintain the volume of mix to kg/m^3I have done replacement in absolute weight method. If I consider 10 kg of fine aggregate with same 25% replacement pond ash my quantity of pond ash will be 2.1675 kg in absolute weight replacement method and fine aggregate is equal to 7.8325 kg.

Table 2: Absolute weight replacement of Pond ash

Sl. no	Fine aggregate(kg)	% of replacement	Fine aggregate(kg)	Pond ash(kg)
2	10	25	7.9368	2.063

Replacement of Pond Ash = FA*(SG of PA/SF of FA)*(%PA/100)

In all the journals the ratio of specific gravity of pond ash to specific gravity of fine aggregate is taken as 1, that is the reason why volume is exceeding kg/m^3 . But the ratio is less than 1 it is equal to 0.8252.

MIX PROPORTIONS

Cement = 339.28 kg
Coarse aggregate = 1272.998 kg
Fine aggregate = 740.877 kg
Water = 190 lit

VOLUMES OF MATERIALS AND SPECIMEN

- Vol. of cubes =0.15×0.15×0.15×6=0.02 m^3
- Vol. of cylinders =(3.14×0.15×0.15×0.3)/4)×3=0.02 m^3
- Total vol. =1.2×(0.02+0.02)=0.04 m^3
- FA =(Total volume × FA)((100-%FA)/100)
- %PA =FA×(SG of PA/SG of FA)×(%PA/100)

Table 3: Mix proportions for each mix

Mix	Cement(kg)	CA(kg)	Water	FA(kg)	Pond ash(kg)
M20 0 PA	14.72	55.23	8.24	32.14	0.00
M20 10 PA	14.72	55.23	8.24	28.93	2.65
M20 20 PA	14.72	55.23	8.24	25.71	4.77
M20 30 PA	14.72	55.23	8.24	22.50	6.36
M20 40 PA	14.72	55.23	8.24	19.29	7.43
M20 100 PA	14.72	55.23	8.24	0.00	15.91

3. RESULTS AND DISCUSSIONS

COMPRESSIVE STRENGTH

- The compressive strength of hardened concrete is that the most vital of all properties of concrete.
- Strength tests don't seem to be created on neat cement paste due to difficulties of excessive shrinkage and succeeding cracking of neat cement
- Casted cubes are kept in curing for 7 & 28 days.
- After removal of cubes they are allowed to lose moisture by exposing to sun.
- Place the required specimen in Compressive Testing machine and apply the load at 35 N/sq.mm/minute.

Compressive strength is the ratio of maximum load applied just before breakage to area of cube.

Table 4: Compressive strength of M20 concrete

Mix	Compressive strength (7 days)N/sq.mm	Compressive strength (28 days)N/sq.mm
M20 0 PA	27.843	31.846
M20 10 PA	29.624	32.549
M20 20 PA	31.105	34.364
M20 30 PA	33.475	39.376
M20 40 PA	31.846	38.363
M20 100 PA	31.105	34.364

SLUMP VALUES

- Vertical settlement of concrete is defined as slump.
- Slump is done to find the workability of concrete. This test measures the consistency of concrete in specific batches.
- The length of cone is 30 cm with top diameter 10cm and bottom diameter 20 cm.
- Slump is of three types
  - Collapse slump
  - Shear slump
  - True slump

Table 5: Slump values of Different Mixes

Mix	Slump (mm)
M20 0 PA	50
M20 10 PA	45
M20 20 PA	40
M20 30 PA	30
M20 40 PA	10
M20 100 PA	0

**PA=Pond Ash

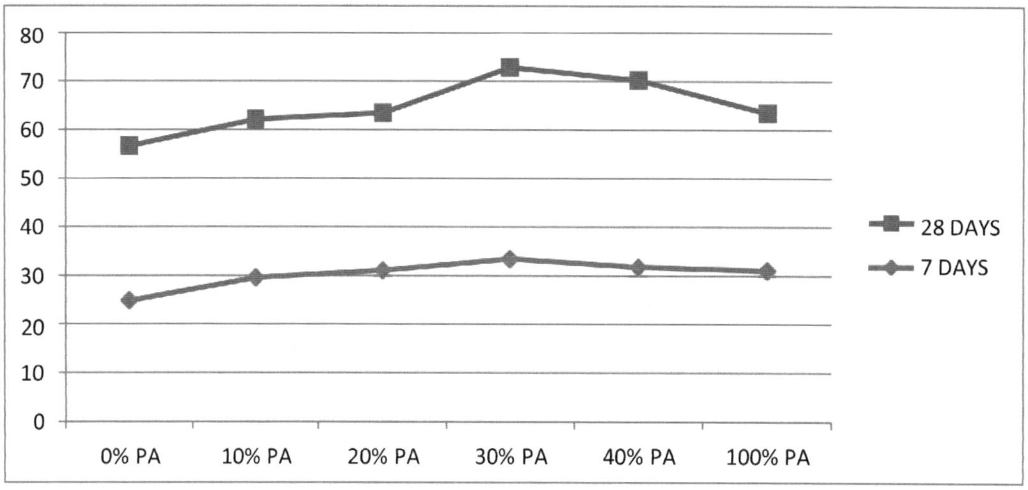

Graph 1: Comparative Compressive strength of M20 Grade concrete

DISCUSSION ON TEST RESULTS:

- With the addition of pond ash there is reduction in slump value of fresh concrete. It is clearly observed from the table 5; slump is reduced from 50 mm to 0 mm.
- The 7days, 28days strength shows that the strength increases from standard Concrete up to the\ addition of 30% replacement of fine aggregate with pond ash. Initially strength of concrete at 0 % PA for 7 days curing is 27.843 N/$mm^2$and it has increased to 33.475 N/mm^2 for 30 % PA for 7 days curing.
- The7days and 28days strength get reduced on further additions of pond ash as sand replacement 30%. At 30 % PA for 7 days curing strength is 33.475N/$mm^2$and it got reduced to 31.846 N/mm^2 for 40 % PA and 31.105 N/mm^2 for 100 % PA for 7 days of curing.
- For 7 days of curing strength for 0 % PA is 27.843 N/mm^2 and for 28 days of curing strength is 31.846 N/mm^2. It is clear that strength of concrete increases with increase in age of concrete.

4. REEFERENCES:

[1] Prof. P.P. Bhangle, Prof. P.M. Nemade, "Study of pond as h (BTPS) "Use as a fine aggregate in cement concrete- Case Study", Shri SantGadge Baba College of Engineering & Technology, Bhusawal, Maharashtra, india.

[2] Lee Bong Chun Kim Jung Sung, Kim Tee Sang and ChaeSeong Tae, "A Study on the fundamental properties of concrete incorporating pond ash in Korea", The 3rd ACF International Conference-ACF/VCA, pp.401-408, (2008)

[3] Arumugam, ilangovan, James Manohar, "A Study on characterization and use of pond ash as a fine aggregate in concrete" International journal of civil and structural engineering Volume 2, No 2, 2011

[4] Agarwal P., Y. Agarwal and S.M. Gupta, 2007, Effect of bottom ash as replacement of fine aggregate in concrete. Asian J. Civil Eng. Build. Housing, 8(1): 49-62

[5] http://teaching.ust.hk/~civl111/CHAPTER3.pdf

[6] http://teaching.ust.hk/~civl111/CHAPTER3.pdf

[7] IS 6491-1972

[8] http://civil-engg-world.blogspot.in/2009/09/quality-of-water-in-concrete-mix.html

[9] IS 2720-PART 3

[10] IS 2386 (Part III)-1963

[11] IS 383-1970

[12] IS 10262-2009

[13] IS 456-2000

IOP Conf. Series: Materials Science and Engineering 1112 (2021) 012030 doi:10.1088/1757-899X/1112/1/012030

A Study on Physico-Chemical analysis of Ground Water and Heavy Metal Analysis of Leachate in Kodungaiyur Landfill Site, Chennai

A.Harinadha Babu[1], Dr.N.Kumara Swamy[2], Dr.S. Krishnaiah[3], M.S.Senthil[4]
[1]Research Scholar, Jawaharlal Nehru Technological University, Ananthapuramu, India
[2]Professor & Dean of Academics, Department of Civil Engineering, Vasireddi Venkatadri Institute of Technology, Guntur, India.
[3]Professor, Department of Civil Engineering, Jawaharlal Nehru Technological University, Ananthapuramu, India.
[4]Research Scholar, University College of Engineering, Anna University, Trichy, India
*Corresponding Author: harinadhababume@gmail.com

Abstract: The Leachate heavy metal samples were analysed for its Physico-Chemical and Heavy metal characteristics to assess a pollution level in soil condition of kodungaiyur dumping yard. Water quality analyses were carried out for its suitability of drinking water condition and compare the result with WHO and BIS standards. Investigation on the accumulation heavy metals like Fe, Pb, Zn, Cu, Cd and Cr in leachate as per APHA methods are carried out. In leachate, the order was Pb>Zn>Cu>Cr>Cd.

Keywords: Heavy Metals, Geo-chemical Index, Enrichment Factor, Contamination Factor, Leachate, Dumping Yard.

Introduction

The accumulation of heavy metals and metalloids in environment is of increasing concern now a day. Potentially harmful metal in nature may come from the bedrock itself and anthropogenic sources like solid and liquid waste deposits from various sources like domestic, industry, agricultural and transport emission etc., (Wilson and Pyatt, 2007). Excessive accumulation of heavy metals in nature results in ground water pollution, soil contamination and has consequences for health effects on living beings and agriculture failure.MSW generated in Chennai includes 68 % of residential waste, 16 % commercial waste, 14% Institutional waste and 2 % industrial waste. The Physico-chemical properties of the MSW generated in Chennai, showed that the majority of the waste is composed of green waste (32.3%) and inert materials (34.7%) *viz.*, stones and glass (CPCB, 2000 and Damodaran et.al., 2003).

Study Area

The Kodungaiyur dumping yard located at the northern part of Chennai city is in operation since 1980. It comes under Perambur Taluk of Chennai City District. Initially the leachate generated from the Chennai city has been dumped in Kodungaiyur dumpsite. Due to urbanization, increase in population, changes in lifestyle and consumption pattern, the problem of water waste management in Chennai has been increased. So two other major, Perungudi in south and Kodungaiyur in north have been proposed and being used as open dumps for disposal of Municipal Water Waste from the Chennai city. Now Kodungaiyur has been used as a dumping ground for municipal waste since 1987. Latitude and longitude of kodungaiyur will be 13° 08' 02" N and 80° 16' 09" E. Elevation of 7 meter above mean sea level. The method of disposal is open dumping and levelling by bulldozer.

The dumping site covered about 30 ha in 1995 and increased to 54.75 ha in 2002 which is twice that of the area in 1995.The dumping area is estimated to be 117 ha in 2009 which is again twice as that of in 2002. Kodungaiyur sewage water treatment plant located adjacent to the dumpsite discharges the sewage water near to the dumpsite. The waste at the site mixes with the sewage water and contaminates it further. The dumpsite lies at 1.5 to 2.0 km from the western side of Buckingham canal and 3 km west of Bay of Bengal coastline. It is situated within a low lying IOC marsh which extends for a length of approximately 10 km from north to south and for a width of 3 to 4 km from west to east that makes the dumpsite always surrounded by water courses. Many industries in and around kodungaiyur such as Madras Fertilizer Limited (MFL), Indian Oil Corporation (IOC), and Madras Refinery Limited (MRL), etc., It is said to be the largest dump yard in Chennai city maintained by Chennai Corporation. It has many harmful effects on residents in the vicinity to the dump yard, in violation of MSW Rules framed by Ministry of Environment and Forests and Climate Change.

Table. 1 Monthly average Weather Statistics of Kodungaiyur

S.No	Months	Temperature			Precipitation
		Normal	Warmest	Coldest	Normal
1	January	24.6°C	28.8°C	20.4°C	1
2	February	26.2°C	30.7°C	21.6°C	1
3	March	28.4°C	33.2°C	23.5°C	0
4	April	30.9°C	35.6°C	26.2°C	1
5	May	32.9°C	38.0°C	27.7°C	2
6	June	32.4°C	37.4°C	27.4°C	7
7	July	30.7°C	35.3°C	26.0°C	9
8	August	30.1°C	34.5°C	25.6°C	10
9	September	29.7°C	34.0°C	25.3°C	10
10	October	28.2°C	31.9°C	24.4°C	12
11	November	26.1°C	29.5°C	22.7°C	11
12	December	25.0°C	28.4°C	21.5°C	6

Source: Madras/Minambakkam weather station: 16 m.a.s.l., 16.0 km away from Kodungaiyur.

Sampling and Methodology

This study collected six leachate samples and ground water samples from the buffer zones of landfill site at the distance interval of 200m to identify the variations in quality of samples. The site is under marshy land, 3 to 4 km away from coastline of Bay of Bengal. The study area experiences tropical climate and features fairly hot most of the months. According to Minambakkam weather station in Chennai recorded the normal maximum and minimum temperatures are 32.9°C (May) and 24.6°C (January) respectively. The samples were collected in a pre-cleaned one litter polyethylene sample container. The container was washed with detergent and soaked with 1% nitric acid. Then sample containers were rinsed with deionised water. For measuring heavy metal concentration, 65% concentrated HNO_3 acid was added to each sample immediately after the collection to bring the pH below 2 to minimize precipitation and adsorption in the container walls (APHA 1998).

Table. 2

Parameter	Instrument Used	Method Used	Method Reference
pH	Digital pH meter	-	APHA (1998)
EC	Digital Conductivity Meter	-	Trivedi and Goel (1986)
TDS	TDS Meter	Evaporation	
COD	Refluxing Assembly	Refluxtitrimetry method	APHA (1998)
BOD	BOD incubator and titration assembly	Winkler azide method	APHA (1998)
Na^+	Flame photometer	Calibration	Flame atomic absorption spectrometer (FAAS)
K^+	Flame photometer	Calibration	
Cl^{2-}	Burette	Titration	APHA (1998)
NO_3	UV-Spectrometer		
Elements	Atomic Absorption Spectrometer		

Assessment of Metal Contamination

In order to assess the heavy metal contamination in the landfill site of the study area. Three quantitative indices namely: Geo-accumulation Index (Igeo), Enrichment Factor (EF) and Heavy Metal Pollution Index (HPI) were utilised.

Geo-accumulation Index

The geo-accumulation Index is used to evaluate the accumulation of contamination in sediments (Muller, 1969).It is defined as:

$$\mathbf{I_{geo} = \log C_n / 1.5B_n}$$

Where,

C_n – Measured Concentration of Element;

B_n – Geo-chemical Background Value;

1.5 – Constant;

The constant 1.5 allowed to predicting the possible natural fluctuations in background data due to litho-logic effect.

Table. 3 Concentration of heavy metals in world surface rock average.

Elements	World Surface Rock Average in ppm	Indian River Sediment in ppm	Undisturbed Sediment Value
Fe	35900	29000	46000
Pb	16	-	20
Zn	127	16	95
Cu	32	28	45
Cd	90	-	0.3
Cr	71	87	90

Source: Martin Meybeck (1979), Subramaniam et al (1987), Turekian and Wedepohl (1961). Values are in ppm.

Table.4The seven grades or classes of profile of the Geo-Accumulation Index proposed by Muller (1981).

S.No	Sediment Accumulation Index (I_{geo})	I_{geo} Class	Pollution Intensity
1	>5	6	Very Strong Pollution
2	>4-5	5	Strong to Very Strong

3	>3-4	4	Strongly Polluted
4	>2-3	3	Moderately to Strongly
5	>1-2	2	Moderately to Polluted
6	>0-1	1	Unpolluted to Moderate
7	<0	0	Practically Unpolluted

Enrichment Factor/ Contamination Factor (EF)

The enrichment factor is a measure to differentiate the sources of heavy metals between lithogenic and naturally occurring Iron (Fe) to normalize the heavy metal distribution (Rule 1986; Naji and Ismail, 2011). The Enrichment factor for Fe-normalization is defined by;

$$EF = (M_{x\,(sample)} / Fe_{x\,(sample)}) / (M_{c\,(shale)} / Fe_{c\,(shale)})$$

Where Mx – Concentration of heavy metals in sourced sample

Mc – Concentration of heavy metals in Undisturbed Sediments;

Fe_x - Concentration of Fe in Spotted Samples;

Fe_c - Concentration of Fe in Average Undisturbed Sediments;

The average shale values used in this study were tabulated in Table No:

Table: 5 The Classification of Enrichment Factor

EF Range	Assessment Category	Result
< 1	I	No Enrichment
1 -2.9	II	Minor Enrichment
3-4.9	III	Moderate Enrichment
5-9.9	IV	Moderately Severe Enrichment
10-24.9	V	Severe Enrichment
25-49.9	VI	Very Severe Enrichment
>50	VII	Extremely Very Severe Enrichment

Contamination Factor

The level of contamination of sediment by heavy metal is expressed in terms of contamination factor (CF) calculated as

$$\text{Contamination Factor (CF)} = S_{ample} / C_{Background}$$

Where, S_{ample} – Concentration of the given metal in River Sediment.

$C_{Background}$ – Concentration Value of the metal equals to the world surface rock average.

Table:6 Hakanson Classification of Contamination Factor

Contamination Factor	Classification
< 1	Low Contamination
1 < CF < 3	Moderate Contamination
1 < CF < 3	Considerable Contamination
CF > 6	Very High Contamination

Heavy Metal Pollution Load Index

Heavy metal pollution index (HPI) is a method of rating that shows the composite influence of individual heavy metal on the overall quality of water. It is used to evaluate overall pollution with respect to heavy metals (Prasad B, Kumari S, 2008). Pollution Load Index of particular site can be estimated using the method proposed by Tomilson et al. (1980).

$$PLI = (CF1 \times CF_2 X\ CF_3 X \ldots \ldots \ldots \times CF_n)^{1/n}$$

Where CF is the contamination Factor and n is the no of metals.

Results and Discussion

Leachate pH level: Leachate is generally found to have **pH** between 4.5 and 9 (T. H. Christensen, P. Kjeldsen, and P. Kjeldse, 20010). The pH of young leachate is less than 6.5 while old landfill leachate has pH higher than 7.5 (A. A. Abbas, G. Jingsong et al, 2009). Initial low pH is due to high concentration of volatile fatty acids (VFAs) (J. Bohdziewicz and A. Kwarciak,). Stabilized leachate shows fairly constant pH with little variations and it may range between 7.5 and 9.

Physio-chemical characteristics of the leachate depend primarily upon the waste composition, depth of waste, moisture content, available oxygen, temperature, co-disposal with inclinator ash, waste processing, age of landfill, toxicity and water content in total waste. The characteristics of the leachate samples collected from the dumpsite has been presented in Table 7. Variation in the conductivity was very high (2360µmhos/cm) and is the main indicator of dissolved inorganic species or total concentration of ions. Though the pH values were within the stipulated standards, the levels can be used to explain the nature of the leachate. Higher pH (> 8) value indicates the alkaline nature of leachate. This shows that the biochemical activity in the landfill was in its final stage and the organic load was biologically stabilized. During the initial stage or acetogenic stage, the pH values are quite low (<7).The maximum pH value of leachate at dumpsite is 7.68. The presence of high BOD (20282 mg/l) and COD (27621 mg/l) indicates high organic strength of leachate. In all six leachate samples sample-1 shows the high value of ammonia nitrogen (2251 mg/l) indicates the decomposition of organic matter. The high TDS value of 28512 mg/l shows presence of inorganic material in the leachate sample.

Table. 7 Physico-Chemical and Heavy metal characteristics of Leachate at landfill sites

S.No	Parameter	L = 200 m	L = 400m	L = 600m	L = 800m	L = 1000m	L = 1200m
		S_1	S_2	S_3	S_4	S_5	S_6
1	pH	7.68	7.33	7.13	7.10	7.04	6.93
2	EC	2596.00	2478.00	2436.48	2419.98	2397.15	2368.80
3	TDS	28512.00	27216.00	2426.76	24577.16	24345.30	2359.35
4	COD	27621.00	26365.50	26848.80	25172.88	24935.40	26103.00
5	BOD	20282.90	19360.95	17695.80	1756.42	1739.85	17204.25
6	Na+	741.40	707.70	597.24	571.34	565.95	580.65
7	K+	1393.70	1330.35	1245.24	1193.56	1182.30	1210.65
8	NH4+	2251.70	2149.35	1888.92	2058.52	2039.10	1836.45
9	NO3	433.40	413.70	368.28	395.38	391.65	358.05
10	Fe	85.80	81.90	70.20	64.66	64.05	68.25
11	Pb	1.33	1.27	1.38	1.38	1.37	1.34
12	Zn	2.20	2.10	1.80	1.80	1.79	1.75
13	Cu	0.86	0.82	0.68	0.73	0.72	0.66
14	Cd	0.04	0.04	0.03	0.03	0.03	0.03
15	Cr	0.42	0.40	0.38	0.35	0.35	0.37

Table. 8 Statistical Analysis of Leachate Characteristics

Para	Min	Max	Average	Median	SD	Kurtosis	Skewness	Variance
pH	6.93	7.68	7.20	7.12	0.33	1.11	0.36	0.07
EC	2368.8	2596	2449	2428	102.50	1.43	0.19	6513
TDS	2359.35	28512	18239	24461	9651.64	5.16	-2.22	153154089
COD	24935.4	27621	26174	25769	1420.24	-0.98	-0.17	1025449
BOD	1739.85	20283	13006	9726	9563.97	-3.25	0.03	77287474
Na+	565.95	741.4	627	584	83.42	-1.47	0.83	5891
K+	1182.3	1393.7	1259	1219	100.13	-0.93	0.54	7185
NH4+	1836.45	2251.7	2037	2048	132.82	-0.63	0.31	24266
NO3	358.05	433.4	393	393	24.61	-0.65	0.51	779
Fe	64.05	85.8	72	67	10.26	-1.64	0.70	84
Pb	1.27	1.38	1.35	1.35	0.05	-1.40	-0.59	0.00
Zn	1.75	2.2	1.91	1.80	0.20	-1.19	0.91	0.04
Cu	0.66	0.86	0.75	0.73	0.07	-1.19	0.78	0.01
Cd	0.03	0.04	0.04	0.03	0.01	-1.58	0.94	0.00
Cr	0.35	0.42	0.38	0.36	0.03	-1.51	0.36	0.00

Table. 9 Results of Geo chemical Index

S.No	Fe	Pb	Zn	Cu	Cd	Cr	Result
S1	-9.29	-4.17	-6.44	-5.80	-11.72	-7.99	Cd > Fe > Cr > Zn > Cu >Pb
S2	-9.36	-4.24	-6.50	-5.87	-11.72	-8.06	Cd > Fe > Cr > Zn > Cu >Pb
S3	-9.58	-4.12	-6.73	-6.14	-12.14	-8.13	Cd > Fe > Cr > Zn > Cu >Pb
S4	-9.70	-4.12	-6.73	-6.04	-12.14	-8.25	Cd > Fe > Cr > Zn > Cu >Pb
S5	-9.72	-4.13	-6.73	-6.06	-12.14	-8.25	Cd > Fe > Cr > Zn > Cu >Pb
S6	-9.62	-4.16	-6.77	-6.18	-12.14	-8.17	Cd > Fe > Cr > Zn > Cu >Pb
Result	**Practically Unpolluted**						

Table. 10 Results of Enrichment Factor

S.No	S1	S2	S3	S4	S5	S6	EF -Quality	Result
Fe	1	1	1	1	1	1	-	Pb>Zn>Cu>Cr>Cd
Pb	36	36	45	49	49	45	Extreme	Pb>Zn>Cu>Cr>Cd
Zn	12	12	12	13	14	12	Significant	Pb>Zn>Cu>Cr>Cd
Cu	10	9.5	7.9	8.4	8.0	7.7	Significant	Pb>Zn>Cu>Cr>Cd
Cd	0.23	0.24	0.21	0.22	0.23	0.21	Minimal	Pb>Zn>Cu>Cr>Cd
Cr	2.5	2.5	2.77	2.77	2.79	2.77	Moderate	Pb>Zn>Cu>Cr>Cd

Table. 11 Results of Contamination Factor

S.No	S1	S2	S3	S4	S5	S6	Classification
Fe	0.002959	0.002824	0.002421	0.00223	0.002209	0.002353	Low Concentration
Pb	0.08	0.08	0.09	0.09	0.09	0.09	
Zn	0.14	0.13	0.11	0.11	0.11	0.11	
Cu	0.03	0.03	0.02	0.03	0.03	0.02	
Cd	0.000444	0.000444	0.000333	0.000333	0.000333	0.000333	
Cr	0.004828	0.004598	0.004368	0.004023	0.004023	0.004253	

IOP Conf. Series: Materials Science and Engineering 1112 (2021) 012030 doi:10.1088/1757-899X/1112/1/012030

Table. 12 Physio chemical Characteristics of Ground Water Samples.

S.No	Parameter	L = 200 m	L = 400m	L = 600m	L = 800m	L = 1000m	L = 1200m
1	pH	6.85	7.05	7.27	7.25	7.19	7.08
2	EC	1615.95	2386.65	2253.30	2702.70	1841.70	2058
3	TDS	1134.00	1446.90	1930.95	2336.25	1614.90	1263
4	Ca^+	65.10	117.60	168.00	147.00	50.40	111
5	Mg^{2+}	33.60	114.45	98.70	117.60	64.05	74
6	TH	304.50	774.90	832.65	861.00	394.80	588
7	K^+	31.50	30.45	72.45	12.60	18.90	48
8	Cl^{2-}	127.05	290.85	469.35	127.05	294.00	669
9	So_4^{2-}	95.55	118.65	90.30	154.35	99.75	177
10	Na^+	204.75	218.40	128.10	201.60	305.55	569
11	NO_3^{2-}	0.33	1.28	1.01	0.72	0.55	0.51
12	F	0.70	0.36	0.89	0.76	0.14	0.41

Parameter	Min	Max	Avg	Median	SD	Kurtosis	Skewness	Variance
pH	6.85	7.266	7.11	7.13	0.16	0.48	-0.97	0.02
EC	1616	2702.7	2143.0	2155	390	-0.59	0.08	152116
TDS	1134	2336	1621.0	1531	448	-0.26	0.76	200645
Ca^+	50.40	168	109.90	114	45	-1.38	-0.17	2072
Mg^{2+}	33.60	117.6	83.83	87	32	-0.80	-0.56	1059
TH	304.50	861	625.98	681	236	-1.98	-0.47	55672
K^+	12.60	72.45	35.70	31	22	0.65	1.00	474
Cl^{2-}	127.05	669.9	329.70	292	209	-0.06	0.82	44031
So_4^{2-}	90.30	177.45	122.68	109	35	-1.04	0.87	1265
Na^+	128.10	569.1	271.25	211	156	3.49	1.78	24489
NO_3^{2-}	0.33	1.281	0.73	0.64	0.35	-0.56	0.68	0.12
F	0.14	0.8925	0.54	0.56	0.29	-1.44	-0.24	0.08

Table. 13 Comparison of Spotted Water with Water Quality Standards

Parameter	Average	BIS (10500) - 1992		WHO 1997
		Desirable Limit	Maximum Limit	
pH	7.11	6.5-8.5	MR	7.0-8.5
EC	2143.0	-	-	750
TDS	1621.0	500	2000	500
Ca^+	109.90	75	200	75
Mg^{2+}	83.83	30	100	30
TH	625.98	300	600	100
K^+	35.70	-		100

Cl^{2-}	329.70	250	1000	250
So_4^{2-}	122.68	200	400	200
Na^+	271.25	-	-	50
NO_3^{2-}	0.73	45	100	-
F	0.54	1	1.5	-

Conclusion

1. The pH level of leachate reveals that the nature of leachate is stabilised one and age old.

Reference

[1] APHA 1975. *Standard Methods for the Examination of Water and Wastewater.* 14thEdition, American Public Health Association, Washington DC.

[2] APHA-AWWA-WPCF 1989. *Standard Methods for the Examination of Water and Wastewater.* American Public Health Association, Washington DC

[3] CPCB (2000) Management of Municipal Solid Waste. Central Pollution Control Board, New Delhi.

[4] Damodharan N., Robinson A., David E. and Kalas-Adams N. (2003) Urban Solid Waste Generation and Management in India. Procs. Sardinia 2003, IX Int... Landfill Symposium, CISA, Cagliari.

[5] Martin JM, Meybeck M (1979). Elemental mass balance of materials carried by major world rivers. Mar Chem. 7:173-206.

[6] Nazi A, Ismail A (2011). Assessment of metals contamination in Klang river surface sediments by using different indexes, Environment Asia; 4(1): 30-38.

[7] Prasad B, Kumari S (2008) Heavy metal pollution index of ground water of an abandoned open cast mine filled with fly ash: A case Study .Mine Water Environ 27:265-267.

[8] Rule JH (1986). Assessment of trace element geochemistry of Hampton roads harbour and lower Chesapeake bay area sediments, Environment Geology and Water Sciences; 8(4): 209-19.

[9] Wilson B, Pyatt FB (2007) Heavy metal dispersion, persistence and bioaccumulation around an ancient copper mine situated Anglesey, UK Ecotoxical Environ Saf 66:224-231.

[10] T.H.Christensen, P.Kjeldsen, and P.Kjeldsen, "Biogeochemistry of landfill leachate plumes," *Applied Geochemistry*, vol. 16, no. 7-8, pp. 659–718, 2001.

[11] WHO, 1984. *International Standards for Drinking Water.*3rdEdn, World Health Organization, Geneva.

[12] WHO 1996. *Guidelines for Drinking Water Quality.* Second Edition, Vol. 2, Health Criteria and Other Supporting Information, World Health Organization, Geneva.

[13] WHO, 1998. *Guidelines for Safe Recreational Water Environments.* Vol. 1, Costal and Fresh waters. Draft for Consultation. World Health Organization, Geneva, WHO/EOS/98.

[14] A.Abbas, G. Jingsong, L. Z. Ping, P. Y. Ya, and W. S. Al-Rekabi, "Review on landfill leachate treatments," *Journal of Applied Sciences Research*, vol. 5, no. 5, pp. 534–545, 2009.
[15] J. Bohdziewicz and A. Kwarciak, "The application of hybrid system UASB reactor-RO in landfill leachate treatment," *Desalination*, vol. 222, no. 1–3, pp. 128–134, 2008.

AUTHOR INDEX